Industrial Ecology

Series Editor

Syed Abdul Rehman Khan, Tsinghua University, Beijing, Beijing, China

Industrial ecology and circular economy is a peer-reviewed book series that focuses on different disciplinary approaches to waste management, sustainable practices & strategies on different scientific, societal, pyschological, technological, economic, governance, and cultural and political aspects of the ongoing and emerging debate. This primary goal of this series is to offer scientists from different school of thoughts and institutions a platform for scientific analysis and debate.

Undeniably, Industrial ecology is a rapidly growing field that systematically examines local, regional and global materials and energy uses and flows in products, processes, industrial sectors and economies. It focuses on the potential role of industry in reducing environmental burdens throughout the product life cycle from the extraction of raw materials, to the production of goods, to the use of those goods and to the management of the resulting wastes. Industrial ecology is ecological in that it (1) places human activity—industry in the very broadest sense—in the larger context of the biophysical environment from which we obtain resources and into which we place our wastes, and (2) looks to the natural world for models of highly efficient use of resources, energy and byproducts. By selectively applying these models, the environmental performance of industry can be improved. Industrial ecology sees corporate entities as key players in the protection of the environment, particularly where technological innovation is an avenue for environmental improvement. As repositories of technological expertise in our society, corporations provide crucial leverage in attacking environmental problems through product and process design.

Adeel Shah · Che Rosmawati Che Mat ·
Alisa Ibrahim · Yu Zhang · Samreen Muzammil

Industrial Ecology

A Fusion of Material and Energy in Green Supply Chain Context

Springer

Adeel Shah
Department of Supply Chain Management
Institute of Business Management
Karachi, Pakistan

Alisa Ibrahim
Arshad Ayub Graduate Business School
Universiti Teknologi MARA
Selangor, Malaysia

Samreen Muzammil
Transformational Entrepreneurship
Centre of Business in Society
Coventry University
Coventry, UK

Che Rosmawati Che Mat
UniKL Business School
Universiti Kuala Lumpur
Kuala Lumpur, Malaysia

Yu Zhang
Department of Supply Chain
School of Economics and Management
Chang'an University
Xi'an, China

ISSN 2730-5775 ISSN 2730-5783 (electronic)
Industrial Ecology
ISBN 978-981-97-3618-8 ISBN 978-981-97-3619-5 (eBook)
https://doi.org/10.1007/978-981-97-3619-5

This Springer imprint is published by the registered company Springer Nature Singapore Pte Ltd.
The registered company address is: 152 Beach Road, #21-01/04 Gateway East, Singapore 189721, Singapore

Dedicated to the late Mr. Shahjehan Syed Kareem, visionary founder and President of Institiute of Business Management, and to my beloved family

Dr. Adeel Shah

Preface

In an interconnected world where goods traverse continents, and production processes span borders, sustainability has become paramount. The intricate dance between industry, environment, and society necessitates a holistic approach—one that transcends traditional silos and embraces a systems perspective. Welcome to the world of Industrial Ecology. A discipline that seeks harmony between human activities and the natural world. This book delves into the heart of industrial processes, supply chains, and their impact on our planet in the era of post-COVID-19. It invites you to explore the intricate relationships that bind production, consumption, waste, and regeneration. As we embark on this journey, we will unravel the threads that weave together industrial systems, ecosystems, and human well-being. From raw material extraction to product design, manufacturing to distribution, and end-of-life disposal to circular economy solutions, each thread contributes to the fabric of our global supply chain and production ecosystem.

The book aims at readers of university students and professionals. The manuscript focuses on educating about the basic principles of industrial ecology and its impact on organizational performance and society. Reading the book allowed the audience to rationalize and select critical first steps to illustrate better strategies. We use quantitative studies and empirically test them to bring meaning to the whole concept of industrial ecology. The book aims to bridge the gap between equivocal and thorough understanding of the subject matter. Besides designing the book to be a stepping stone in the science of industrial ecology, we hope our audience can understand the challenges in the post-COVID-19 era.

The chapters educate the readers to embrace a system's perspective and recognize that every action has repercussions. Thus, it enables an understanding of optimizing resource use, minimizing waste, and creating closed-loop systems wherever possible. We will delve into the life cycle of products, assessing their environmental impact from the cradle to the grave and the hidden costs embedded in our everyday items while educating new sustainable strategies that businesses of the future would implement, like cradle to cradle in developing their product and services. The whole concept of industrial ecology thrives on collaboration between stakeholders and transitional stages. We shed light to educate businesses, policymakers, and communities working together to create regenerative systems. Resilience and adaptability are also important features for businesses; thus, climate change and its impact causing disruptions of the value chain also looms and should be considered for supply chain adaptation and how industrial symbiosis can play an essential role in creating resilience.

Our effort is not to design or produce an academic exercise, but it is a call to action. Whether you are a student, a practitioner, or a concerned citizen, you hold a thread in this intricate tapestry. Let us weave a future where industry and ecology dance in harmony—a future where supply chains nourish rather than deplete.

No particular background knowledge is required to read the book; anyone with basic English language skills can easily understand the main crux. We have used the spiral learning technique to impregnate the whole concept of industrial ecology to the book's reader. A few chapters where the empirical study is used to provide logical solid backing would require external online help to understand the analysis and discussions of the results. However, the concepts in each chapter are easily relatable to different industries, thus providing coherence and clear understanding. The book introduces industrial ecology to expose the readers to definitions from known authors, providing history and its relation with United Nations sustainability goals (UN SGDs) and green supply chain while commenting on the post-COVID-19 era. In Chap. 1. Chapter 2 introduces the current practices, methods, and tools used in industrial ecology. Chapter 3 provides an overview and implementation of industrial ecology applications by studying organizations from developing countries' manufacturing and services industries.

Chapter 4 builds an understanding of the importance of the digital supply chain, while case studies from the Pharma and healthcare industry are brought forward for experiential learning in Chaps. 5 and 6. The Reduce, Reuse, and Recycle (3Rs) are explained in Chap. 7, along with understanding organizational intention to use the whole concept as sustainable development by testing a hypothesis in a quantitative study under Chap. 8. Social economic benefits derived from industrial ecology are brought forward in Chap. 9, and the same is discussed with the help of exploratory research conducted in Chap. 10 to understand the actual social, economic benefits derived in the textile industry, which is the most established sector globally with product consumption by people in all seven continents.

Chapter 11 touches on the most important topic under the limelight, climate change and its relation with industrial ecology; a case study is brought forward from a sourcing company looking for sourcing options in future to prepare a supplier selection strategy in the best suited location. Moving forward, Chap. 12 in the book discusses industrial ecology in the food supply chain and makes themes to improve industrial ecology advancements through industrial experts. Finally, the book's closure starts with Chap. 13, which discusses the impact of COVID-19 on industrial ecology. The topic is understood by put forward the case from an automated industry. Finally, Chap. 14 concludes the book by addressing the challenges of industrial ecology in the current manufacturing sector.

Karachi, Pakistan	Adeel Shah
Kuala Lumpur, Malaysia	Che Rosmawati Che Mat
Selangor, Malaysia	Alisa Ibrahim
Xi'an, China	Yu Zhang
Coventry, UK	Samreen Muzammil

Contents

Abbreviations

A&F	Abercrombie and Fitch
AAB	American Association of Blood Banking
AGVs	Automated Guided Vehicles
AI	Artificial Intelligence
AMRs	Autonomous Mobile Robots
APICS	Association for Supply Chain Management
APIs	Application Programming Interfaces
APS	Advance Planning and Scheduling
AR	Augmented Reality
ASCM	Association for Supply Chain Management
AVE	Average Variance Extracted
BDA	Big Data Analytics
BI	Business Intelligence
BIPOC	Black, Indigenous, and People of Color
BPCS	Business Planning and Control System
BYOB	Bring Your Shopping Bag
C2C	Cradle to Cradle
CAIC	Computer Numerical Control
CAP	College of American Pathologists
CCS	Carbon Capture and Storage
CDMOs	Contract Manufacturing and Development Organizations
CDP	Carbon Disclosure Project
CE	Circular Economy
COVAX	COVID-19 Vaccines Global Access
CPFR	Collaboration Planning, Forecasting, and Restocking
CPS	Cyber-Physical Systems
CRCP	Climate Risk Country Profile
CRM	Customer Relationship Management
CSC	Circular Supply Chain
CSCP	Certified Supply Chain Professional
CSR	Corporate Social Responsibility

CTP	Conformance to Plan
CTRP	Conformance to Release Plan
CTS	Conformance to Supply
DC	Distribution Center
D-EMRP	Domestic Manufacturing Enterprise Resource Planning
EIP	Eco-Industrial Parks
EOQ	Economic Order Quantity
EPA	Environmental Protection Agency
EQAS	External Quality Assurance System
ERP	Enterprise Resource Planning
ESG	Economic, Social, and Governance
ETAs	Estimated Time of Arrival
ETP	Effluent Treatment Plant
EUETS	European Union Emissions Trading System
Evs	Electric Vehicle
FDA	Food and Drug Administration
FEA	Finite Element Analysis
FSC	Forest Stewardship Council
GBG	Green Banking Guidelines
GDP	Gross Domestic Product
GHG	Green House Gas
GMP	Good Manufacturing Practices
GOP	Government of Pakistan
GPN	Green Purchasing Network
GPP	Green Public Procurement
GPS	Global Positioning System
GRI	Global Reporting Initiative
GSCM	Green Supply Chain Management
GTFS	Green Technology Financing Scheme
HEIs	Higher Education Institutions
HR	Human Resource
HYVs	High Yielding Varieties
ICT	Information and Communication Technologies
IE	Industrial Ecology
ILO	International Labor Organization
IoT	Internet of Things
IPCC	Intergovernmental Panel on Climate Change
IPOs	Initial Public Offerings
IPS	Indoor Positioning Systems
IR 4.0	Industrial Revolution 4.0
IS	Industrial Symbiosis
ISBT	International Society of Blood Transfusion
ISRI	Institute of Scrap Recycling Industry
IT	Information Technology
ITA	International Trade Administration

JIT	Just in Time
KPIs	Key Performance Indications
LCA	Life Cycle Assessment
LCM	Life Cycle Management
LEED	Leadership in Energy and Environmental Design
MENGO	Malaysian Environmental NGOs
MES	Manufacturing Execution System
MFA	Material Flow Analysis
MIT	Massachusetts Institute of Technology
ML	Machine Learning
MPMA	Malaysian Plastics Manufactures Association
MSC	Marine Stewardship Council
MS-DOS	Microsoft Disk Operating System
NGO	Non-Profit Organizations
OEE	Overall Equipment Effectiveness
OEKO-TEX	Textiles Testing for Harmful Substances
OEMs	Original Equipment Manufacturers
ORION	On-Board Integrated Optimization and Navigation
PC	Personal Computer
PDA	Personal Digital Assistant
PEPA's	Pakistan-Environmental-Protection-Act 1997
PET	Polyethylene Terephthalate
PI	Pharmaceutical Industry
PLS	Partial Least Square
PLS-SEM	Partial Least Square Structural Equation Modeling
PPEs	Personal Protective Equipments
PPP	Profit People and Plant
PV	Photovoltaic System
QR Code	Quick Response Code
R&D	Research and Development
RFID	Radio Frequency Identification
RFQs	Request of Quotations
RH	Relative Humidity
RLNG	Regasified Liquefied Natural Gas
ROHS	Restriction of Hazardous Substance
ROI	Return on Investment
SAC	Public Private Partnership
SAFE	Coffee and Farmer Equity
SAP	Systems, Applications, and Products in Data Processing
SBP	The State Bank of Pakistan
SBTI	Science-Based Target Initiative
SC	Supply Chain
SCM	Supply Chain Management
SCS	Supply Chain Sustainability
SDGs	Sustainable Development Goals

SFI	Star Fades International
SME	Small and Medium Size Firms
SMS	Short Message Service
SOPs	Standard Operating Procedures
SSCF	Sustainable Supply Chain Foundation
SSCM	Sustainable Supply Chain Management
TBL	Triple Bottom Line
TIF	Thalassemia International Federation
TMS	Transportation Management System
UKM	Universiti Kebangsaan Malaysia
UN SDG	United Nations Sustainable Development Goals
UN	United Nations
UNEP	United Nations Environment Program
UNIDO	United Nations Industrial Development Organization
UNISDR	United Nations Office of Disaster Risk Reduction
USGBC	United States Green Building Council
VF	Vanity Fair Mills
VIF	Variance Inflation Factor
VMI	Venders Managed Inventory
VOC	Volatile Organic Compound
WEEE	Waste Electrical and Electronic Equipment
WFG	World Federation of Hemophilia
WHO	World Health Organization
WWF	World Wildlife Fund

Chapter 1
Introduction to Industrial Ecology

The 21st Century's most significant threat to planet Earth is humanity's industrial revolution, which caused drastic climate change that affects current and future generations. Besides humankind, other branches of bio-organisms are at risk of extinction, too. Recently, policymakers have been captivated by the potential ramifications of climate change and other environmental concerns on economies and communities, owing to the indisputable significance of these matters. Due to regulations and consumer awareness, organizations are increasingly pressured to do business sustainably. There is an emphasis on sustainability and the necessity to better the flow of material and energy in the industrial ecosystem to meet the economic, social, and legal concerns of all stakeholders (investors, customers, policymakers, and the general public). Many organizations are ignorant of industrial ecology to achieve sustainability and desire to increase profits at the cost of the environment, which has specifically caused global warming and changes in biodiversity.

1.1 Introduction

IE has been the focus of the past few years, but a lot is still to be researched in the context of economics (Yu & Zhang, 2021) and resilience (Kendall & Spang, 2020). Recently, there has been considerable focus on climate change and other environmental concerns, which policymakers find intriguing due to the potential ramifications for economies and populations. Organizations are under more and more pressure to do business sustainably in both developed and developing economies. For organizations to be sustainable, Industrial Ecology (IE) can significantly help businesses by preventing wastages that impact the environment, thus preventing or slowing climate change (Awan, 2022) and reversing all the ill effects. The existing literature provides many definitions of IE, and an important one is brought forward

A. Shah et al., *Industrial Ecology*, Industrial Ecology,
https://doi.org/10.1007/978-981-97-3619-5_1

by Jelinski et al. (1992) to suggest that it is a new manufacturing concept that efficiently supplies material for achieving sustainability. The nebulous explanations and definitions of the subject make it difficult to differentiate between IE and sustainability (Hicks, 2022). To better understand the concept of IE, Table 1.1 illustrates a set of definitions by different authors over the past years.

In pursuit of understanding, we can divide IE into three phases;

1. It investigates how the parts and pieces of industrial and natural systems interact.
2. To introduce fundamental ideas as practical approaches for material and energy flows for industrial metabolism.
3. Providing policy to create a roadmap for sustainable development.

The IE is coupled with sustainable supply chain methodologies to improve performance. The efforts for IE to achieve sustainability in recorded history started in the 18th Century in a German company for the continuity of their business related to woods and forests. Sustainability deviates from the noun "sustenance," which means "which one retains." As mentioned earlier, the term sustainability was used for the first time in the 18th Century by German Hans Carl von Carlowitz in his book Silvicultura Economic (Spindler, 2013). According to Schmithüsen (2013), the sustainability concept was first developed in central Europe to conserve forestry. In the 1330s, it was mentioned in common law for Germany that woodcutting should be moderate and should not impact the perpetuity of this industry. This was done through different principles, like trees with fruits should not be cut, and forests near the villages and towns should be there to provide shade to the people and animals. Trees will be planted and should be protected from the al grazing to sustain themselves. In France, this term was used in the Ordonnance de Brunoy, which was related to waterways and forests.

Table 1.1 Definitions of industrial ecology

Authors	Definitions
Frosch and Gallopoulos, (1989)	'Transformation of conventional production activities of value-adding raw material to produce finished goods and discard waste, to an integrated production ecosystem: an industrial ecosystem similar to biological ecosystems'
Graedel (1996)	'A systematic study to understand the collaboration of human activities and the environment. In the context of industrial application, industrial ecology pursues improving the flow and consumption of virgin material to produce a final product and dispose of waste during the process'
Korhonen (2000)	'Commonly defined as management of material in manufacturing, also focusing on flow material and energy extracted by the enterprise and co-partners. It also focuses on the waste and emissions produced during the process and its environmental impact
Allenby (2006)	'A multifaceted discourse that pursues to comprehend evolving behavior of human and natural systems and its complications'
Vujanović et al. (2022)	'An organic industrial system which imitates a biological system'

Further, historical evidence is brought forward that, in 1346, King Philippe V made a law that the owners of rivers and forests would make arrangements for the waterways and forests to be sustained in perpetuity. A book entitled Sylva: A Discourse of Forest Trees and the Propagation of Timber in His Majesty's Dominions was published by John Evelyn in 1664 and presented to the King, the royal society, and the general public in Britain. This was reprinted many times in the 17th Century to encourage the planting of trees (Schmithüsen, 2013). Spindler (2013) suggests that the book published by the Royal Society in 1664 was on the conservation, growing, and use of wood in perpetuity. The book advises enterprises to cut the forest to the extent they can regrow through planned forestation projects to sustain this industry; in the true sense, it related to the understanding of sustainability, which means that resources are used to satisfy current needs in such a way that future needs can be met without any compromise.

However, IE became popular research (Frosch & Gallopoulos, 1989). The concept has been evolving since the 20th Century; according to Erkman (2001), in the 1950s, there was a concept named "end of pipe," which was to clean or treat your waste at the end of the process. Though a good step, it is not sufficient for the current global issues due to the existence of industries in the ecosystem. To overcome the problem, IE tries to resolve the issues of humans and the environment to minimize ill actions on the ecosystem and to control industrial global problems.

The environmental hazards caused by the Industrial Revolution are global. To correct the trajectory of growth and industrial development (United Nations, 2023) has proposed seventeen sustainable development goals that need to be achieved by all signatory nations. In developed countries, laws are stringent and are correctly implemented, unlike in developing countries.

1.2 History of Industrial Ecology

Industry ecology (IE) is a pragmatic approach to sustainability. An investigation into the correlation between physical surroundings and living organisms is called ecology. IE is the set of tools, perspectives, and principles taken from the concept of ecology, and it analyses the industrial impact on the people surrounding it and the environment. IE gives the perspective of process optimization. Initial work was done only on the material flow of the system, but currently, only energy flow is also being focused on (Lowenthal & Kastenberg, 1998).

According to Erkman (2001), industrial ecology was initially considered an oxymoron, which means industry and ecology were considered as there is no relationship between them. These are proposed to each other like bittersweet is an oxymoron. In the 1950s, a concept named "end of pipe" was developed after awareness of environmental pollution caused by the industry. According to this concept, companies started treating their waste streams to reduce or control the pollution caused by their processes. This was the start of realizing that the industrial process impacts the environment and the surrounding biosphere. Industrial ecology refers to all human activities at manufacturing sites, hotels, and other service industries. Industrial Ecology is an effort to address the question, "How can the concept of sustainable development be made operational in an economically feasible way" (Erkman, 2001). IE emerged after it was evident that the end-of-pipe concept is inadequate to solve the environmental problem and will be expensive in the long run. There are three perspectives regarding IE on which more or less all researchers agree and are as follows:

A. It encompasses a thorough, methodical, and cohesive examination of the interconnections among every element of the industrial economy and the biosphere in which it operates.
B. It correlates with the biophysical foundation of human activities.
C. It examines technological dynamics, specifically the evolution of clusters of critical technologies over an extended period. It is an essential component in facilitating the shift from the present unsustainable industrial system to a sustainable industrial ecosystem in the future.

IE looks at the industrial system with complete focus. It does not only talk about pollution and environmental issues but also the technology economy and other issues involved in managing a business. It talks about the optimum usage of scarce resources. Different authors frequently used The term Industrial ecology after the 1970s literature. This expression resurfaced at the beginning of the 1990s, initially in a group of industrial engineers connected with the National Academy of Engineering in the US. There was a practice to write on a particular topic every September. In 1989, the special issue on which they wrote was "Managing Planet Earth." In this issue, Robert Frosch and Nicholas Gallopoulos, engineers at General Motors, wrote on "Strategies for manufacturing." This focus was on the idea that production systems can be designed that may have less impact on the environment than the current processes; this led them to the term industrial ecosystem. They thought the human population was growing exponentially; conversely, the natural resources were depleted due to high consumption. So, the process should be designed so that the products to be sold and used should be maximum, and waste should be minimized. In 1994, White introduced the idea that, as in IE, we study the flow of energy and material in the industrial sector, and this should also involve the impact of consumer activities (Allenby & Richards, 1994). The International Society for Industrial Ecology was established in the early 21st Century (2001), and the Journal of Industrial Ecology (MIT Press) was established in 1997.

IE is an emerging concept that deals with environmentally sound manufacturing and material and energy consumption. This is due to the limited resources available, and wastes should be in a quantity that the environment can bear without affecting future needs (Van Berkel et al., 1997). In 2006, according to Allenby (2006), IE was a system-based approach that was a multidisciplinary concept to understand the human-natural system interaction.

Industrial metabolism was established in 1988 and is a fundamental concept in IE. It relates to the impact of human activity on the environment and the utilization of natural resources. IS, or Industrial Symbiosis is the lifeblood of the IE discipline. Its inception can be traced back to biological principles and the established symbiotic relationships observed in nature, wherein two or more inconsequential classes of living organisms exchange substances, energy, or data for the benefit of all (Saavedra et al., 2018).

Four challenges in the IE framework need to be addressed.

1. Systematic use of waste and by-products of the process
2. Minimize the loss caused by the dispersion of fertilizers and pesticides
3. Reduce the material flow in the process and increase the efficiency
4. Reduce or avoid the use of hydrocarbon fuels in the system

Coastal cities have more population than non-coastal cities. This is because these provide the gateway to other countries and act as nodes in the international supply chain. Most industries exist in these cities, and urbanization near these plants is meant to provide job opportunities. Plants near the ports are primarily refineries, chemical plants, coal processing facilities, and renewable energy sources. These industrial hubs produce 25% of the world's primary production. So, implementing IE in these cities is more critical as these industrial areas have most of the material and energy and produce most of the waste (Cerceau et al., 2014).

The industrial revolution has increased pollution, and the cost of disposal through incineration disposals in landfills is too much and involves natural resources, so it is not entirely sustainable. Industrial ecology states that one industry's waste should be used by another, which will own the utilization of natural resources and eliminate the cost of disposing of the trash. Eco-industrial parks (EIP) have been developed to work on this phenomenon and reduce waste generation. EIP is an actual practical approach to the application of industrial ecology. The most well-known is in Denmark Kalundborg. This concept was built in the 1960s but was not developed with it then. Now, a benchmark for all over the world to compare the EIPs. This included refineries, pharmaceutical companies, and power plants. Another idea in the name of the environmentally balanced industrial complex uses similar industries that can use each other's wastes and raw materials, which saves the storage and transportation of waste/raw material, which also requires natural resources.

Previously, there was an approach called "cradle to grave," which meant that every material produced had a limited life and was disposed of as waste. Companies are working on "built-in obsolescence," meaning they design their products so they must waste some time selling a new version to the same customer. Then McDonough and Braungart proposed a concept named "cradle to cradle," which states that products should be patterned to allow their material to be reused infinitely. That is a sustainable approach. The product flow based on the cradle-to-cradle approach is shown in Fig. 1.1 for easy understanding.

Industrial ecology somewhat talks about the cyclic flow of materials, not the linear approach like cradle to grave. In 2001 the International Society for Industrial Ecology was established at Yale University. In 2018, this society was registered as a nonprofit organization in the Netherlands. They started the Journal of Industrial Ecology. In

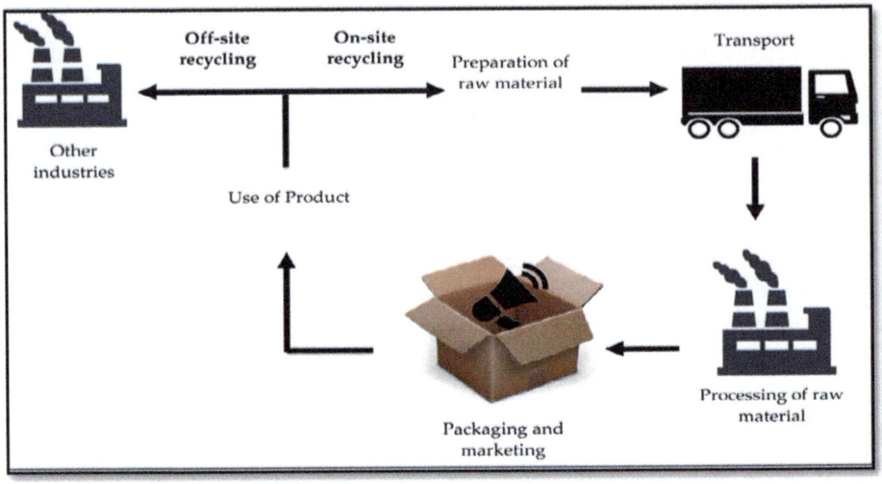

Fig. 1.1 Cradle to cradle approach (McDonough & Braungart, 2010)

2003, the first Ph.D. was offered in industrial ecology, and then after two years, it was taught in the master's program as well (ISIE, 2022).

1.3 The Intriguing World of Industrial Ecology

IE is a subfield of sustainable research that evaluates the stages of service and product production from the standpoint of the natural world, intending to imitate a natural system through resource conservation and reuse. Industrial ecology is relatively new and challenging for engineers, scientists, and policymakers. Frosch and Gallo Poulos' 1989 Scientific American article "Strategies for Manufacturing" gave rise to the contemporary discipline of IE, also known as the "science of sustainability." On the other hand, historical allusions to IE extend back to the early 1970s. Industrial ecology is interdisciplinary, making offering a clear and widely acknowledged description challenging.

The biological comparison, the application of systems views, the importance of technological progress, the involvement of businesses, eco-efficiency and dematerialization, and forward-looking research and practice are some of the fundamental components that define this phenomenon, i.e., Industrial Ecology. There are several analytical techniques and resources available for each of these sub-topics. The most significant characteristics and costs of the fundamental elements—particularly those that are pertinent to bio-degradable and sustainable energy—are covered in the subject (Hariyani & Mishra, 2021).

IE is the study of the systemic interactions that occur between the environment, economy, and society. The approach underscores the utilization of technology to

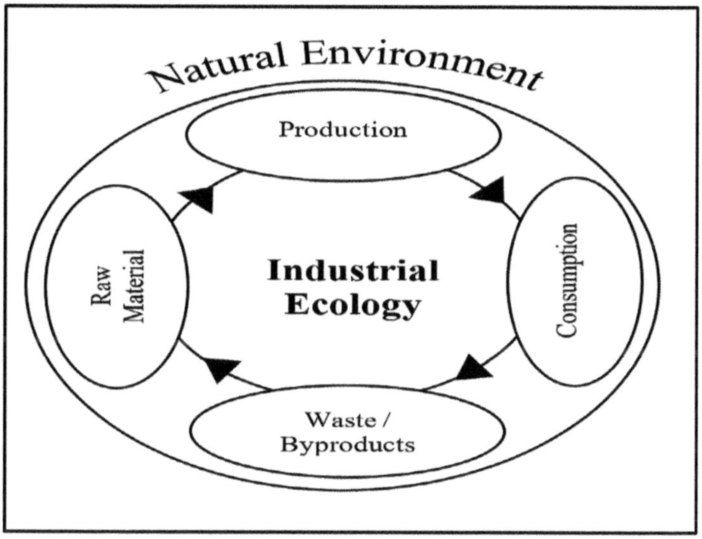

Fig. 1.2 The basic concept of industrial ecology (Celik, 2013)

lessen adverse environmental impacts and find the right balance between ecological stewardship and human advancement while recognizing the significance of socioeconomic factors in pursuing these objectives. The examination of industrial ecology often involves the assessment of resource utilization, recycling, and exchange with the environment (via extraction and emissions) (Al-Thani & Al-Ansari, 2021). These assessments focus on various scales and levels, encompassing municipalities, eco-industrial zones, nations, and the worldwide economy. Originating from the notion that ecosystems and economic systems (including production processes) are comparable is the term "industrial ecology." Industrial ecology incorporates many ecological concepts, including but not limited to the cycles of matter and energy, highly adaptable systems, and environmental networks (Fig. 1.2) (Bruel et al., 2019).

Industrial ecology has advanced since its inception thanks to the creation of databases, more complex mathematical modeling, cooperation between natural, physical, and social scientists, and the expansion of theory independently and in conversation with other related subjects. Meanwhile, industrial ecology significantly impacts environmental management and policy, offering insights on various topics, including climate change, biodiversity loss, water, and more. Despite its young, breadth, and interdisciplinary nature, industrial ecology may claim several subfields within its purview.

1.4 Importance of Industrial Ecology

IE results from cleaner manufacturing, eco-industrial parks, and life cycle analysis. Given the ongoing economic, cultural, and technical progress, IE represents a deliberate and logical approach humanity can adopt to achieve and sustain sustainability. The concept advocates for the collaborative evaluation of an industrial ecosystem alongside its adjacent systems instead of its isolated assessment. The objective is to maximize the efficiency of the entire materials cycle, encompassing raw materials, finished goods, components, finished products, obsolete products, and final disposal. It is imperative to optimize resources, energy, and capital, among other factors (Baumann & Lindkvist, 2022).

Understanding how industrial systems—such as a factory, eco-region, or national or international economy—interact with the biosphere is the goal of industrial ecology. An "ecosystem" built on resources and infrastructural capital rather than on natural capital, similar to how different components of industrial systems interact with one another, may be seen through the lens of natural ecosystems (Sun & Wang, 2021). It encourages sustainable design by capitalizing on the notion that waste does not exist in natural systems. Finding ways to reduce modern consumerism's ecological impact economically is crucial to developing sustainable solutions that protect future generations' resources (Sullivan et al., 2018).

One of Natural Capitalism's four goals is to protect industrial ecology. The other three are to save energy and materials and to rethink how linked international trade markets carefully and product stewardship interactions work as a service economy. This method stops people from buying things morally because they don't know what's happening far away (Singh & Basak, 2018). It also suggests a political economy that puts much value on natural capital and needs more educational capital to build and run each industrial ecosystem.

IE also studies societal concerns and how they relate to environmental and technological systems. Through this holistic viewpoint, IE acknowledges that different components cannot be regarded in isolation and that solving issues requires understanding the linkages between these systems (Belaud et al., 2019). Modifications made to one component of a more extensive system frequently spread to affect changes made to a different component. Therefore, you can only comprehend an issue if you consider its components in light of the total. Based on this concept, IE adopts a systematic thinking perspective when examining environmental challenges (Bahers et al., 2020).

As we already said, industrial ecology is an idea about the environment created by academics to help businesses better manage the environment. Industrial ecology tries to make environmental sustainability and industrial operations work together so that neither hurts the other. The cradle-to-cradle idea says that this method should be used to create manufacturing processes that reduce waste and harmful substances in materials. The "cradle-to-cradle" idea can be seen as building on the "industrial ecology" idea. L. W. Jelinski says that IE is "a new way of designing goods and processes for industry and using environmentally friendly ways to make things."

This concept says an industrial system should work with other systems instead of by itself (Lybæk et al., 2021).

1.5 Industrial Ecology and UN SDGs

The international community accepted the Sustainable Development Goals (SDGs), which address global difficulties in education, health, economic security, social equity and justice, and environmental concerns, in September 2015. The SDGs, which are a component of a larger 2030 Agenda that builds on the Millennium Development Goals established in 2000, were raised by the UN as a model for sustainable development on a worldwide scale. The Sustainable Development Goals (SDGs) entered force on January 1, 2016. Although they are not legally enforceable, they provide a way for nations to organize to reduce poverty, combat climate change, and ensure fair livelihoods for all individuals.

According to Rweyendela (2022), the 2030 Agenda proposes 169 targets for global transformation by 2030 in addition to 17 SDGs. Assuring sustainable utilization and production patterns is a requirement of the 12th SDG (United Nations 2023). Under this goal, the following targets are set: 12.2 (efficient use and sustainable management of natural resources), 12.3 (halving per capita global food waste at the consumer and retail levels and reducing food losses along supply chains), 12.4 (environmentally sound management of all wastes and chemicals throughout their life cycle), and 12.5 (reducing waste generation through prevention, reduction, recycling, and reusing). We must substantially reduce our environmental impact to meet them by rethinking how we produce and consume goods and resources to increase resource efficiency and decrease waste. Although significant efforts are being made to persuade governments to support the SDGs, and many are showing enthusiasm for doing so, application gaps still exist, and many countries are still figuring out their implementation strategies.

1.5.1 Five-Node Resource Nexus Approach

According to many experts, land should be included in a resource nexus approach because of its numerous vital environmental roles. Soil and other natural resource formation, as well as the production, management, and purification of water and the provision of resources for livelihood and development, all depend on the land.

Figures 1.3 show how land functions as an input into all other categories and how it is vitally linked to water. Land must be considered while defining appropriate nexus approaches for implementation purposes in SDGs 2 (Zero Hunger), 7 (Affordable and clean energy), 11 (Sustainable cities and communities), 13 (Climate action), and 15 (Life on land). Regarding land and food, functioning landscapes that offer water-based ecosystem services for agricultural output are crucial to the sustainability of food systems. The food system is necessary for achieving SDG 2, as it requires input from all other resources and has numerous critical interconnections. In contrast, it is possible to have biomass as a dimension on its own to emphasize its significance for bioenergy, food, and bio-based goods from forestry. Figure 1.3 represents biomass as a component of energy, food, and land, highlights the interconnections, and places more focus on the systems of provision than the resources themselves. Further, based on Fig. 1.3, include materials for at least four reasons, in keeping with reference:

1. The other resource categories and materials are closely related. The provision of water (SDG 6), urbanization, and the production and distribution of energy (SDG 7) are all significantly impacted by base metals, vital minerals, and building minerals (SDGs 9 and 11). Mineral fertilizers are also used to produce food (SDG 2).
2. As per the material flow analysis methodology, non-energy, non-renewable resources account for around 50% of the physical units of natural resource use in most industrializations.
3. Throughout the 2000s, the costs of acquiring and processing materials in manufacturing industries accounted for around 40% of total production costs.
4. Lastly, base metals and fertilizers have damaging environmental effects, including greenhouse gas (GHG) emissions 23 and land and water degradation. Nexus research is an excellent way to determine the best mineral inputs for agriculture.

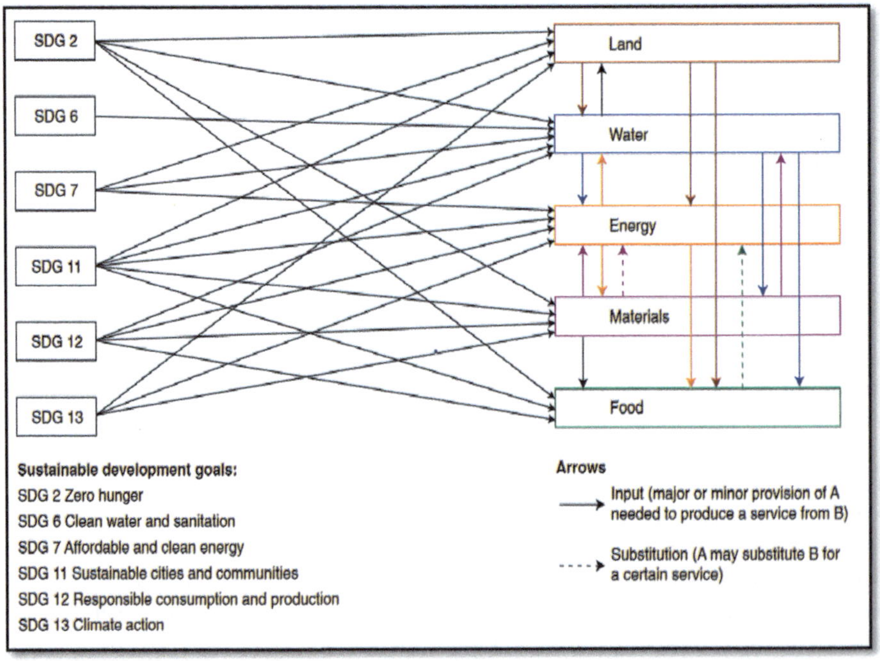

Fig. 1.3 Five-node resource nexus approach (Bleischwitz et al., 2018)

The benefits of consistency, comprehensiveness, and policy relevance come with a five-node nexus. It recognizes broader environmental roles that land, water, and soil play in regulating and stabilizing the inflow of resources from other sources, like biomass for food production. It targets SDGs 2, 6, and 7, maintaining the benefits of other nexus approaches, and it may be more effective than other ways of tackling SDGs 9, 11, 12, and 15 than other approaches. An easy understanding of the relationship and interaction of SDGs and significant environmental resources like land, water, energy, materials, and food is added in Fig. 1.3.

1.6 Industrial Ecology and Triple Bottom Line (TBL)

Hourneaux et al. (2018) suggest that to measure the sustainable performance of companies, there is a model named Triple Bottom line (TBL), which is Profit, People, and Planet (PPP). This model deals with performance measurement in three aspects: economic or traditional profit, sociological impact, and planet impact due to the industrial processes of the companies. If companies don't consider the planet and people and only focus on the profit part, then they are miscalculating the cost of doing business. The term TBL was offered by John Elkington in 1994, and the idea was that companies should give equal importance to all three aspects: the economic, social, planet, and environmental aspects of their business.

According to this, the company should be responsible towards people, including their employees, and treat them responsibly. This includes paying them so they can live and at least follow the minimum wage rule. Please provide them with safe working conditions, including the safe process design provision of personal protective equipment (PPEs) and proper refresher training. If the company pays well and offers a better workplace and culture, more people will be inclined to work with them, even with slightly fewer financial benefits. So, this will ultimately increase the economic benefit along with social responsibility. The company should dispose of all waste responsibly, including all the waste emitted from its processes, like gaseous emission, effluent, and soil contamination. It should be as low as possible through proactive approaches, and waste should be disposed of in a manner provided by the environmental protection agency (Investopedia). According to the Harvard Business Review, seventy-one percent of industrial emissions are attributable to one hundred energy companies (Miller, 2020).

On the contrary, not just the top companies should work on it to change the climate positively. Every organization should work on this to protect the planet. The reason for climate change is the pollution caused by the Industrial Revolution.

Previously, companies always thought about their shareholder value maximization. However, the trends are changing, and companies are now discussing shareholder value, including their employees, shareholders, and community. The organization has partnered with nonprofit organizations to improve their community as well. Fair hiring in a good working environment also means caring about the people.

Harvard Business School professor Henderson (2021) says companies can "do well and do well" simultaneously. According to her, "Doing well" means the company is performing well in terms of financial targets, and "Doing good" means solving social and environmental problems. She further suggests that earning profit is vital to work on the social and environmental side. Working on sustainability is profitable in the long term; it might be seen as an increase in the cost of doing business in the short term. Similarly, McKinsey & Company (2023) suggests that companies with elevated environmental, social, and governance (ESG) ratings outperform in the medium and long term.

It will facilitate the business in the long run in the following ways:

- It will protect your brand name and reduce the risk of defamation. Brand revitalization costs so much to the organization. It will save the company from front-page scandals regarding social or environmental issues.
- Millennials are more willing to pay for sustainable products, and sales of sustainable products have increased by 20% in 5 years from 2014 to 2019.

Research shows that the company can earn profit through increased sales by adopting the triple bottom line concept. In 2018, surveys conducted by Nielsen revealed that 48% of American consumers were open to modifying their consumption patterns to lessen their environmental impact. As the US is the biggest consumer market, the consumer can drive this change quickly, and companies can maintain profit while working on TBL. The findings of this study were validated in 2018 when sales of sustainable, fast-moving consumer goods reached an estimated $128.5 billion.

1.6.1 Industrial Ecology and Green Supply Chain

Industrial ecology (IE) is a cutting-edge strategy for achieving sustainable development. It attempts to reduce waste production and maximize natural resources and energy use. The study of industrial ecology examines how people may continue to live within the desired carrying capacity in the face of ongoing cultural, economic, and technical development. As suggested in theory, all industrial systems must be seen not in isolation from other systems but rather in collaboration with them (Hariyani & Mishra, 2021).

IE aims to maximize the whole material life cycle, from raw materials to completed goods, components, waste products, and final disposal. This impacts the green supply chain and has everlasting ecological benefits for SC firms. The definition suggests that industrial ecology entails the optimization of the entire material life cycle, including the components and wastes that result from each stage of material production and process, from the extraction of raw materials to the final product. Industrial ecology's primary tool is cleaner production, which means looking at how industrial systems and ecological systems relate to and affect each other. It allows us to figure out how different industries hurt the environment and make plans to lessen these effects.

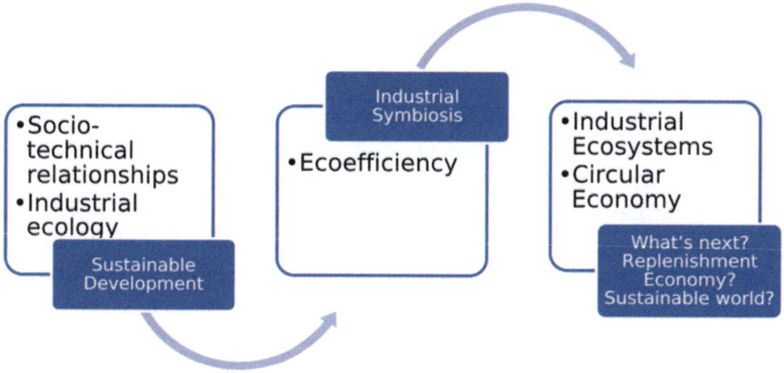

Fig. 1.4 Oughton et al. (2022) progression framework

In conclusion, IE's primary goal is to promote sustainable development, which will lead to compliance with environmental laws meant to protect the environment. Industrial ecology aims to change industries such that they mimic natural ecosystems, where some life uses any material or energy that is accessible. Utilizing techniques like life cycle analysis, the managerial strategy of IE examines how industry and the environment interact (LCA) (Han et al., 2021). Conversely, the technical approach entails implementing innovative product and process design strategies, such as cleaner production and eco-industrial parks. Because of industrial ecology, many economic theories have been created to make resources more useful by minimizing waste and maximizing their benefits. The notion of the circular economy, which gained popularity in the 2010s, is the culmination of these principles. IE now offers scientifically sound methodology, equipment, and strategies for understanding and implementing circular economy principles. Figure 1.4 to understand the progression from industrial ecology to the circular economy while asking an open-ended question as to what would next develop from the circular economy.

1.7 Post-COVID-19 and Industrial Ecology

The COVID-19 (Corona Virus Disease), a pandemic brought on by the SARS-CoV-2 virus, has wreaked havoc on the lives and livelihoods of individuals and resulted in numerous fatalities and significant economic losses. All countries have been impacted by the global pandemic, which has exposed hidden socioeconomic issues like environmental management, social fairness, and healthcare systems. Over 559 million instances of The COVID-19 pandemic have been officially confirmed, with a global toll exceeding 6.36 million lives lost. The epidemic has affected almost all nations. The pandemic and related response strategies provide scientists with a rare opportunity to reexamine socioeconomic networks and their connections to long-term ecological resilience (Besson, 2022). The socioeconomic ramifications encompassed a fall in the demand for goods and services, disruption of the supply chain operations, unemployment, and income loss.

Recently, regulatory authorities have realized that biological hazards affect environmental health and constitute a long-term, widespread impediment to socioeconomic factors. The effects of such pandemic occurrences on various state sectors, particularly those that directly support economic stability, are typically not immediately apparent but have adverse long-term effects. As the possibility of health loss breeds a sense of fear among the populace, these effects permeate the whole industrial sphere and have a long-lasting impact on regional supply chains and national marketplaces.

Developing economic recovery strategies founded upon sustainability and inclusiveness principles is critical in these unprecedented times. The policy packages' objectives so far vary from nation to nation. Some people have prioritized the health sector to lessen the strain it bears while simultaneously coping with the adverse effects their economies are experiencing. Trillions of dollars in economic stimulus have been made available worldwide as part of the extraordinary reaction to the COVID-19 epidemic, and there have never been more clamorous requests for a recovery that is in line with other global concerns. Beyond the epidemic, many see a unique chance to create a strong, low-carbon economic rebound.

The "circular economy" is a paradigm of production and utilization that highlights the sharing, reusing, renting, recycling, and repairing used goods for the longest time possible. Such measures extend the useful life of products. It pertains to the reduction of waste. Circular economy principles aim to separate economic activity from the depletion of finite resources. Its sturdy structure provides environmental, financial, and societal benefits. A framework for systems-level solutions to challenges like biodiversity loss, pollution, waste, and climate change, the circular economy is an example of such an approach.

The world, the definition of sustainability, and the practice of sustainable development have all changed due to the current COVID-19 situation. This enormous task must be viewed as a chance for researchers from all fields and disciplines to collaborate to address the problem posed by human interference with natural ecosystems. The COVID-19 problem and the repercussions of climate change have made it clear that our connection with nature and our economic system is not sustainable. If nothing drastically changes, humans will soon require another planet to call home. On the other side, this severe crisis has allowed us to rebuild a greener and more durable world by altering how we live, create, consume, and relate to nature.

Conventional definitions of the circular economy idea place a substantial weight on lowering the utilization of raw materials, goods, and components and recycling and reusing raw resources. One of the most apparent global problems, global warming, cannot be resolved via unilateral, bilateral, or regional solutions. The circular economy is one of the conceptual tools to meet this problem, and policies must be used globally. In reaction to the COVID-19 epidemic, countries have spent unlimited amounts on economic stimuli that are being made accessible worldwide. Recovery objectives are more important than ever, given other global issues.

Because of this, the phrase "build back better" has become widely used to refer to the chance to create a robust, low-carbon economic rebound.

To achieve this, societies must take decisive action that prioritizes defending national economies during crises to prepare the path for a more extensive economic transformation that is more robust to upcoming global dangers. The circular economy is a practical way to realize this goal and is more important than ever since it offers a chance for a future that is both inclusive and resource-wise. This is a chance for both areas to embark on a change that reflects the ambition required to work toward the SDGs and to adapt to a future with scarce resources.

1.8 Industrial Ecology in Developing Countries

Most materials used in modern society have been linear: they are dug up, processed, used, and then discarded. When newspapers and a few other materials started to be gathered for potential reuse in the 1980s, people began to feel apprehensive about this method. However, these acts were not part of a larger strategy, and they only slightly altered the philosophy used in the linear materials (Harper & Graedel, 2004). When Robert Frosch and Nicholas Gallopoulos of the General Motors Research Laboratories published an article for Scientific American titled "Strategies for Manufacturing" in 1989, conceptual thinking underwent a paradigm shift.

By investing time, energy, and resources in a resilient economy and utilizing the concepts and tools of IE, developing nations can guarantee that any enhancements they implement to the well-being of ecological systems do not compromise the quality of life for their citizens. According to Chiu and Yong (2004), polluting now and cleaning up later is an outdated philosophy that will only worsen a country's ecology and quality of life while increasing the cost of remediation work in the future.

Given the severe resource availability limits that developing nations face, IE is just as important in emerging countries as in industrialized ones. The main draw for international sponsorship of solar energy in underdeveloped nations is the potential to promote a clean power source and raise minimal living standards. Various sources of funding are available for developing countries. Solar-related initiatives are specially funded by several multilateral and bilateral aid organizations, which have benefited the industry. PV systems were installed in distant villages in nations like Indonesia and the Philippines as part of large projects costing tens of millions of dollars. An expanding trend is the use of microfinance, such as revolving money, to support sustainable regional businesses that can offer PV systems at a reasonable price.

1.9 Industrial Ecology in Developed Countries

The theoretical foundation of IE is the shift from a linear to a cyclical mode of material utilization in product manufacturing. An evolving approach for assessing industrial activities in relation to their environmental impacts is called industrial ecology (IE). Three primary pillars may be identified in industrial ecology. First, it examines the interactions between the parts and elements of both industrial and natural systems. Second, IE introduces fundamental ideas as practical approaches, including those of energy and material flows and industrial metabolism.

Sustainable development is unquestionably not a straightforward, well-established strategy. Different interest groups conflict with which social and environmental problems should be addressed first and how to solve them. According to earlier analyses of industrial ecology in developing nations, IE provides an overview framework, or a kind of panoramic view, within which specific local issues could be addressed with workable solutions. Solutions based on IE can simultaneously tackle several connected problems, resulting in advantages for all parties involved.

However, most interactions between human and natural capital lack directness; instead, they are mediated by manufactured capital, which takes the shape of massively constructed infrastructure, industrial production facilities, and communication technologies. Sustainability science has paid relatively little attention to this element up to this point. Manufactured capital is the complete physical stock of things humans make and procedures. Buildings, transportation, water, energy, waste infrastructure, industrial production facilities, and any long-lasting consumer and production commodities, such as machinery, vehicles, aircraft, and computers, are included. Industrial metabolism is defined as the socially coordinated continuous movement of material and energy from and to the environment required to reproduce manufactured capital.

Manufacturing companies positioned in clusters have a variety of chances to generate value thanks to industrial ecology. Aggregating businesses with comparable waste and by-product streams offers the potential to focus and minimize trash collection expenses in one place. Attaining a critical mass of waste via geographic concentration offers prospects for encouraging the co-location of enterprises engaged in waste recycling, subsequent sale to cluster members, or utilization of recycled water, heat, flare gas, or virgin materials. This might lower the costs of utilities like water and energy and material delivery for each company. The economies of scale for reprocessing are reduced in smaller and more scattered industrial operations due to quantity and transportation costs, necessitating the transportation of many waste products to other centers for processing or disposal. Thus, industrial ecology can provide opportunities to add value to industrial output through collaborations and commercial networks that naturally arise in clusters. In many nations, this opportunity is only now starting to be realized (Roberts, 2004).

1.10 Summary

An essential new planning instrument, industrial ecology promotes sustainable industrial development. Industrial Ecology entails implementing novel systems, practices, and management techniques that incentivize industries to recycle their waste materials and emissions instead of disposing of them in landfills, waterways, atmosphere, or other disposal facilities. The presented guidelines of UN sustainable development goals endorse global standards for organizing and promoting the growth of industrial ecology. Further knowledge and understanding are required in industrial ecology and its practical implementation, as the IE framework presents several obstacles that must be surmounted.

It is evident from the COVID-19 outbreak and the effects of climate change that neither our economic system nor our relationship with nature is sustainable. If nothing changes drastically, humanity will soon relocate to another planet. While certain developing nations have come to recognize the advantages of IE approaches and have implemented its principles to foster sustainable community development, others have yet to grasp the significance of this methodology. Developing countries gain from IE the formulation of policies and strategies that promote sustainable development.

References

Allenby, B. (2006). The ontologies of industrial ecology? *Progress in Industrial Ecology, an International Journal, 3*(1–2), 28–40.

Allenby, B. R., & Richards, D. J. (1994). *The greening of industrial ecosystems.* National Academy Press Washington.

Al-Thani, N. A., & Al-Ansari, T. (2021). Comparing the convergence and divergence within industrial ecology, circular economy, and the energy-water-food nexus based on resource management objectives. *Sustainable Production and Consumption, 27*, 1743–1761.

Awan, U. (2022). Industrial ecology in support of sustainable development goals. In *Responsible consumption and production* (pp. 370–380). Springer.

Bahers, J.-B., Tanguy, A., & Pincetl, S. (2020). Metabolic relationships between cities and hinterland: A political-industrial ecology of energy metabolism of Saint-Nazaire metropolitan and port area (France). *Ecological Economics, 167*, 106447.

Baumann, H., & Lindkvist, M. (2022). A sociomaterial conceptualization of flows in industrial ecology. *Journal of Industrial Ecology, 26*(2), 655–666.

Belaud, J.-P., Adoue, C., Vialle, C., Chorro, A., & Sablayrolles, C. (2019). A circular economy and industrial ecology toolbox for developing an eco-industrial park: Perspectives from French policy. *Clean Technologies and Environmental Policy, 21*(5), 967–985.

Besson, T. (2022). *Artificial closed ecosystems, life support systems & industrial ecology.*

Bleischwitz, R., Spataru, C., VanDeveer, S. D., Obersteiner, M., van der Voet, E., Johnson, C., Andrews-Speed, P., Boersma, T., Hoff, H., & Van Vuuren, D. P. (2018). Resource nexus perspectives towards the United Nations sustainable development goals. *Nature Sustainability, 1*(12), 737–743.

Bruel, A., Kronenberg, J., Troussier, N., & Guillaume, B. (2019). Linking industrial ecology and ecological economics: A theoretical and empirical foundation for the circular economy. *Journal of Industrial Ecology, 23*(1), 12–21.

Celik, B. (2013). Exploring sustainable development and its interpretation in the built environment. *Journal of Sustainable Development*, 6. https://doi.org/10.5539/jsd.v6n12p83

Cerceau, J., Mat, N., Junqua, G., Lin, L., Laforest, V., & Gonzalez, C. (2014). Implementing industrial ecology in port cities: International overview of case studies and cross-case analysis. *Journal of Cleaner Production, 74*, 1–16.

Chiu, A. S. F., & Yong, G. (2004). On the industrial ecology potential in Asian developing countries. *Journal of Cleaner Production, 12*(8–10), 1037–1045. https://doi.org/10.1016/j.jclepro.2004.02.013

Erkman, S. (2001). Industrial ecology: A new perspective on the future of the industrial system. *Swiss Medical Weekly, 131*(37–38), 531–538.

Frosch, R. A., & Gallopoulos, N. E. (1989). Strategies for manufacturing. *Scientific American, 261*(3), 144–153.

Graedel, T. E. (1996). On the concept of industrial ecology. *Annual Energy and the Environment Review, 21*(1), 69–98.

Han, F., Feng, Z., Wang, C., Yang, N., Yang, D., & Shi, F. (2021). Interweaving industrial ecology and ecological modernization: A comparative bibliometric analysis. *Sustainability, 13*(17), 9673.

Hariyani, D., & Mishra, S. (2021). Organizational enablers for sustainable manufacturing and industrial ecology. Cleaner Engineering and Technology, 100375.

Harper, E. M., & Graedel, T. E. (2004). Industrial ecology: A teenager's progress. *Technology in Society, 26*(2–3), 433–445. https://doi.org/10.1016/j.techsoc.2004.01.013

Henderson, R. (2021). Reimagining capitalism in a world on fire: Shortlisted for the FT & McKinsey Business Book of the Year Award 2020. Penguin UK.

Hicks, A. L. (2022). The role of community-based learning in teaching about industrial ecology and sustainability in the context of engineering education: A case study from the field. *Journal of Industrial Ecology, 26*(3), 1136–1146.

Hourneaux Jr, F., da Silva Gabriel, M. L., & Gallardo-Vázquez, D. A. (2018). Triple bottom line and sustainable performance measurement in industrial companies. Revista de Gestão.

ISIE. (2022). History—International Society for Industrial Ecology—ISIE.

Jelinski, L. W., Graedel, T. E., Laudise, R. A., McCall, D. W., & Patel, C. K. (1992). Industrial ecology: Concepts and approaches. *Proceedings of the National Academy of Sciences, 89*(3), 793–797.

Kendall, A., & Spang, E. S. (2020). The role of industrial ecology in food and agriculture's adaptation to climate change. *Journal of Industrial Ecology, 24*(2), 313–317.

Korhonen, J. (2000). *Industrial ecosystem: Using the material and energy flow model of an ecosystem in an industrial system* (Issue 5). University of Jyväskylä.

Lowenthal, M. D., & Kastenberg, W. E. (1998). Industrial ecology and energy systems: A first step. *Resources, Conservation and Recycling, 24*(1), 51–63.

Lybæk, R., Christensen, T. B., & Thomsen, T. P. (2021). Enhancing policies for deployment of Industrial symbiosis—What are the obstacles, drivers and future way forward? *Journal of Cleaner Production, 280*, 124351.

McDonough, W., & Braungart, M. (2010). *Cradle to cradle: Remaking the way we make things.* North Point Press.

McKinsey. (2023). Environmental, social, and governance (ESG).

Miller, K. (2020). *The triple bottom line: What it is & why it's important.* Harvard Business School Online.

Oughton, C., Kurup, B., Anda, M., & Ho, G. (2022). Industrial Symbiosis to circular economy: What does the literature reveal for a successful complex industrial area? *Circular Economy and Sustainability, 2*(4), 1317–1344.

Roberts, B. H. (2004). The application of industrial ecology principles and planning guidelines for the development of eco-industrial parks: An Australian case study. *Journal of Cleaner Production, 12*(8–10), 997–1010. https://doi.org/10.1016/j.jclepro.2004.02.037

Rweyendela, A. G. (2022). Getting closer to SDG12: Incorporating industrial ecology principles into project EIA. *Journal of Environmental Planning and Management, 65*(6), 953–974. https://doi.org/10.1080/09640568.2021.1974361

Saavedra, Y. M. B., Iritani, D. R., Pavan, A. L. R., & Ometto, A. R. (2018). Theoretical contribution of industrial ecology to circular economy. *Journal of Cleaner Production, 170*, 1514–1522.

Schmithüsen, F. J. (2013). *Three hundred years of applied sustainability in forestry.* Working Papers/Forest Policy and Forest Economics Department of Forest Sciences. International Series, 2013(1).

Singh, A., & Basak, P. (2018). Economic and environmental evaluation of municipal solid waste management system using industrial ecology approach: Evidence from India. *Journal of Cleaner Production, 195*, 10–20.

Spindler, E. A. (2013). The history of sustainability the origins and effects of a popular concept. In sustainability in tourism (pp. 9–31). Springer.

Sullivan, K., Thomas, S., & Rosano, M. (2018). Using industrial ecology and strategic management concepts to pursue the sustainable development goals. *Journal of Cleaner Production, 174*, 237–246. https://doi.org/10.1016/j.jclepro.2017.10.20

Sun, Y., & Wang, N. (2021). Development and correlations of the industrial ecology in China's Loess Plateau: A study based on the coupling coordination model and spatial network effect. *Ecological Indicators, 132*, 108332.

United Nations. (2023). THE 17 GOALS | Sustainable Development. https://sdgs.un.org/goals

Van Berkel, R., Willems, E., & Lafleur, M. (1997). The relationship between cleaner production and industrial ecology. *Journal of Industrial Ecology, 1*(1), 51–66.

Vujanović, A., Puhar, J., Čolnik, M., Plohl, O., Vidovič, T., Volmajer Valh, J., Škerget, M., & Čuček, L. (2022). Sustainable industrial ecology and environmental analysis: A case of melamine etherified resin fibres. Journal of Cleaner Production, 369, 133301. https://doi.org/10.1016/j.jclepro.2022.133301

Yu, X., & Zhang, Y. (2021). An economic mechanism of industrial ecology: Theory and evidence. *Structural Change and Economic Dynamics, 58*, 14–22.

Chapter 2
Current Practices in Industrial Ecology

Industrial ecology presents an innovative approach that holds promise for achieving sustainability in resource-based cities and industrial settings. By imitating the principles of natural ecosystems, such as optimizing material and energy flows, industrial ecology facilitates the transformation of industries towards symbiosis. It creates a circular economy with minimal waste. One critical element of industrial ecology is its potential to promote a shift to low-carbon industries, thereby contributing to efforts to mitigate the adverse environmental impacts of industrial production processes. Moreover, adopting principles such as "industrial metabolism" and "technological food webs" provides industries with tools to aid in transitioning toward more sustainable practices. Collaborative efforts between public and private actors are crucial for building a future sustainable society, particularly in green transport initiatives.

2.1 Introduction

The effectiveness of industrial sustainability policies is evident through indices like the Industrial Environmental Sustainability Index, underscoring the significance of cooperation in driving environmental progress. The incorporation of cleaner production methods along with closed processes into industrial operations shows substantial promise for achieving sustainable development goals. These approaches contribute to preserving human health while ensuring ecological health and demonstrate potential financial gains while reducing environmental degradation. Furthermore, embracing life-cycle assessment and net energy analysis within this context enables comparative studies on material usage, which can significantly impact pivotal decisions leading toward more environmentally friendly alternatives mentioned earlier. Overall, industrial ecology presents a transformative framework that integrates economic, social, and environmental considerations to achieve sustainable development. The principles of industrial ecology offer a comprehensive approach to achieving sustainability

A. Shah et al., *Industrial Ecology*, Industrial Ecology,
https://doi.org/10.1007/978-981-97-3619-5_2

by optimizing material and energy flows in industrial processes and transitioning to a circular economy. This transformation is crucial for resource-based cities, as it allows symbiosis between different industries, minimizing waste and fostering low-carbon practices.

Industrial ecology proposes a shift towards sustainable development by emphasizing the use of resources, preserving human and ecological health, and promoting environmental equity. Implementing cleaner production and closed processes within industries has shown promising results, reducing environmental degradation and contributing to financial gains. Industrial ecology acts as an agent of change by providing the necessary principles and tools to foster a more sustainable society. By incorporating concepts like "industrial metabolism," "closed industrial ecosystem," and "industrial symbiosis," industries can make significant strides toward sustainability. Furthermore, life-cycle assessment and net energy analysis enable comparative studies, guiding industries towards environmentally friendly alternatives. Industrial ecology is also crucial in developing green transport and environmental policy.

Recently, the pressing need to address environmental impact and the escalating demand for raw materials and natural resources have underscored the importance of industrial ecology. By embracing the principles and practices of industrial ecology, industries can transition from a linear trend of industrialization to a circular model that maximizes the use of products and manufacturing waste as industrial materials. By implementing industrial ecology, cities can achieve a more aesthetically and functionally appealing environment with minimal waste and align with sustainability principles. This shift in industrial practices reflects a broader global movement towards achieving sustainable planning policies and fostering a more sustainable future. As we delve further into industrial ecology, it becomes evident that its potential impact on the economy needs to be articulated more explicitly.

While strides have been made in linking natural and industrial systems, the translation of these connections to economic systems and their interaction with business studies and industrial investment decision-making remains under-explored. The economic implications and potential opportunities that industrial ecology presents need to be more thoroughly investigated to provide a comprehensive understanding of its wider impact. Moreover, the need for cooperation between public and private actors in building a sustainable society, particularly in the context of green transport, cannot be overstated. The successful implementation of environmental policy hinges on this collaboration, and the Industrial Environmental Sustainability Index stands as a testament to the importance of such partnerships in driving industrial sustainability policies across different regions and countries.

Incorporating industrial ecology, cleaner production methods, and closed processes into industrial operations fosters sustainability and presents substantial financial gains. Moreover, the interconnectedness of different industries through industrial symbiosis promotes efficient resource utilization, reduced environmental impact, and improved economic outcomes. This interconnectedness is pivotal in fostering sustainable development and preserving human and ecological health.

Furthermore, applying life-cycle assessment and net energy analysis as tools in industrial ecology facilitates comparative studies of material and energy conversion processes. It serves as a foundation for informed decision-making towards sustainable practices. These tools are invaluable in guiding industrial processes toward more sustainable and efficient models, thereby contributing significantly to the overarching goals of sustainable development. In conclusion, the potential of industrial ecology to drive sustainable transformation in industrial settings and cities cannot be underestimated. By further exploring its economic implications, fostering collaboration between public and private actors, and embracing the principles and tools it offers, industries can pave the way for a more sustainable and prosperous future. The increasing urgency to address environmental impact and the growing demand for natural resources has highlighted the significance of industrial ecology. Embracing its principles allows industries to transition from a linear model to a circular one, optimizing product use and reducing waste. Implementing industrial ecology leads to aesthetically pleasing cities with minimal waste and aligns with sustainability principles. This shift reflects a global movement towards sustainable planning policies.

The economic implications of industrial ecology need more exploration, including how it interacts with business studies and investment decision-making. Collaboration between the public and private sectors is crucial for building a sustainable society, especially in green transport initiatives. Successful environmental policy implementation relies on such partnerships, as seen through the Industrial Environmental Sustainability Index across different regions. Incorporating industrial ecology into operations results in financial gains while promoting sustainability. Interconnectedness among industries through concepts like industrial symbiosis encourages efficient resource utilization, reduced environmental impact, and improved economic outcomes—a vital aspect of sustainable development.

Tools like life-cycle assessment enable comparative studies of material conversion processes essential for informed decisions toward sustainability—contributing significantly to sustainable development goals. Industrial ecology has substantial potential to drive sustainable transformations within industries by exploring its economic impacts, fostering collaboration between public/private actors, and embracing its tools—an avenue leading towards an environmentally friendly future.

2.2 Closed-Loop in Industrial Ecology

Industrial ecology focuses on shifting from linear production models to circular systems, where waste is reused as a valuable resource for other operations. This transition reflects the patterns found in natural ecosystems and helps decrease waste and improve efficiency, aligning with sustainability goals. Industrial ecology seeks to establish self-contained industrial systems that mimic biological ecosystems by repurposing the waste outputs of one process as inputs for another. The process of re-engineering is expected to lead to substantial decreases in resource use and waste generation.

Industrial ecology prioritizes the shift from linear production models to circular systems, in which waste is reused as a valuable resource for other operations. This transition replicates the patterns found in natural ecosystems and aids in reducing waste and improving efficiency, aligning with sustainability goals. Industrial ecology seeks to establish self-contained industrial systems that mimic biological ecosystems by repurposing the waste outputs of one process as inputs for another. The re-engineering process is expected to substantially decrease resource consumption and waste generation while enhancing overall efficiency.

2.2.1 The Conceptual Framework of Closed-Loop Systems

In industrial ecology, the emulation of natural ecosystems is a foundational principle. This discipline seeks to model industrial processes and systems based on nature's regenerative and self-sustaining characteristics. The envisioned ideal is an interconnected, cyclical industrial framework where waste products from one process are repurposed as valuable inputs for another. This methodology fosters a continuous material flow within the industrial ecosystem, significantly diminishing resource consumption and waste production, thus harmonizing industrial activities with sustainability principles.

The role of closed-loop systems in sustainability is central to the sustainability paradigm within industrial ecology, which is the implementation of closed-loop systems. These systems are pivotal in reducing resource consumption and waste generation, catalyzing cost efficiencies, enhancing operational effectiveness, and yielding substantial environmental benefits. This approach necessitates a collaborative ethos among industrial entities, mirroring the complexity and interconnectivity observed in natural ecosystems while simultaneously promoting a culture of continuous innovation and improvement.

2.2.1.1 Technological Integration and Resource Optimization

The advancement of closed-loop systems is inextricably linked to adopting innovative technologies. This involves optimizing the use of resources per unit of product—particularly energy and water—and amplifying recycling efforts while mitigating toxic emissions. Transitioning from traditional open-loop systems, characterized by a linear resource-to-waste trajectory, to closed-loop systems where waste is reincorporated as an input represents a fundamental paradigm shift. Integrating industrial ecology principles into industrial practices signifies a substantive stride toward sustainability and environmental stewardship.

2.2.1.2 Holistic Approach and Ecological Interconnectivity

Industrial ecology recognizes the imperative of analyzing material and energy flows emanating from industrial activities and the need to foster a symbiotic relationship between industrial processes and ecological systems. This holistic approach is aligned with the broader science of sustainability, supporting the reduction of environmental impact and leading towards a more sustainable, interconnected industrial ecosystem.

2.2.1.3 Conclusion: The Path Towards a Sustainable Industrial Ecosystem

Adopting closed-loop systems within the industrial ecology framework is crucial for evolving a sustainable, interconnected industrial ecosystem. Industrial processes are reimagined to minimize resource consumption and waste generation by emulating the efficient resource recycling found in natural ecosystems. These systems' interconnected, cyclical nature ensures that waste outputs are transformed into valuable inputs, establishing a continuous material and energy loop.

As industries increasingly embrace the principles of industrial ecology, we anticipate a shift towards autonomous and regenerative industrial systems. This transformation is poised to yield significant sustainability and environmental responsibility advancements, marking a new era in industrial evolution. This chapter, in its essence, encapsulates the transformative potential of closed-loop systems in industrial ecology, delineating a path towards a more sustainable and environmentally conscious industrial future.

2.2.2 *Types of Closed-Loop Systems*

2.2.2.1 Industrial Symbiosis in Kalundborg

Industrial Symbiosis in Kalundborg, Denmark, is frequently referenced as a classic illustration of industrial symbiosis, an integral element of closed-loop systems. The interconnected industrial network in Kalundborg includes various facilities such as a power plant, an oil refinery, and a pharmaceutical plant. By-products from one industry are utilized as raw materials by another. For example, surplus heat from the power plant is used to warm local homes, and fly ash from the same facility is employed in cement production; this exemplifies the practical application of closed-loop principles. With the expanding global focus on sustainability and environmental stewardship, there is growing interest in adopting closed-loop systems within industrial ecology. This shift towards emulating the efficient recycling of resources found in natural ecosystems is essential for fostering a more connected and regenerative industrial ecosystem. A crucial aspect of closed-loop systems lies in their capacity to reduce resource consumption and waste generation while establishing a continuous flow of materials and energy within industrial processes. Moving away from linear production models toward interconnected cyclical systems allows waste outputs to be transformed into valuable inputs for other methods. This contributes significantly to enhancing overall efficiency and sustainability within the industrial ecosystem.

Additionally, adopting closed-loop systems results in environmental advantages and fosters innovation and cooperation across different industrial sectors. By embracing fresh technologies and reducing inputs per product unit, industries can notably improve their resource efficiency and reduce their environmental footprint. This collaborative effort aligns with the concept of industrial ecology, highlighting the interconnectedness and interdependence of industrial processes within the ecological system. Moreover, the comprehensive approach of industrial ecology emphasizes preserving ecological equilibrium through closed-loop processes, thereby supporting sustainability as a primary objective. Recognizing the flow of materials and energy from industrial activities and integrating these principles into operational practices enables industries to make significant progress toward environmental stewardship. As the industrial landscape continues to develop, incorporating closed-loop systems alongside principles of industrial ecology will undeniably play a crucial role in transforming industry procedures for a more sustainable future. Through resource efficiency initiatives, waste reduction strategies, and sustainable design approaches, industries can advance towards establishing a circular and regenerative industrial ecosystem.

2.2.2.2 Apple's Focus on Closed-Loop Supply Chain

Apple Inc. has been committed to establishing a closed-loop supply chain that utilizes renewable resources or recycled materials in product construction. The company's advanced material recovery lab, Daisy, can dismantle iPhones to retrieve valuable materials like aluminum and cobalt, which are reused in new products or responsibly recycled for other purposes (Apple, 2020). This strategy reduces the need for new materials and supports environmental conservation by promoting resource efficiency and waste reduction. With the growing global emphasis on sustainability and environmental stewardship, there is increasing interest in implementing closed-loop systems within industrial ecology. This transition toward replicating the effective resource recycling seen in natural ecosystems is essential for creating a more interconnected and regenerative industrial ecosystem.

Additionally, the adoption of closed-loop systems not only leads to environmental advantages but also fosters innovation and cooperation across different industrial sectors. By embracing new technologies and reducing inputs per product unit, industries can significantly improve their resource efficiency and lessen their environmental footprint. This collaborative effort aligns with the concept of industrial ecology, highlighting the interconnectedness and interdependence of industrial processes with the ecological system. Moreover, the comprehensive approach of industrial ecology emphasizes preserving environmental balance through the closed-loop process, thereby supporting sustainability as a primary goal. By recognizing the flow of materials and energy from industrial activities and integrating these principles into practices, industries can make significant progress toward environmental stewardship. As the industry evolves, incorporating closed-loop systems and principles from industrial ecology will play a crucial role in transforming industrial processes for a more sustainable future. Industries can move closer to establishing a circular and regenerative industrial ecosystem through efforts focused on resource efficiency, waste reduction, and sustainable design choices.

2.3 Methods and Tools of Industrial Ecology

Industrial ecology encompasses various tools, methodologies, and approaches for analyzing and optimizing industrial systems while considering their environmental impact.

2.3.1 Material Flow Analysis (MFA)

The methodology known as Material Flow Analysis (MFA) holds significant importance in industrial ecology. It systematically assesses material flows and stocks within a specific system, considering spatial and temporal factors. The utilization of this tool is crucial in various domains, including environmental management, urban planning, and resource stewardship, as it enables a comprehensive comprehension of material interactions and their potential ecological consequences.

The versatility of MFA is shown in its capacity to be applied to a wide range of systems, encompassing individual processes, industry sectors, and products, as well as more prominent entities such as cities, regions, and even the world stage. MFA examines various materials, such as elements, compounds, generic materials, and specific products, to support multiple industrial ecological investigations. Furthermore, MFA can be classified into two main categories: static analyses, which capture a particular moment, and dynamic analyses, which track material changes over time. These categories provide researchers with the flexibility to handle different study requirements.

2.3.1.1 The Fundamental Procedures in Material Flow Analysis

System Definition: The first phase of this process involves clearly defining the system's limits and determining the research's temporal span. This establishes the basis for a cohesive and targeted investigation of MFA.

Identifying Relevant Materials: Identifying materials relevant to the investigation is a critical stage that guarantees a focused and pertinent analysis following the study's objectives.

Measuring the Quantity of Flows and Stocks: Currently, the primary emphasis is quantifying materials being introduced, removed, and accumulated inside the system. This process may entail using direct measurement or estimation methodologies, which serve as crucial data sources for further analysis.

Examining the Findings: The outcome of Material Flow Analysis (MFA) involves analyzing data to derive significant insights into material movement, consequences, and potential environmental effects. This process aids in making well-informed decisions within the field of industrial ecology.

2.3.1.2 Practical Applications of MFA

MFA in Waste Management and Recycling Policy: The application of Multi-Factor Analysis (MFA) is particularly relevant in waste management and recycling, as it facilitates the comprehension and standardization of waste statistics and recycling processes. Although standardizing data and quantifying flow can be difficult, MFA provides vital insights that aid in the development and execution of successful policies in waste management.

MFA in Material Efficiency and Scenario Analysis: Another notable application is the utilization of MFA to enhance material efficiency and conduct what-if scenario analyses. MFA empowers stakeholders to make informed decisions by meticulously studying material flows, paving the way for optimized resource utilization and reduced environmental footprint.

2.3.1.3 Conclusion

The Material Flow Analysis as a fundamental tool in industrial ecology. It offers a systematic approach to understanding intricate material interactions within different systems. This technology's application spans various fields, highlighting its importance in promoting sustainable management practices and supporting industrial ecology's objectives.

2.3.2 Life Cycle Assessment (LCA)

Life Cycle Assessment (LCA) is a crucial tool in industrial ecology that offers a thorough assessment of the environmental effects across the whole lifecycle of a product, starting with the extraction of raw materials (cradle) to its disposal or recycling (grave). LCA provides a comprehensive assessment of the environmental effects of a product over its entire life cycle, avoiding a narrow focus on individual stages and promoting a more sustainable approach to product design, production, and disposal.

2.3.2.1 Phases of Life Cycle Assessment

Goal and Scope Definition: The initial stage of this process is clearly describing the product or service being examined, developing a functional framework for comparison, and specifying the scope of the study. This phase establishes the Life Cycle Assessment (LCA) trajectory, guaranteeing precision and pertinence in the evaluation's goals and limitations.

Inventory Analysis: The assessment quantifies energy consumption, raw material usage, and environmental emissions at every stage of the product lifetime. Subsequently, the gathered data is correlated with a process flow chart, directly connecting to the specified functional foundation and furnishing a comprehensive record of inputs and outputs.

Impact Assessment: In this step, the collected resource utilization and emissions data are classified and measured into separate impact categories. The categories mentioned above can be assigned additional weights to calculate a unified aggregate score, which concisely represents the environmental effects across various dimensions.

Interpretation: The concluding stage integrates the inventory analysis and impact assessment results, formulating valuable judgments and recommendations. The interpretation step is significant in converting data into practical information, facilitating stakeholders' decision-making, and implementing targeted environmental enhancements.

2.3.2.2 Practical Applications of LCA

LCA in Achieving Net Zero and Sustainable Practices: In today's industrial environment, Life Cycle Assessment (LCA) plays a crucial role for organizations aiming to achieve net zero emissions and implement sustainable operational procedures. By incorporating Life Cycle Assessment (LCA) into their strategic planning, firms acquire a comprehensive framework for assessing and improving their environmental performance. This framework is essential for generating well-informed sustainability reports and formulating effective strategies.

LCA and Circular Economy: Life Cycle Assessment (LCA) is crucial in providing in-depth insights into the environmental aspects of circular economy plans. Through evaluating the lifecycle effects of products and systems, Life Cycle Assessment (LCA) offers significant insights that facilitate the shift from linear to circular models, focusing on enhancing resource efficiency and minimizing waste.

EPA's Greenhouse Gas Equivalencies Calculator: An example of implementing Life Cycle Assessment (LCA) is the Environmental Protection Agency's (EPA) Greenhouse Gas Equivalencies Calculator. This tool enables organizations to measure and compare their products' greenhouse gas emissions, promoting more environmentally conscientious decision-making.

2.3.2.3 Conclusion

Life Cycle Assessment (LCA) is a fundamental component of industrial ecology, providing a systematic and all-encompassing framework for comprehending and addressing the ecological consequences associated with products and activities. LCA, with its systematic stages and adaptable uses, enables stakeholders to effectively negotiate the intricacies of environmental sustainability, hence facilitating advancements towards industrial practices that are more resilient and environmentally friendly.

2.3.3 *Industrial Symbiosis*

Industrial symbiosis, a fundamental idea in industrial ecology, embodies the basic principles of a circular economy by converting the waste or byproducts generated by one industry into the primary resources for another. Industrial symbiosis exemplifies the combination of environmental care with economic and social prosperity, mirroring the cyclical and waste-free nature of natural ecosystems.

2.3.3.1 Mechanisms of Industrial Symbiosis

Industrial symbiosis revolves around the strategic cooperation between industries to effectively utilize each other's leftover outputs or energy flows. This collaboration, which can occur between nearby enterprises or over larger geographical areas, promotes a network of sharing resources amongst industries, resulting in mutual economic, social, and environmental advantages.

2.3.3.2 Implementing Industrial Symbiosis

Identifying Collaborative Partnerships: The first step in developing an industrial symbiosis network is identifying industries with complementary waste and resource needs. This will help identify prospective opportunities for symbiotic relationships.

Analyzing Symbiotic Potential: After identifying a partner, a comprehensive analysis is undertaken to assess the technical feasibility and economic viability of repurposing waste streams. This analysis aims to ensure that the symbiotic relationship is sustainable and advantageous for all stakeholders.

Establishing Symbiotic Exchanges: Once a mutually beneficial connection has been established, the subsequent step is to create and execute the necessary infrastructure and procedures to share and utilize resources effectively.

Monitoring and Optimization: Continuous monitoring is an essential aspect of evaluating the effectiveness of symbiotic exchanges. At the same time, constant optimization is necessary to ensure the long-term viability and enhancement of the industrial symbiosis network.

2.3.3.3 Examples of Industrial Symbiosis

Kalundborg Industrial District, Denmark: A quintessential example of industrial symbiosis in action is Denmark's Kalundborg Industrial District, where a network of material, thermal, and energy exchanges between industries has evolved since the 1960s, demonstrating the long-term viability and benefits of such collaborations.

Eco-industrial Parks: The concept extends to creating eco-industrial parks, where firms are located together to promote networks that share resources, minimize waste and emissions, and enhance the community's well-being and the environment.

2.3.3.4 Conclusion

Industrial symbiosis is significant in industrial ecology since it presents a practical strategy for attaining sustainability by replicating closed-loop processes in natural environments. Through this approach, industries can surpass conventional linear models and transition towards a more integrated, efficient, and sustainable framework of production and consumption. The preceding chapter has comprehensively explained the core principles, research approaches, and notable instances of industrial symbiosis, emphasizing its crucial significance in the shift toward a circular economy.

2.3.4 Other Approaches

Other Approaches, such as socioeconomic metabolism for national economies, urban metabolism for cities, and the analysis of critical socioeconomic factors such as consumer behavior, business models, and public policy.

Socioeconomic Metabolism: Socioeconomic metabolism pertains to energy and material exchange dynamics within a given society or economy. This approach facilitates comprehension of resource utilization and waste generation within communities and the impact of social and economic factors on these phenomena. A recent example is a study that systematizes major socio-metabolic research traditions that study the biophysical basis of economic activity, such as urban metabolism, the multiscale integrated assessment of societal and ecosystem metabolism, biophysical economics, material and energy flow analysis, and environmentally extended input–output analysis.

Urban Metabolism: This concept examines the movement of energy and materials within urban areas. Understanding resource consumption patterns, waste generation, and potential strategies for enhancing sustainability in urban areas is beneficial. For instance, a recent study analyzed the urban metabolism in China's megacities, tracking the lifecycle of material stock across over a hundred distinct end-use types and 12 categories of materials.

Consumer Behavior Analysis: This field of research examines consumers' decision-making processes regarding the acquisition, utilization, and disposal of items or services, as well as how these behaviors manifest their beliefs, attitudes, and social influences. A recent example is a comprehensive guide that explores the intricacies of customer behavior analysis, delves into real-world customer behavior analysis examples, and equips readers with insights on consumer behavior analytics.

Business Models: Business models refer to firms' strategic approaches to generate and acquire value. They establish a company's operational framework, revenue generation strategies, and customer, supplier, and partner interactions. A recent example is a Forbes article that discusses the digital era's seven most successful business models, including the ad-supported, freemium, and subscription models.

Public Policy: Public policy encompasses the deliberate acts and choices undertaken by governmental entities to attain specified societal objectives. It covers a diverse array of concerns, including, but not limited to, healthcare, education, housing, and environmental preservation. A recent example is the list of Malaysian Government Policies, which includes policies on foreign affairs, national education, national transport, shared prosperity vision, security, public order, and many others.

2.3.5 Conclusion

These concepts and approaches are all interconnected and play crucial roles in shaping our societies and economies. They help us understand the complex dynamics of how we use resources, how we interact with our environment, and how we can make these processes more sustainable and equitable.

2.4 Industrial Ecology—New Perspective

The emerging area of industrial ecology presents a novel viewpoint that questions and redefines conventional industrial methods by promoting a mutually beneficial connection between industry and the environment, reflecting the recurring and interconnected characteristics of ecological systems. This novel methodology highlights the transition from the traditional linear economy, characterized by a linear model of 'take, make, dispose of,' to a circular economy. In the circular economy, waste is effectively eradicated, resources are regenerated, and the entire lifecycle of products is optimized to minimize environmental impact.

The fundamental idea of industrial ecology revolves around converting waste materials into valuable resources, a concept exemplified by industrial symbiosis. An illustrative instance can be observed in the municipality of Kalundborg, located in Denmark. Within this context, a complex network of interactions among diverse organizations, such as a power station, an oil refinery, and a pharmaceutical firm, is a tangible manifestation of the pragmatic implementation and advantages associated with industrial ecology. The power plant's waste heat is utilized to heat residential properties and a nearby fish farm. Additionally, fly ash and sludge are transformed into primary resources for construction and agricultural purposes, illustrating the implementation of a closed-loop system.

In addition to waste reuse, industrial ecology promotes the development of innovative methods to mitigate the environmental impact associated with manufacturing activities. This encompasses the progression of environmentally friendly technologies, utilizing sustainable energy sources, and adopting methodologies like life cycle assessment (LCA) to examine and mitigate the ecological consequences of goods across their entire life cycle.

The adoption of industrial ecological principles carries significant economic repercussions. By using these techniques, companies can attain substantial cost reductions, decrease their dependence on unpredictable raw material markets, and improve their brand image among consumers who are becoming more environmentally aware. Moreover, this transition facilitates the emergence of novel business models and employment prospects within industries prioritizing recycling, remanufacturing, and sustainable product design.

However, to fully harness industrial ecology's capabilities, it is imperative to go beyond mere technological advancements. This demands a fundamental transformation in the mindset of industries and consumers, reevaluating the criteria for determining value and achievement in business and establishing supportive policy frameworks that encourage sustainable practices and promote collaboration among different sectors.

Essentially, industrial ecology entails a fresh viewpoint and a fundamental reconceptualization of how industrial processes might function sustainably within the ecological boundaries of the Earth. This strategy not only addresses the negative impacts of industrial activity but also presents a means to establish a robust and prosperous economic structure that aligns with the principles of environmental sustainability.

Reference

Apple Inc. (2020). *Environmental progress report*. Retrieved from https://www.apple.com/environment/pdf/Apple_Environmental_Progress_Report_2020.pdf

Chapter 3
Industrial Ecology Implementation in Manufacturing and Service Sector: A Context of Developing Country

This chapter studies two cases conducted based on thorough research on industrial ecology. In the following sections, an As-is and To-be scenario is created by taking the textile industry and, more specifically, a denim garment company as a base for this research to propose global best practices to be implemented. For a better understanding, a study of the service sector is included in this chapter, which compares Pakistan's banking sector with the international sustainable practice. The second case envelops the manufacturing sector, which entails the Pakistan apparel industry. The chapter delves into initiatives the local industries took to support green supply chains, especially post-COVID-19, where the supply chain has taken a new route and changes. A denim garment factory and the banking sector of Pakistan have been taken as samples to apprehend the implementation of industrial ecology in developing countries. The selected samples are studied to check the implementation of industrial ecology: two retail customers in the USA for the manufacturing sector and the international bank sector of Hong Kong for the services sector based on their sustainable objectives for writing the chapter exploratory methodology based on the available data through online sources and interview of the industry professional. The result suggests that the COVID-19 impact halts sustainability and industrial ecology progress. Lastly, the chapter suggests recommendations that would help developing countries in the manufacturing sector improve their sustainability and industrial ecology targets progress.

3.1 Introduction and Literature Review

As per the literature review, it is evident that Germans were the first to use sustainability for the forest industry in the eighteenth century. In that era, it was just for the continuity of their business related to woods and forests. But now, the need to work on sustainability has substantially increased due to climate changes, global warming,

A. Shah et al., *Industrial Ecology*, Industrial Ecology,
https://doi.org/10.1007/978-981-97-3619-5_3

rising air, water, and soil pollution, and disturbance in the overall ecosystem. In the 1950s, there was a concept named "end of pipe," which was to clean or treat your waste at the end of the process. It is a good step but not sufficient for the current global issues due to the existence of industries in the ecosystem. So, industrial ecology is the need of the time to minimize the impact of our actions on our surrounding environment and to control global issues. In developed countries, laws are stringent and implemented correctly compared to those in developed countries. There is a need to compare the sustainability work done in developed and underdeveloped countries like Pakistan. For analysis, a case study of the manufacturing and service sectors is used for benchmarking. A comparison is required between developed and underdeveloped countries so that companies can better understand the current trends in the manufacturing and production processes and buildings. Based on the outcome, suggestions can be given to the company to improve its processes. Companies should have a tool to measure their sustainability and industrial ecology performance. This will provide an easy comparison and be used to publish their sustainability reports, which will be helpful in the business growth and awareness of different stakeholders. Currently, companies in underdeveloped countries face serious financial challenges in implementing policies for industrial ecology as implementation increases the cost of doing business. However, government commitment is also required to implement sustainability effectively. The research objective wraps around the UN SDG framework to fill the above gaps. The research identifies the difference in progression between developed and developing countries. Further, it also elaborates on how the COVID-19 pandemic impacts industrial ecology and green practices.

3.2 Research Methodology Approach

This study follows a qualitative approach, and the research philosophy for qualitative research will be interpretivism. It entails researchers interpreting the main components of the survey and combining human interest into the study. "Interpretive researchers, therefore, assume that access to reality (whether inherent or socially constructed) is restricted to instruments, shared meanings, language, and consciousness." The development of interpretive philosophy was precipitated by the critique of positivism in the social sciences. This philosophy is about the study based on qualitative analysis, which examines and focuses on the critical issue for better generalization. This establishes the differentiation between the two distinct methodologies. The inductive approach is a departure from the general to achieve specific goals (Bryman, 2012). As a result, the research focus can be established after gathering information. While this may appear to be the juncture where novel theories are formulated, it is also possible that upon further analysis, the data may be discovered to be consistent with an existing theory.

The research is based on industrial ecology and its importance in recent times. It requires an in-depth study of the manufacturing industry of Pakistan and its counterpart in the developed nation regarding their commitment, progress, and challenges in the sustainability framework and industrial ecology. This study will focus on the problem in detail to solve it better. This research strategy focuses on one or more cases of any area of interest, which is examined systematically to develop extending a theoretical framework. An exploratory design focuses on exploring one case in considerable depth. The exploratory approach applies to both qualitative and quantitative analysis. The present study employs a qualitative analysis approach using a single exploratory design. This research endeavor will assess the local practices implemented by a prominent garment manufacturing company in Pakistan. Its objective is to generate an AS-IS and TO-BE scenario concerning global practices. Upon examining several tangible instances of supply copresences in the real world, it is possible to deduce that the case-study design methodology could prove advantageous in scrutinizing diverse facets of the procedure. However, although numerous scholars have criticized exploratory research for lacking rigor, it can be a valuable method for conducting thorough analyses, provided it is undertaken systematically and thoroughly documented. A solitary design exhibits numerous prospective benefits. One of the primary benefits is a more profound comprehension of a specific viewpoint regarding the causes and mechanisms of events.

The time horizon of this study is cross-sectional as the data will be collected from the industry at one point in time from different locations and not in various periods from the exact location; that is called longitudinal. This is done to compare the other places or areas at the same point and compare them correctly. The data collection method for this research is interviews with the focal person of the company selected and a site visit to understand the company's current standing and challenges entirely so that the findings of this study can suggest the solutions available to their challenges in the research by comparing it with the developed countries' industry.

3.3 Pakistan Garment Manufacturer

The garment manufacturing company under study is a family-owned business with a rich history of evolution and growth since 1949, from a small retail store to 16 manufacturing units in Karachi, Pakistan. The company is famous worldwide for its innovation and sustainability in the jeans manufacturing and design process. The organization can manufacture 30 million garments, 108 million meters of fabric, and 88 million pounds of yarn annually.

The garment manufacturing company is the denim industry powerhouse in Pakistan and is working on PPP, a triple bottom line. It is working on innovation in design and manufacturing process, automation to effectively utilize the resources, save the electricity and water consumption for sustainable development, and to have quality consistency, people; it is the first fair trade certified company, they are equal opportunity employer and provide transportation to the employees and lunch as well to maintain the working environment, their building are USGBC (United States Green Building Council) LEED certified. LEED certification is an internationally renowned emblem of leadership and sustainability accomplishment. This includes climate change, human health, biodiversity, water resources, green economy, community, and natural sources usage and impacts. They are working deep-heartedly, from sustainable design to the sustainable manufacturing process. They are working on cradle-to-cradle (C2C) instead of cradle-to-grave vision.

The company is working on renewable energy sources as part of its industrial ecology policy. It has a 100 MW wind energy plant and an 8.5 MW Solar system in the installation and commissioning phase in collaboration with an internal energy company of Korean origin. Further, it is also working on Hydro plants to reduce the carbon footprints, and per unit energy consumption in production is being monitored to track this down. This is being reduced through design innovation and efficient automated plants. The company is working on Industry 4.0 technologies, and this automation has reduced the process's water and energy consumption requirements. This has also resulted in zero defects and consistent quality. This reduces the water and energy consumption by 70%.

The garment manufacturing company ethically sources all the materials, including organic cotton and other fibers like polyester and viscose. It produces 9 million meters of finished product per month and has extensive dye ranges in Asia with 12 dips. It has a production capacity of 30 million garments per year.

The company has a virtual showroom so people can understand its design philosophy. It has an experience center in Dubai and works as a hub for design rethinking. It also has Soho Incubator, NYC, which is also for innovation and co-creation with the help of partners, like customers and brands. It also has the SFI-Los Angeles, a design studio and denim laundry. The garment manufacturing company's vision is "championing the denim revolution of the future to transform lives and change the world for the better." Further, to achieve the vision, the company's mission is "to sustain our reputation as one of the top manufacturers in the world by striving for excellence in each function of our business."

Per its vision and mission statements, the company has high aims, wants to be world-class, compete with the companies of developed countries, and wants to change the whole world for the better.

Its core values include the following:

- Authenticity
- Inclusivity
- Force for good
- Planet Positive

These values are the source of their innovation and brand name, which means they work on continuous improvement; their innovation comes through inclusivity, that is, they involve their customer and partners in their innovation side to improve the design and process further to reduce their impact on the environment.

3.3.1 Sustainable Synergies: Forging Green Collaborations in the Garment Industry

The garment manufacturing company studied has set the highest standards for them, and they want to surpass or implement all the standards of local and international nature. It does not just want to work on fulfilling basic compliance requirements but exceed them. To scale up its sustainability work, it is collaborating with world-class organizations already working on sustainability and following all the guidelines and rules, which are way more stringent than those in underdeveloped countries like Pakistan. They do not just want to make economic profits but to be responsible organizations worldwide. Some green collaborations of the company are mentioned below in the form of the organization or institute logo in Fig. 3.1. In 2020, it was the only company that promised to follow the UN's business goal of 1.5 C. It was the world's first denim company certified by Cradle 2 Cradle.

The company's Cotton Initiative is a unique project, a fashion-sustainable cotton program. WWF implemented this on the ground and is located in a city in Pakistan, Rahim Yar Khan. This covers the complete cotton supply chain as an ecosystem and is verified by the Control Union Pakistan.

The garment manufacturing company knows that we are in an ecosystem, and each of our steps has a consequence or potential when we take on the society we live in and the environment outside us. It is aware of climate change due to human activities. Balance in these activities is vital for sustainability. The global average temperature is increasing at a faster pace than in the past, and this is due to the industrial pollution/emissions being generated. It is working continuously to clean, save, and protect the environment. Since the company started, it has collaborated closely with all the stakeholders. It gives importance to all the stakeholders, including

Fig. 3.1 Green collaborations of the Garment Manufacturing Company (privacy concerns prevented the citation of this figure)

employees, suppliers, community, and customers, so they can continuously improve all aspects, whether quality or design.

3.3.2 *Renewable Energy Firm for Production*

The garment manufacturing company has an energy firm, a wholly-owned subsidiary, which produces the cleanest and most sustainable energy through wind turbines, solar plants, and hydropower plants. The wind plant produces 100 MW and has zero occupational incidents. This lightens up the 50,000 households through national grid integration. Two solar projects of 25 MW and 50 MW are in the final phase in Sukkur (Pakistan). Two hydropower plants are being processed in the province of Kyber Pukhtoonkhua, and they have a cumulative capacity of 100 MW.

3.3.3 *USA Market*

Pakistan's biggest denim customer is mainly the US, one of the leaders in implementing sustainable development goals. In Developed countries, the US is one of the largest textile raw material producers and importers of textile and apparel consumer products. As part of the exploratory study, we have taken retail brands' UN SDGs targets as benchmarks for Pakistani garment manufacturers to see progress in developing countries for industrial ecology and sustainable goals. Abercrombie and Fitch (A&F) is a US retail brand imported from the manufacturer. The A&F has chalked out its sustainable goals based on UN SDGS (A&F, 2022a, b). A&F has selected their sustainable goals till 2025 out of 17 sustainable goals. A&F is aligned with 10 SDGs out of 17, and they have tremendously worked on this. The goals chalked by A&F are illustrated in Table 3.1. The retail giant targets for 2025 are given below (Table 3.1). In August 2020, A&F Company updated the status of all three goal areas, detailing accomplishments, tasks in progress toward completion, and those that were underway. Each of A&F Company's 2025 sustainability targets, which affect one or more SDGs, was disclosed in August 2019 (Table 3.2).

Rainwater harvesting by the suppliers is done to reduce fresh groundwater consumption and wastewater volumes. Some factories work on zero liquid waste by recycling, treating, and reusing water. VF is working on circulatory to reduce waste and material consumption for the sustainability of the business. Products will be designed in a way that they will be "Made to last." That these will be recycled so many times.

Table 3.1 A&F targets for 2025 (A&F, 2022a, b)

Product goals	Global home office goals	Global store goals
• 100% responsible down standard • 30% water reduction in denim production	• Engage/Educate associates • Recycle 100% of cardboard, E-waste, denim and fabric scraps • Reduce print By 25% compared to the 2017 baseline	• Reduce print By 105 globally year over year
• Train 25,000 more workers on human trafficking • 25% better cotton	• Train 100% relevant associates on human trafficking • 50% waste reduction	• Recycle 100% of hazardous waste in all domestic stores • Recycle 100% of polybags in participating stores
• Expand capacity building programs to 50,000 more workers • 100% responsible wool standard wool • 50% canopy supported innovative viscose • 25% recycled polyester		

3.3.4 *Interviewing the Focal Person*

This study gathered data through interview sessions with selected organization focal persons for the sustainability projects and compliance department, and it will be presented in its summary. As per the interview, the studied company works on almost all the UN SDGs related to their industry and is working on 13 SDGs out of 17. The main drivers of working on sustainability are customer requirements and management commitment to the planet, earth, and society. Around 25,000 families are being fed through their factories. Every sustainability initiative revolves around these goals. For example, the Cotton Initiative covers seven SDGs: no poverty, reduce inequalities, Gender inequality, Clean Water, Industry innovation and infrastructure, responsible consumption and production, and climate action. They are aggressively working on these issues throughout the organization. They are an equal opportunity employer and have raised the number of female workers where possible per the job role. They are 30% female and 70% male workers; they conduct regular sessions with female workers to motivate them for education and growth. They educate workers on managing their households and families in these difficult times. The company also talks about their physical and mental health as well. The company provides them with free pick-and-drop service and on-site meals to achieve their gender equality targets.

Their primary customers are the US and Europe region, approximately 70% and 30% in share, respectively. So, the researchers are comparing them with the US industry to comply with customer requirements and increase their sales. The

Table 3.2 Sustainability targets of A&F 2025 Align with UNSDGs (A&F, 2022a, b)

VF goals	Made for change goals and targets	Status/Progress through FY2020
	People	
	Gender parity in leadership positions by 2025	38% female leaders, a 3% improvement since 2018
	Attain pay equity for associates, sponsored athletes, and influencers by 2024	New goal
	Integrate training at all levels that foster inclusion and address unconscious bias by 2020	375+ directors and above trained
	Slate 50% diverse candidates (women, BIPOC, individuals, veterans, and individuals with disabilities) when hiring or promoting associates	New goal
	Aspire to achieve 25% representation of black, indigenous, and people of color (BIPOC) representation within our director and above population by 2030 in the US	New goal
	VF associates will contribute 1 million hours to local communities by 2025	New goal
Supply chain goals	Double supplier diversity spending by 2025	36,000 cumulative hours tracked
	Focused supplier factories are required to establish a mechanism by 2025 for the prevention and reporting of gender-based violence	New goal
	Enhanced well-being for one million laborers and their communities by 2025 and two million individuals by 2030	290,315 reached, representing 130,000 more people since 2018
	By 2025, every factory within the supplier scope will be improving and growing industry-leading health and safety initiatives	New goal
	By 2026, none of the VF supply chain's employees will be paid	New goal
	By 2030, the representation of supply chain personnel within the designated scope should be increased by establishing gender-proportional workplace committees	New goal
Science-based targets	*Planet*	
	Reduce absolute Scope 1 and 2 greenhouse gas (GHG) emissions by 55% by 2023 (2017 baseline year)	67,826 MT CO^2e, a 17% reduction
	The target is to decrease absolute Scope 3 greenhouse gas emissions from upstream transportation and purchased goods and services by 305 percent by 2030 (base year set at 2017)	3,700,000 MT CO^2e, no change

(continued)

Table 3.2 (continued)

	By 2020, we aim to attain a state of zero waste at every owned and operated distribution center (DC)	72% of our DCs are zero-waste, a 17% improvement since 2018
	By 2025, 100% renewable energy must be utilized in all our owned and operated facilities	29% renewable energy, a 17% improvement since FY 2017
	By 2030, all single-use plastics and non-essentials shall be eliminated from VF's direct operations and sponsored events	New goal
	Product	
	Produce traceability maps for one hundred of the most recognizable products of our brands by the end of 2021	44 maps completed since CY2018
	By 2021, tanneries audited by the leather working group will process all footwear leather	100% goal achieved
	By 2025, 50% of polyester will be derived from recycled materials	24% recycled polyester, a 12% reduction from FY2018
	By 2025, all cotton acquired by VF will be sourced from the United States Australia, or managed under a third-party sustainability scheme for cotton cultivation	75% grown in the US, Australia, or under a third-party sustainability scheme
	Eliminate and restrict 100% of unwanted chemicals or substances using the innovative CHEM-IQ^{SM} program from VF's supply chain by 2025	New goal
	Eliminates single-use plastic packaging by 2025	New goal
	100% trace five critical materials utilized by VF throughout our supply chain by 2027	New goal

company's business model is B2B, business to business; they work for the world's top denim brands like H&M and Levi's.

Their current energy mix includes K-Electric, RLNG, Gas, Diesel, and solar power. They are targeting a net zero carbon footprint and zero discharge. The deadline for the zero discharge is by 2030. They have installed effluent treatment plants in their facilities, and they recycle the water instead of draining it into the waste lines. Their per day recycling is around 1.5 million gallons of water. The company is working on solar, wind, and hydropower projects to mitigate the effect of their business's carbon footprint.

Pakistan is the major player in world denim production, and the company is the leader in the market.

From the government, they need support in the following aspects,

- Control the cost of electricity to be competitive globally; otherwise, this share can be lost to India or Bangladesh.
- Control the daily wages by controlling the inflation.
- Farmers need to be aware of cotton production as cotton is one of the bottlenecks in the textile industry. As per the installed capacity of manufacturing units, the cotton requirement is 18 million bales per year. Last year, production was just 7–8 million bales. The remaining cotton was imported at a higher rate than the local cotton. So, the government can support by subsidizing the farmers for cotton production to motivate them to increase the export bill and boom this textile industry. The price of imported cotton is usually 10–15% higher than local cotton.
- The government can also go for the textile park with the related industries side by side to reduce the logistic cost and improve collaboration in the industry. This park should have ETPs for the small sector as the initial cost of ETP is not feasible for the smaller units.

They also work on materials' traceability through blockchain, a new technology. This tag will share data on all the processes, utilities, and workers involved in its supply chain. Per the management's commitment to sustainability, all the projects are on track and will meet the deadline. The currently available data on American and garment manufacturing companies are available through online sources, their published sustainability reports, and interviews with the focal person from the company.

All of them are working on the UN SDGs, and they have set the timelines for them along with the action plans. Their goals are common as they all work on less water and energy consumption, reducing polybags that are not readily decomposable and polluting the environment. As per the current scenario, there are stringent laws compared to those in Pakistan, which is one reason for shifting the manufacturing business from the US to Asian regions. However, the company works on world-class standards, has UN bodies certification, and works with international organizations on sustainability and human development. Like US companies, the company also has LEED-certified buildings certified by the UN Green Building Council. The company is also working on the labor laws and providing them with a good working environment. The company is also stringent in its child labor and harassment policies. VF aimed to utilize 100% renewable energy sources by 2025; in 2022, they are at 29%.

This shows they are behind the target if their year-on-year performance is considered. This might be due to COVID-19, which has negatively impacted business growth in the last two years. VF also aimed to "zero waste at their distribution centers by 2020," but they could not achieve this and have reached 72% waste reduction. They are also unable to reduce the GHG emissions in their upstream transportation; to date, there is no such improvement, and the target is to reduce it by 30% by 2030. Their goal was to reduce their GHG greenhouse gas emissions by 55% by 2030, reaching a 17% reduction. They had a goal to give 1 million hours to the local community by 2025, and to date, only 36,000 h have been tracked. This was due to covid-19 lockdown. A&F has achieved its target of water consumption reduction by 30%, recycling cardboard, educating and engaging associates, and reducing its carbon footprint by 25%. They have also set new targets for 2025, and their progress data will be shared in the following sustainability report. The company has an energy requirement of around 90 MW, and 285 MW of renewable energy units are installed all over Pakistan. The garment manufacturing company has around 28,000 tons yearly carbon footprint. The garment manufacturing company has reduced its water consumption by 19% compared to their baseline year 2018. The garment manufacturing company has some strategic goals of net zero carbon footprints by 2050. In 4 years, they have achieved a 43% reduction compared to 2018.

Significant drivers of sustainable practices in the industry are consumer requirements, global pressure, government regulations, and brand pressure on the upstream supply chain. Out of these, consumers are also the barrier to sustainable practices. Most want retailers and manufacturers to work on sustainable practices but are unwilling to pay. When discussing their purchase decision, they think organic cotton-made products are costly and old-fashioned, so they will not buy them. In this decision, the main issue is the lack of awareness about the products, their composition, and their carbon footprints. So, companies need to work on this to educate the consumer by focusing on market-driven sustainable strategies. These will enable them to promote the greener product and tell the audience how it is better than regular textiles from an environmental perspective. Companies should also focus on process-driven strategies, which state that processes should be environmentally efficient and consume less water and other material or energy resources. Companies can replace old machines with new ones that are more energy-efficient and can produce higher yields, reducing costs (Desore & Narula, 2018).

3.3.5 Interview Summary

Industrial ecology is an emerging field that is old enough, but its implementation is boosting nowadays. Now, people are aware of their influence on the environment and society. Responsible companies are working worldwide, and the garment manufacturing company is one of the great names in Pakistan's Denim industry. Despite a lack of resources and exposure to new technologies in Pakistan and a lack of infrastructure for sustainable development, the company is competing with the whole world in the Denim industry. The government should focus on the sustainability infrastructure so that other SMEs can work on sustainability. The following are some suggestions for improvement, specifically for the garment manufacturing company and the overall Pakistan denim industry.

- The company should evaluate its old energy and water-intensive products and gradually replace them with new and efficient machines with incredible speed, product quality, and environmentally efficient production capability.
- The company should work on the traceability of its supply chain; this can be done through blockchain technology. OEKO-TEX is an organization that issues ecolabels based on the company's performance and production process. One of

the labels is MADE IN GREEN, which ensures the complete traceability of the product, shows the details of the material and energy used, and provides the consumer that this labeled product is tested for harmful substances. It is safe to wear and ensures that it is produced under safe environmental conditions for the workers. This label can be in the form of a QR code, which can be scanned easily through our smartphones and shows the details of the product produced and the process through which it passes. The same is followed by the company's Turkish competitor Orta in the name of the blue app. This is their designed app, which shows their product details on the scan. This is being used for the product's life cycle assessment and to calculate the environmental impacts of the denim produced.
- The company should work on the product's reverse logistics and return the ability to reuse the denim again, as their Turkish competitor is working on it with the brand's collaboration. As per Orta sustainability literature:
  - Orta works with 20 ladies from Aksu, a small village in Kahramanmaras, Turkey. These women remove labels, seams, and metals from used jeans.
  - After removing accessories from the jeans, these are dispatched to the following plant location at Gaziantep, Turkey, for shredding.
  - Then, these fiber-shredded materials are converted into yarns to make recycled fabrics.
  - This after-use of jeans recycling is difficult because the company needs access to many used jeans.
  - Orta collaborates with various brands on a project-by-project basis for this endeavor. For instance, Levi's donated Orta used consumer jeans for recycling, which yielded four new fabrics comprised of 20% recycled fiber. Additionally, two of the fabrics are reinforced with recycled polyester.

3.3.6 Actionable Steps

The garment manufacturing company should work on this; this will reduce material consumption and environmental pollution.

- Cotton is a very water-consuming crop, and the textile process is the primary water user and a significant contributor to worldwide pollution. The company is working on all the factors. However, awareness among the farmers regarding cotton's new production methods in the world, which are more sustainable in nature, consume less water, produce high yield, and avoid pesticides and fertilizers, requires further focus.
- The company should work on the rainwater harvesting technique, which they can use for their processes after the treatment, gardening, and other functions like washingrooms, etc. There are two main rainwater harvesting techniques: surface runoff and rooftop rainwater harvesting (C, 2018). The company can use both of them as required. This will save water from being salty as it will ultimately

reach the ocean. Karachi also suffers from the rainwater drainage system due to the poor sewerage system. So, it will be a CSR project as well. Karachi's average annual rainfall was 174.6 mm from 1981 to 2010. And now it is increasing year by year. In 2020, on August 27, the record rainfall in Karachi was 223.5 mm. So, the garment manufacturing company should work on this as a leader in this sustainability initiative as their Bangladeshi competitors are doing this already. Bores can be made for rainwater, saving groundwater instead of sending it to ocean water, as it is not drinkable after spending much money on it. So, the company can bore into the water channels sides so that water can seep into the ground. This will also save water and the destruction caused by this in society.

- There is a lack of collaboration among Denim manufacturers in Pakistan; the government of Pakistan should focus on this to increase their revenue. Companies should learn good practices from others. Companies can also collaborate on more significant projects and get their payback in a shorter period. Through this collaboration, companies can get the advantage of economies of scale, and contracts can be made at a lower cost for transportation or export from Pakistan.
- The company also should work on the whole upstream and downstream supply chain as it is the key driver of its upstream supply chain and can implement excellent and responsible practices at its suppliers. As it is now part of international standards, you should implement the laws on your supplier as much as possible to make the supply chain responsible.
- The garment manufacturing companies can use the sea for solar panel installation as they need space to install the panels. The government should promote this initiative as China has already done this to reach a net zero emission target by 2030 (Shen, 2022).
- The government should work on the Eco-industrial parks, which can provide utilities to the small industries and effluent treatment plants, mainly as its cost is so high that an average SME cannot afford an ETP of his own, which pollutes the environment and ultimately impacts the society.
- The government should work on the new industrial zones and promote the related industry to be on the nearby premises, through which they can reduce transportation and inventory costs. This will also increase the responsiveness in the supply chain.

3.4 Industrial Ecology in the Services Sector of Pakistan

Today's corporate world associates "green" with providing environmentally friendly goods and services. Green banking, which refers to lending and depository products and services that banks offer to their regular and potential customers that are environmentally friendly, has been one of the rising themes in the financial sector over the past ten years. "green banking" is already thriving in developed and developing Asia. The State Bank of Pakistan (SBP) recently launched the Guidelines for Green

Banking in 2017, based on PEPA's advice (Pakistan-Environmental-Protection-Act, 1997).

These green banking guidelines aim to introduce environmentally friendly loan products and establish green banking services related offices in all commercial banks. Since commercial banks must implement GBG within a year of its issuance, there is no prior experience to draw upon in Pakistani commercial banks as they begin from blank. To that end, there was a discussion with senior specialized bankers from various commercial banks. Electronic banking products are currently provided and utilized by banks in both the public and private sectors. These products include e-statements, online transactions, debit and credit cards, credit on phone products, e-investment, mobile banking, and internet banking. However, this notion is relatively novel in Pakistan, specifically concerning the Green Banking Guidelines that the SBP issued. These guidelines mandate commercial banks to offer more environmentally sustainable products and serve as intermediaries between reductions. Green banking implementation by Pakistani commercial banks is justified based on economic benefits, CSR, environmental consideration, and sustainability, according to a 2015 SBP concept paper. Presently, the government is working to modify its product portfolio to incorporate a more significant proportion of low-carbon and environmentally sustainable energy sources. Numerous banks have implemented green building practices and green IT infrastructure and discontinued paper-based banking in pursuit of environmental sustainability. Financial and state regulators advise that paperless banking adheres as closely as possible to CSR and environmentally responsible practices (Becker et al., 2015).

"green banking" refers to financing goods and services that are more ecologically friendly and have lower carbon emissions. To ensure sustainability and the efficient conversion and use of natural resources, the banking sector could contribute to the greening of the economy by financing green vehicle projects, solar panel installation, and green building initiatives. As the primary source of funding, green banking has become a hot topic now throughout the world. Banks are scrutinized and given rigorous instructions by authorities to ensure compliance with implementing low-carbon projects and green financing. The Green Bank also functions as a commercial bank, and its funding comes from deposits from the general public and philanthropic donations or changes made by international organizations. The issues with adopting green banking, "Greening the Banking System," a survey report by the United Nations Environment Programme, identifies the following as critical issues in the financial sector: policy uncertainty and externalities, regulatory alignment, developing business cases, skill, capacity, and culture, and the availability and quality of E&S information (UN Environment Annual Report, 2016).

The Green Banking Guidelines (State Bank of Pakistan, 2017) by SBP also list a few factors that prevent the adoption of GBG, including behavior and attitude toward environmental concerns, changes in market conditions, indirect risk, direct and reputational risk, climate change, legal and regulatory risk, site, and sector. Nevertheless, these risk cans are unbelievable, as seen in applying environmental risk management modules, environmental risk planning tools, and ecological risk rating techniques. The adoption of electronic banking products such as PC banking, internet banking, mobile banking, EFT, ATMs, e-statements, online transactions, debit and credit cards, credit on phone products, and e-investments has been discussed in numerous studies. Although Pakistan is considered a developed country in the telecom industry, this concept is relatively new (Javeria et al., 2019). The Financial Stability Review on green banking underlines the following elements: green banking should be adopted by Pakistani commercial banks for four key reasons: corporate social responsibility (CSR), environmental concern, sustainability, and economic benefits (Becker et al., 2015). Many banks embrace green buildings, green IT infrastructure, and banking practices to reduce their use of environmental resources. The implementation of paperless banking complies to the fullest extent with CSR and ecological sustainability recommendations made by the financial and state regulators (Becker et al., 2015), reducing paper-based.

Most commercial banks are engaged in implementing green banking policies; however, there is not much research available in Pakistan that examines the factors that prevent the implementation (Pakistan Infrastructure Housing and SME Finance Department, 2017). The significance of "green banking" is being emphasized on a global and national scale, and its implementation calls for the utmost care to be more considerate of environmental sustainability while complying with legal requirements.

3.4.1 Banking Sector in Developed Country (Hong Kong)

The Hong Kong retail banking sector is the main subject of this study. Seventy of the top 100 banks worldwide have operations in Hong Kong, making it a global financial hub (Hong Kong Monetary Authority—Annual Report, 2016). According to a worldwide ranking of IPOs, Hong Kong came in #1 (The Wall Street Journal's News, 2015). After Tokyo, it boasts Asia's second-largest stock exchange (Clark Billings Obituary (1949–2016)—Liverpool, NY—Syracuse Post Standard). Furthermore, China, the second-largest economy in the world behind the United States, has looked to the financial sector in Hong Kong as a model.

Banks failed to address the balance of interests of various stakeholders during the 2008 financial crisis, instead placing an unbalanced emphasis on top management's gain (Karkowska, 2019). The crisis forced banks to reevaluate their use of "unsustainable" business strategies (Yip & Bocken, 2018). To become competitive again after the crisis, banks must repair their reputations and reduce expenses. However, it is impossible to completely rule out greenwashing due to the difficulty in measuring and evaluating the influence on personal sustainability performance and objectives see Table 3.3.

Table 3.3 Value analysis of the eight financial sustainable business model archetypes (Yip & Bocken, 2018)

Model	Value proposition	Value creation and delivery	Value capture	Difference from mainstream banking business model
1. Increase the effectiveness of materials and energy	Less resource-intensive services produce less waste and emissions and provide comparable functionality	The innovation of the internal operational process is the main focus	Profits rise as costs are decreased through improved operational effectiveness	Instead of taking a piecemeal approach, banks want to execute a bank-wide strategy to enhance material and energy efficiency
2. Replace with digital methods	One can enhance the flexibility of a business concerning cost, speed, convenience, and precision, as well as reduce environmental impact, by integrating electronic methods into the service delivery process	Innovation in service delivery design, such as delivery channels, ultimately enhances the speed, comfort, affordability, and accuracy of service provision to customers	Revenue is increased by giving clients additional convenience, which could lead to more frequent transactions. Manpower and associated costs can be reduced to save money	Banks continue to innovate with digital processes for customer engagement to reduce or do away with the old branch network through FinTech
3. Strengthen sufficiency	Solutions that aim to lower demand (previously typically exaggerated) by accurately assessing client demands, lowering financial product mis-selling, and lending moral hazard. The reward system and customer connections are the main points of emphasis	This may entail increasing the share of fixed compensation paid to frontline sales workers, boosting need-based selling through proper product pairing, and promoting prudent financing	Customers' loyalty and contentment could rise, which might result in increased sales Reduced compliance risk lowers the likelihood of regulatory fines. Customers receive the products they need in the exact amount and quality, capturing the social benefit	Banks abandon the strategy of "selling more" and replace it with premium services and products that precisely meet the demands of their clients

(continued)

Table 3.3 (continued)

Model	Value proposition	Value creation and delivery	Value capture	Difference from mainstream banking business model
4. Assume a stewardship position	Services that are provided with the intention of actively and sincerely engaging stakeholders to ensure their long-term well-being. By better engaging customers, broader stakeholder benefits frequently become a significant component of the values argument	Ensure that partners and actions are concentrated on providing stakeholders with well-being. The worth chain is sure to provide social or environmental advantages	Stewardship tactics can build brand value, reduce costs, and guarantee future business The satisfaction of stakeholders generates long-term business gains. For instance, employees who are healthy and content might take fewer sick days and work more	Banks frequently use a shared value approach in addition to standard CSR initiatives to leverage and advance their primary business, such as hiring retired employees on a part-time basis
5. Creating value for all	I Innovations in goods and services that target underserved markets or high-risk customers generate inclusive value	Process innovation is essential to reducing the risks involved, for instance, when managing the hazards of SME financing by adopting methods like credit scoring and portfolio management	Growth in market share. When customers grow more lucrative, client loyalty may help secure more company prospects. Additionally, included is societal gain	Banks benefit by assisting the less fortunate, who may become devoted clients. Additionally, if the risk is reasonable, it means more significant business in the short term

(continued)

Table 3.3 (continued)

Model	Value proposition	Value creation and delivery	Value capture	Difference from mainstream banking business model
6. Repurpose for the environment and society	It offers financial services that meet consumers' demands and generate societal and environmental advantages	In addition to being experts at delivering financial services to this specific market niche and gaining economies of scale, banks are utilizing sustainability as a consideration for selecting consumers	Only offer banking services to disadvantaged people and environmentally and socially responsible businesses, including "positive screening" against these criteria	Banks should focus primarily on sustainable enterprises when segmenting their clientele rather than only employing the present harmful screening method
7. Continuity in loan approval	To weed out unsustainable businesses, combine innovation in the loan approval process with sustainability standards (i.e., negative screening)	The cost of capital and possible bad debt could be decreased if the sustainability risk is reduced	We are loaning to clients who pose little or no sustainability risk This may directly cut off financial support for businesses with adverse sustainability effects	Banks utilize a more comprehensive system to eliminate unsustainable borrowers across all dimensions, encompassing corporate and individual clientele
8. Financially sound products	For investors/ savers and debtors, products are developed for the asset and liability sides	Product innovation promotes sustainable development and brings up new markets for sustainable finance	Product innovation makes it possible to participate in sustainability, which may give both savers and borrowers a platform to follow financial gain in sustainable companies	Banks give consumers access to new financial platforms to meet the market's demands for sustainable enterprises

3.4.2 Banking Sector of Developed Countries

The banking sector should focus on digital financing and operations management rather than the physical availability of documents. The banking industry in Pakistan is at a point when the bank that adopts green banking initiatives first will profit from competitive prevention. Additionally, banks have significant influence over various stakeholder groups, which can contribute to a genuinely green future for the nation. Banks' adoption of green banking can significantly reduce environmental issues that endanger a nation's ability to prosper. The lack of academic study on the implementation of green banking further contributes to the importance of this research field. Following are the suggestions for banking sector policymakers of Pakistan in the context of industrial ecology and sustainability:

- According to SBP Green Banking Guidelines-Scope and Implementation Challenges (2017), the three major issues affecting Pakistan's economy are the fisc, energy, and climate change. Determining the breadth of green financing in Pakistan would be pointless without knowing the relationship between the energy economy and the environment. Around 5.9 trillion rupees were spent on Pakistan's budget for the 2018–19 fiscal year, of which the energy sector accounted for 3.7 trillion. The circular debt of Rs. 1.9 trillion in the power sector of Pakistan as of December 2019 is mainly attributable to high system losses on the part of energy distribution corporations and the use of imported fossil fuel, specifically imported oil, which constituted 40% of the nation's primary energy consumption in 2019. This serves as evidence of the exorbitant energy costs in Pakistan.
- The government should work on the Eco-industrial parks, which can provide utilities to the small industry and effluent treatment plant, mainly as its cost is so high that an average SME cannot afford an ETP of his own, which pollutes the environment and ultimately impacts the society.
- The government should work on the new industrial zones and promote the related industry to be on the nearby premises, through which they can reduce transportation and inventory costs. This will also increase the responsiveness in the supply chain.

References

A&F. (2022a). *Environmental, social and governance goals.*

A&F. (2022b). *A&F Cares/Sustainability/Corporate Goals, Environmental, Social, Governance | Abercrombie & Fitch.*

Becker, F. G., Cleary, M., Team, R. M., Holtermann, H., The, D., Agenda, N., Science, P., Sk, S. K., Hinnebusch, R., Hinnebusch A. R., Rabinovich, I., Olmert, Y., Uld, D. Q. G. L. Q., Ri, W. K. H. U., Lq, V., Frxqwu, W. K. H., Zklfk, E., Edvhg, L. V, Wkh, R. Q., Fatemi, H., et al. (2015). Financial stability review. *State Bank of Pakistan, 7*(1), 37–72.

BizVibe. (n.d.). *Top 10 largest textile producing countries, largest textile exporters 2020.*

Bryman, A. (2012). *Social research methods.*

C, R. (2018). *Methods of rainwater harvesting | how to save rain water | Mettur diary.*

Clark Billings Obituary. (1949–2016). *Liverpool, NY—Syracuse Post Standard* (n.d.). Retrieved August 23, 2022, from https://obits.syracuse.com/us/obituaries/syracuse/name/clark-billings-obituary?pid=180670557

Desore, A., & Narula, S. A. (2018). An overview on corporate response towards sustainability issues in textile industry. *Environment, Development and Sustainability, 20*(4), 1439–1459.

Hong Kong Monetary Authority—Annual Report 2016. (n.d.). Retrieved August 23, 2022, from https://www.hkma.gov.hk/eng/data-publications-and-research/publications/annual-report/2016/

Javeria, A., Siddiqui, S. H., Rasheed, R., & Pakistan, K. (2019). Towards green banking in Pakistan: Problems, players and prospects. *Pakistan Journal of Social Sciences (PJSS), 39*(2), 363–376.

Karkowska, R. (2019). Business model as a concept of sustainability in the banking sector. *Sustainability, 12*(1), 111. https://doi.org/10.3390/SU12010111

Pakistan-Environmental-Protection-Act. (1997). The Pakistan code-law and justice division.

Pakistan Infrastructure Housing & SME Finance Department. (2017). *Green banking guidelines.* IH&SMEFD Circular No 08 Dated October 2017, 1–18.

SBP Green Banking Guidelines-Scope and Implementation Challenges. (n.d.). Retrieved September 5, 2022, from https://dailytimes.com.pk/601325/sbp-green-banking-guidelines-scope-and-implementation-challenges/

Shen, Q. (2022). *China's Shandong to build massive solar panel farms out to sea—Bloomberg.* In news.

State Bank of Pakistan. (2017). Retrieved August 21, 2022, from https://www.sbp.org.pk/smefd/circulars/2017/C8.htm

The Wall Street Journals' News Archive for July 15, 2015. (n.d.). Retrieved August 23, 2022, from https://www.wsj.com/news/archive/2015/07/15

UN Environment Annual Report 2016. (n.d.). Retrieved August 23, 2022, from https://www.unep.org/annualreport/2016/index.php

Yip, A. W. H., & Bocken, N. M. P. (2018). Sustainable business model archetypes for the banking industry. *Journal of Cleaner Production, 174*, 150–169.

Chapter 4
Understanding Digital Supply Chains in the Context of Industrial Ecology

Digital supply chains represent a transformative approach to traditional supply chain management, integrating advanced digital technologies to enhance efficiency, transparency, and sustainability. In industrial ecology, digital supply chains facilitate the transition to a circular economy by tracking and optimizing resource flows, minimizing waste, and promoting the reuse and recycling of materials. Companies can gain real-time insights into their operations through big data analytics, the Internet of Things (IoT), artificial intelligence (AI), and many other digital technologies, allowing for more informed decision-making and fostering a closed-loop system that mirrors natural ecosystems. This integration of digitalization within supply chains not only drives economic growth but also aligns with the principles of environmental sustainability, ensuring that the needs of the present are met without compromising the ability of future generations to meet their own needs.

4.1 Planning 4.0

Planning 4.0 is an innovative strategy for supply chain planning that utilizes digital technology to improve efficiency, flexibility, and reactivity in the supply chain ecosystem. Planning 4.0 utilizes cutting-edge technology like artificial intelligence (AI), machine learning (ML), big data analytics, and the Internet of Things (IoT) to facilitate smarter and automated decision-making across the supply chain. This paradigm change is crucial in the current dynamic corporate environment, where conventional planning methodologies may have difficulties keeping up with the intricacies and uncertainties of global supply chains. An essential element of Planning 4.0 is its emphasis on the visibility and integration of real-time data. Organizations may collect real-time data about their supply chain activities by utilizing IoT sensors, RFID tags, and other data-capturing devices. Having this detailed degree of insight

A. Shah et al., *Industrial Ecology*, Industrial Ecology,
https://doi.org/10.1007/978-981-97-3619-5_4

allows for improved monitoring of inventory levels, manufacturing status, transportation routes, and other essential factors, resulting in more precise demand forecasts and resource allocation.

In addition, Planning 4.0 places great importance on utilizing predictive analytics and AI-powered forecasting models to identify future demand trends and potential disruptions in the supply chain. By examining past data, market patterns, and external influences like geopolitical occurrences or natural calamities, these sophisticated algorithms can produce more precise predictions and proactive tactics for minimizing risks. Adopting a proactive strategy is essential to reduce stockouts, excess inventory, and other expensive inefficiencies in the supply chain. Another fundamental tenet of Planning 4.0 is optimizing networks and fostering cooperation. Instead of working alone, supply chain participants are urged to exchange data, knowledge, and resources within a cooperative ecosystem. Cloud-based platforms and digital markets enhance communication efficiency and stakeholder coordination, resulting in quicker decision-making and improved supply chain visibility. This collaborative method promotes the ability to recover and adjust quickly to shifting market circumstances and client requirements.

Furthermore, Planning 4.0 facilitates flexible and responsive planning procedures that may respond promptly to unexpected interruptions or opportunities. By employing scenario planning and simulation tools, organizations may replicate different hypothetical situations and assess the possible consequences on their supply chain operations. This allows companies to establish contingency plans and reaction tactics beforehand, guaranteeing the uninterrupted operation of their firm under unpredictable circumstances.

Planning 4.0 improves operational efficiency and promotes sustainability, industrial ecology, and ethical sourcing practices in the supply chain. Using data analytics and blockchain technology, organizations can identify the origins of raw materials, monitor carbon emissions, and ensure compliance with environmental and social responsibility norms. Transparency in supply chain processes fosters consumer and stakeholder confidence while diminishing the environmental impact. In addition, Planning 4.0 provides supply chain experts with sophisticated decision support tools and visualization approaches. Interactive dashboards, data visualization tools, and augmented reality interfaces offer user-friendly methods to analyze intricate supply chain data and get valuable insights. This facilitates expedited decision-making and enhances well-informed strategic planning, helping organizations maintain a competitive edge in the current dynamic business landscape. In addition, Planning 4.0 enables the implementation of agile and adaptable production methods, such as additive manufacturing (3D printing) and robots. These technologies facilitate the creation of goods as needed, allowing for customization on a large scale and manufacturing that is spread out. This results in shorter waiting times and lower transportation costs while improving the quality of products and promoting innovation. By adopting these digital manufacturing approaches, organizations may more effectively fulfill client demands for rapidity, personalization, and environmental responsibility.

The digital function fosters ongoing enhancement and originality in the supply chain by utilizing the notion of digital twins. A digital twin is a computer-generated model accurately representing a physical object, procedure, or arrangement. It allows for continuous monitoring, analysis, and improvement in real time. Organizations may enhance their supply chain operations by generating digital replicas of their processes, enabling them to pinpoint inefficiencies, test process improvements, and model the consequences of modifications before implementation.

Ultimately, Planning 4.0 represents a significant transformation in managing supply chains, driven by incorporating digital technology and insights derived from data analysis. This innovative methodology enables organizations to quickly adjust to the intricacies of the current business environment, which is marked by fast-paced transformations, uncertainties, and worldwide interconnections. Companies may improve their agility, resilience, industrial ecology, and sustainability by adopting real-time visibility, predictive analytics, collaborative networks, and adaptive planning procedures. This will allow them to provide more value to consumers and remain competitive in the digital era. In addition, Planning 4.0 promotes a culture of ongoing enhancement and originality in supply chain operations. Organizations may pinpoint inefficiencies, streamline processes, and achieve strategic enhancements by implementing digital twins and utilizing sophisticated decision-support technologies. By using these digital capabilities, organizations may keep ahead of changing client expectations, reduce risks, and take advantage of chances for expansion and differentiation in dynamic market conditions. In summary, Planning 4.0 has the potential to completely transform supply chain management by facilitating more intelligent, more effective, and more adaptable operations. By utilizing cutting-edge technology and data analytics, organizations may achieve higher levels of adaptability, durability, and environmental responsibility, positioning themselves for success in an ever-growing digital and linked world.

4.1.1 Big Data

Big data has become vital in the digital supply chain, providing unparalleled optimization, innovation, and competitiveness prospects. Big data in the supply chain encompasses the large amounts of structured and unstructured data produced throughout all stages. This comprises data from several sources, including IoT devices, RFID tags, corporate systems, social media platforms, and external databases. By integrating and analyzing various data types, organizations may obtain more profound insights into their supply chain operations, consumer behaviors, market trends, and external influences. This process eventually facilitates informed decision-making and strategic planning. A crucial use of big data in the digital supply chain involves demand forecasting and predictive analytics. Organizations may enhance their demand forecasting and predict future demand patterns by utilizing previous sales data, market trends, and other pertinent elements. Organizations may use advanced predictive analytics techniques, such as machine learning and artificial intelligence, to uncover concealed patterns, correlations, and anomalies in data. This improves the precision and dependability of demand estimates. By utilizing this, organizations may effectively optimize their inventory levels, manufacturing schedules, and distribution methods, reducing stockouts, excess inventory, and interruptions in the supply chain.

Moreover, utilizing big data analytics allows for immediate and continuous monitoring of supply chain activities, granting organizations enhanced clarity and authority over their procedures. Organizations may utilize IoT sensors, GPS tracking devices, and other data-capturing technology to actively monitor the real-time movement and status of commodities within the supply chain. Organizations can spot bottlenecks, delays, and inefficiencies by providing detailed visibility. This enables timely interventions and adjustments to the processes. Real-time analytics support proactive risk management by promptly notifying organizations of possible interruptions or deviations from the intended course, allowing them to take remedial steps before issues worsen.

Furthermore, utilizing big data analytics enables the enhancement of supply chain efficiency and the allocation of resources by finding potential areas for reducing costs, improving performance, and optimizing processes. By examining extensive data sets from several sources, organizations may detect inefficiencies, redundancies, and opportunities for enhancement within their supply chain procedures. This may involve optimizing logistics routes, warehouse configurations, manufacturing timetables, and supplier partnerships to save expenses and maximize operational effectiveness. Organizations may utilize advanced optimization algorithms to assess various situations and trade-offs, which aids in making well-informed decisions and attaining superior results.

In addition, big data analytics empowers organizations to improve customer service and happiness by comprehending and predicting client demands and preferences more efficiently. By examining consumer data, including purchase history, browsing behavior, and feedback, organizations can customize their services, suggest appropriate items, and provide a smooth customer experience across various platforms. This enhances client loyalty and retention and stimulates revenue growth and competitive advantage in today's customer-centric industry. Moreover, big data analytics enhances supply chain risk management by allowing organizations to proactively discover, evaluate, and reduce risks and uncertainties. By examining past data, market patterns, and external influences, organizations can detect possible hazards, such as interruptions in the supply chain, political instability, environmental catastrophes, and changes in demand. Organizations may utilize this capability to create backup plans, expand their pool of suppliers, and execute strategies to reduce risks, minimizing the adverse effects of interruptions on their operations and ensuring uninterrupted business operations.

Furthermore, utilizing big data analytics is essential in managing supplier relationships as it allows organizations to evaluate supplier performance, discover possibilities for cooperation, and efficiently address risks. Organizations may assess supplier capabilities, monitor performance indicators, and pinpoint areas for enhancement by examining supplier data, including delivery times, quality metrics, and compliance records. Organizations may establish strategic alliances with suppliers, engage in favorable negotiations, and optimize their supply chain networks to improve resilience and competitiveness. In addition, big data analytics enhances industrial ecology and ethical sourcing practices in the supply chain by enabling organizations to more efficiently manage and analyze environmental, social, and governance (ESG) parameters. By analyzing data about carbon emissions, energy consumption, trash generation, and labor practices, organizations may find ways to diminish their environmental impact, enhance supply chain transparency, and guarantee compliance with legal mandates and industry benchmarks. This not only improves the reputation and value of the company but also mitigates risks related to environmental and social concerns in the supply chain.

Moreover, big data analytics allows organizations to enhance their inventory management and distribution strategies by conducting more efficient analyses of demand patterns, seasonality, and market trends. Organizations may minimize holding costs, stockouts, and excess inventory by utilizing sophisticated forecasting models and optimization algorithms to optimize inventory levels, safety stock levels, and reorder points. By implementing this, organizations may optimize their cash flow, minimize inventory expenses, and elevate customer service standards, eventually leading to increased profitability and a competitive edge in the market. Furthermore, big data analytics enhances supply chain transparency and cooperation by enabling organizations to efficiently exchange data, knowledge, and assets with their supply chain counterparts. Organizations may establish a cohesive and transparent supply chain ecosystem using cloud-based platforms, blockchain technology, and data-sharing protocols. This enables stakeholders to access up-to-date information, cooperate on common objectives, and collaboratively make choices based on facts. This improves the supply chain's synchronization, dexterity, and promptness, allowing organizations to swiftly adjust to shifting market conditions and client requirements.

Moreover, big data analytics empowers organizations to enhance quality management and product traceability by closely monitoring and tracking the complete product lifetime, from acquiring raw materials to the final disposal at the end of its useful life. Organizations may utilize data from sensors, RFID tags, and other traceability technology to ascertain the origins of products, monitor quality indicators, and guarantee adherence to legal requirements and industry standards. Implementing this measure improves the product's safety, decreases the likelihood of recalls, and fosters trust among customers and stakeholders. Ultimately, it enhances the reputation of the brand and promotes customer loyalty.

Incorporating big data analytics into the digital supply chain environment represents a noteworthy advancement in supply chain management methodologies. This technique utilizes various data sources to provide insights, improve operations, and increase decision-making in the supply chain ecosystem. By using big data analytics, organizations may enhance their understanding of their supply chain operations, allowing them to detect inefficiencies, predict variations in demand, and manage risks more effectively. Furthermore, the utilization of big data analytics allows for the creation of predictive and prescriptive models, empowering organizations to make proactive modifications and optimize the allocation of resources in real-time, hence enhancing overall efficiency and customer satisfaction. Moreover, applying big data in the digital supply chain facilitates the promotion of cooperation and openness among supply chain participants. By sharing data and integrating, stakeholders may have immediate access to up-to-date information, facilitating smooth coordination and decision-making processes. This cooperative strategy not only enhances the efficiency of the supply chain but also reinforces the connections between partners, fostering increased confidence and alignment of goals.

Moreover, with big data analytics, organizations may bolster the resilience of their supply chains by detecting possible interruptions and formulating backup strategies to minimize their consequences. This guarantees the uninterrupted functioning of operations, even when confronted with unexpected obstacles. Incorporating big data analytics into the digital supply chain marks a significant change in supply chain management approaches. Organizations may obtain a competitive advantage in today's fast-paced business climate by utilizing the large amounts of data created across the supply chain. This allows them to extract valuable insights and foster innovation. However, fully harnessing the power of big data in the digital supply chain necessitates a systematic approach to data governance, infrastructure investment, and talent development. To fully leverage the advantages of big data analytics in the digital supply chain, organizations must give utmost importance to data security, privacy, and ethical concerns. They should also promote a culture that emphasizes making decisions based on data.

4.2 Product or Process Innovation

In digital supply chains, innovation has emerged as a fundamental element for improving efficiency, cutting expenses, and boosting customer satisfaction. This innovation may be categorized into two main groups: product and process. Both play crucial roles in enhancing digital supply chains, increasing their resilience, agility, and customer-centricity.

4.2.1 Product Innovation in Digital Supply Chain

Product innovation in the digital supply chain involves the creation of novel or greatly enhanced products or services that are introduced to fulfill changing market needs or use new technology. This includes incorporating intelligent technology into products, such as Internet of Things (IoT) devices that offer real-time tracking functionalities. These innovations increase the product's value and allow enterprises to distinguish themselves in competitive marketplaces. The emergence of 3D printing technology is a prime example of product innovation that has significant consequences for digital supply chains. On-demand production facilitates the reduction of inventory costs and lead times.

Furthermore, it allows for the customization of products to meet the particular requirements of individual customers on a large scale, which was previously expensive and intricate. Blockchain technology demonstrates the advancement of product innovation in digital supply networks. Blockchain enhances transparency and confidence among supply chain actors by offering a decentralized and tamper-proof ledger. This is especially advantageous in industries where the origin and genuineness are vital, such as pharmaceuticals and luxury items.

4.2.2 Process Innovation in Digital Supply Chain

Process innovation, however, specifically aims to improve or develop new procedures inside the supply chain. This includes implementing digital technology to optimize processes, enhance information dissemination, and minimize inefficiencies. The objective is to improve the responsiveness and adaptability of supply networks to fluctuations in demand or market conditions. Artificial Intelligence (AI) and Machine Learning (ML) are leading the way in revolutionizing digital supply chain processes. These technologies provide predictive analytics, which can precisely anticipate demand, optimize inventory levels, and detect disruptions in advance. By adopting a proactive strategy, the risks are minimized, and the supply chain operation is particular to be smoother. The use of digital twins signifies a notable advancement in process innovation. A digital twin is a computer-generated model replicating a physical supply chain, enabling real-time simulation, analysis, and monitoring of supply chain activities. This technology facilitates the identification of obstacles, assesses the consequences of alterations in the supply chain, and enables the preparation for future expansions or adjustments without jeopardizing natural resources. Cloud computing has transformed the digital supply chain by offering scalable and adaptable data storage and processing resources. Companies can effectively handle substantial amounts of data from several sources across the supply chain, improving stakeholder decision-making and communication. Integrating IoT devices into supply chain activities enables immediate tracking and monitoring of items, machines, and vehicles. Having this level of visibility allows for more accurate inventory management, decreases the likelihood of theft or loss, and enhances the overall security of the supply chain.

4.2.3 Combining Product and Process Innovation

The combination of product and process innovation in the digital supply chain results in a business model that is more dynamic, efficient, and focused on the needs of the consumer. An example of this is the utilization of intelligent packaging that incorporates Internet of Things (IoT) technology. This enhances the product's functioning (product innovation), inventory management, and logistics (process innovation).

Utilizing digital technology, agile supply chain management strategies provide quick adjustments to market demand or supplier circumstances. This agility is achieved by implementing adaptable and responsive supply chain methods and rapidly introducing new or modified products, which are the outcomes of both process and product innovation.

Furthermore, industrial ecology has emerged as a primary area of emphasis in advancing digital supply chains. Companies are employing both product and process innovation to reduce their environmental footprint. This encompasses the creation of items specifically designed to facilitate recycling processes, as well as the utilization of artificial intelligence to enhance the planning of transportation routes to minimize carbon emissions.

The convergence of product and process innovation inside digital supply chains represents a revolutionary period for industries worldwide. Introducing modern technology into goods and services allows unique chances to stand out and increase value in competitive marketplaces. At the same time, process improvements significantly improve the operational elements of supply chains, making them more efficient, responsive, and able to adjust to the quickly changing market dynamics. This dual strategy addresses immediate goals of efficiency and customer happiness and aligns with long-term objectives like industrial ecology and the capacity to withstand upheavals. As we anticipate the future, the continuous development of digital technology holds the potential to accelerate innovation in digital supply chains. The continuous integration of AI, IoT, blockchain, and other emerging technologies will provide possibilities for new product and process innovations. This transformation will not only change the way organizations oversee their supply chains but also the way they generate value for consumers and society as a whole. The future success of supply chains will depend on the seamless integration of product and process innovations, utilizing digital transformation to address future issues effectively.

4.3 IT System—MES and ERP

In the intricate landscape of the digital supply chain, IT systems play a pivotal role in orchestrating seamless operations, enhancing efficiency, and fostering innovation. Among these systems, Manufacturing Execution Systems (MES) and Enterprise Resource Planning (ERP) systems are fundamental, serving distinct yet interconnected functions. Their deployment and integration are crucial for businesses aiming to optimize production processes, ensure data accuracy, and improve decision-making capabilities.

4.3.1 Manufacturing Execution Systems (MES) in the Digital Supply Chain

MES are computerized systems used in manufacturing to track and document the transformation of raw materials to finished goods. They provide operators, managers, and decision-makers real-time information about production activities, facilitating immediate adjustments to enhance efficiency and productivity. MES bridges the gap between the shop floor and corporate management by providing detailed insights into all aspects of production. One of the core benefits of MES is its ability to improve product quality. By monitoring production processes in real-time, MES can identify deviations from quality standards immediately, allowing for quick corrective actions. This reduces waste and rework and ensures the final products meet the required specifications and customer expectations. MES also enhances production efficiency by optimizing manufacturing schedules and resources. It can adjust real-time production plans based on various factors, including machine availability, labor capacity, and material supply. This flexibility is crucial in responding to market demands and minimizing downtime. In terms of compliance and traceability, MES plays a vital role. It can track the use of materials and components throughout the production process, ensuring compliance with industry standards and regulations. This traceability is essential for industries where safety and quality are paramount, such as pharmaceuticals, food and beverage, and aerospace.

4.3.2 Enterprise Resource Planning (ERP) Systems in the Digital Supply Chain

ERP systems integrate various business processes into a unified system, from procurement and supply chain management to sales and finance. By consolidating data from across the organization, ERP systems provide comprehensive insights into business operations, supporting strategic planning and operational decision-making. ERP systems enhance supply chain visibility by offering a holistic view of operations, including inventory levels, supplier performance, and customer demand. This visibility is critical for effective supply chain management, enabling companies to anticipate and respond to supply chain changes proactively. ERP systems facilitate better financial management by integrating economic and operational data. They allow real-time expenses, revenues, and profitability tracking by product, department, or business unit, supporting more informed financial decisions and strategic planning. ERP systems also support scalability and flexibility in business operations. As companies grow and evolve, ERP systems can adapt to changing business needs, adding new functionalities or integrating with other systems to extend their capabilities. This adaptability is crucial for businesses looking to expand their operations or enter new markets.

4.3.3 Integration of MES and ERP for a Seamless Digital Supply Chain

Integrating MES and ERP systems is pivotal in creating a seamless digital supply chain. MES provides detailed, real-time data from the shop floor, which can enhance the accuracy and timeliness of information within the ERP system. Conversely, ERP systems can feed strategic and planning data back to MES, aligning production activities with broader business objectives. This integration facilitates improved demand forecasting and inventory management. By leveraging real-time production data from MES, ERP systems can provide more accurate forecasts, enabling better inventory optimization and reducing the risk of stockouts or excess inventory. Moreover, the MES-ERP integration supports agile manufacturing processes. It allows for quicker adjustments to production schedules based on changes in demand or supply chain disruptions, ensuring that manufacturing operations are efficient and responsive to market conditions.

Integrating Manufacturing Execution Systems (MES) and Enterprise Resource Planning (ERP) systems represents a strategic imperative in the modern digital supply chain landscape. Together, they form a robust foundation for enhancing operational efficiency, ensuring product quality, and facilitating strategic decision-making across the enterprise. This integrated IT framework empowers businesses to respond swiftly to market dynamics, optimize production processes, and maintain competitive advantage in their respective industries by bridging the gap between the shop floor and corporate strategy. As companies continue to navigate the complexities of global supply chains, the role of MES and ERP systems in fostering transparency, agility, and resilience cannot be overstated. Looking forward, the evolution of digital technologies, including the Internet of Things (IoT), artificial intelligence (AI), and machine learning (ML), promises further to augment the capabilities of MES and ERP systems. The seamless flow of real-time data and insights across all levels of the organization will enable even more precise forecasting, more intelligent resource allocation, and enhanced operational flexibility. This continuous innovation in IT systems is not just transforming the digital supply chain. However, it also redefines the essence of manufacturing and business operations, paving the way for a future where adaptability and efficiency drive sustained success.

4.4 Procurement 4.0

Procurement 4.0 signifies a transformative change in how procurement is conducted in the digital supply chain, which aligns with the more comprehensive Industry 4.0 concept that emphasizes incorporating digital technology into all facets of corporate operations. The use of modern technology is revolutionizing the way organizations handle their sourcing and purchasing activities. This transformation aims to improve efficiency, lower expenses, and establish more robust supplier relationships. The following is a comprehensive analysis of Procurement 4.0 and its impact on the digital supply chain.

4.4.1 Foundations of Procurement 4.0

Procurement 4.0 is a notable advancement in the digital supply chain, aligning with the more significant Industry 4.0 revolution that incorporates sophisticated digital technology into production and logistics. This procurement transformation is marked by a transition towards automation, data-driven decision-making, and linked supplier networks. It significantly changes how organizations handle their sourcing and purchasing activities. Procurement 4.0 is built around the smooth incorporation of technologies like the Internet of Things (IoT), artificial intelligence (AI), machine learning (ML), blockchain, and sophisticated analytics into procurement operations. These solutions provide immediate visibility, anticipatory analytics, and enhanced efficiency across the procurement lifecycle, encompassing supplier selection, contract administration, and order fulfillment.

An essential pillar of Procurement 4.0 is the prioritization of data and analytics. Data is a valuable asset in the digital supply chain, influencing decision-making and strategic planning. Advanced analytics and artificial intelligence algorithms are utilized to analyze large volumes of data to detect trends, predict demand, evaluate supplier risk, and optimize procurement strategies. Using data in decision-making empowers procurement professionals to enhance their decision-making process, save expenses, and manage risks by proactively predicting market fluctuations and adapting procurement strategies appropriately. Procurement 4.0 also involves digitizing procurement procedures, which is a crucial feature. This is the shift from manual, paper-based methods to automated, digital ones, simplifying operations and improving effectiveness. E-procurement systems and cloud-based solutions streamline the process of requesting quotations, conducting auctions, managing contracts, and invoicing electronically. This results in faster processing times and enhanced precision. Digitalization improves transparency and collaboration in the supply chain, allowing organizations to establish closer ties with suppliers and partners.

Blockchain technology enhances Procurement 4.0 by providing a safe and transparent method for conducting transactions and exchanging information inside the supply chain. Blockchain mitigates the possibility of fraudulent activities, boosts the capacity to track and verify transactions in real time, and improves adherence to regulatory requirements by establishing a decentralized ledger. This is especially advantageous in overseeing contracts and upholding the authenticity of supply chain documents, ranging from material certification to delivery confirmation, thereby fostering stakeholder confidence. IoT technology in Procurement 4.0 enables the immediate monitoring and management of the supply chain. Internet of Things (IoT) devices can monitor the whereabouts and state of items throughout transportation, oversee stock levels in warehouses, and even anticipate the maintenance requirements of industrial equipment. This connectivity not only boosts the efficiency of operations but also improves the agility of the procurement function in response to changes in supply chain dynamics. It ensures that materials and components are supplied correctly and in the best possible condition.

Procurement 4.0 in the digital supply chain uses advanced technology and data-driven insights to transform procurement procedures. Organizations may attain unparalleled efficiency, transparency, and cooperation levels by harnessing the power of AI, IoT, blockchain, and analytics. In the current digital age, organizations may enhance their ability to negotiate the intricacies of global supply chains and gain a competitive edge by incorporating these fundamental principles into their procurement practices, promoting sustainable growth.

4.4.2 Strategic Sourcing and Supplier Selection

Procurement 4.0 has significantly changed strategic sourcing and supplier selection in the digital supply chain, focusing on improving efficiency, adaptability, and resilience. This innovative approach utilizes cutting-edge digital technology to rethink conventional procurement procedures, allowing organizations to take proactive measures in response to market fluctuations and interruptions in the supply chain. Procurement 4.0's strategy for seeking and selecting suppliers combines big data analytics, artificial intelligence (AI), and machine learning (ML) capabilities. These technologies enable procurement experts to analyze large quantities of data to get insights into supplier performance, market trends, and risk factors. This ensures a more educated and strategic approach to making sourcing decisions.

Artificial intelligence and machine learning algorithms are vital in improving supplier selection procedures. AI can forecast supplier dependability, fiscal soundness, and adherence to industrial ecology y and ethical criteria by analyzing past data. By utilizing this predictive skill, procurement teams may effectively anticipate and address potential hazards while simultaneously establishing a supplier base that is both financially efficient and in line with the organization's long-term strategic objectives and values. In addition, AI-powered solutions can automate mundane procurement operations, such as issuing RFQs (Request for Quotations) and assessing bids. This simplifies the sourcing process and lets procurement experts concentrate on strategic decision-making. Utilizing digital platforms and e-sourcing technologies is a crucial component of Procurement 4.0. These systems enable online assessments of suppliers, electronic auctions, and cooperative discussions, enhancing the sourcing process's transparency and competitiveness. Digital sourcing platforms also have a worldwide reach, providing access to a larger group of suppliers and promoting a more dynamic and competitive supplier ecosystem. By using extensive and current information, this digital method decreases the time and expenses related to sourcing and improves the quality of supplier selection.

Blockchain technology enhances strategic sourcing and supplier selection by providing a secure, transparent, and unchangeable record of transactions and supplier relationships. This technology can authenticate items, guarantee contract adherence, and monitor suppliers' ethical and environmental practices. Blockchain technology improves trust and transparency in the supply chain, allowing organizations to make better sourcing decisions and establish more robust and dependable supplier relationships.

Procurement 4.0 has revolutionized the process of strategic sourcing and supplier selection in the digital supply chain. Organizations may enhance their procurement process's efficiency, transparency, and resilience by strategically using artificial intelligence (AI), machine learning (ML), digital platforms, and blockchain technologies. These developments facilitate improved risk management and cost reduction and aid in pursuing innovation and industrial ecology goals. To stay ahead in the ever-changing digital world, organizations must adopt Procurement 4.0 strategies. This is essential for maintaining a competitive edge and adjusting to the fast-paced nature of global supply chains. Improved Supplier Relationships Procurement 4.0 highlights the significance of establishing robust and cooperative supplier partnerships. Digital platforms facilitate smooth communication and information exchange between customers and suppliers, promoting transparency and confidence. This collaborative strategy aids in reducing risks, enhancing product quality, and promoting innovation through suppliers' networks.

4.4.3 Risk Management and Compliance

Procurement 4.0, in the context of the digital supply chain, employs advanced technologies to manage risks and ensure effective compliance with modern procurement environments. Using sophisticated digital tools and processes revolutionizes conventional risk management systems, providing unparalleled capacities to discover, evaluate, and reduce risks. The digital transformation relies on artificial intelligence (AI), machine learning (ML), big data analytics, and blockchain technology. Each component is crucial in improving the transparency, efficiency, and resilience of procurement processes. Artificial intelligence and machine learning technologies are leading the way in this transformation, offering the capability to anticipate and address possible hazards with enhanced precision and efficiency. Artificial intelligence (AI) can detect patterns and forecast disruptions that might impact the supply chain by analyzing extensive datasets, including supplier histories, market trends, and geopolitical events. By having this capacity to predict, organizations may take proactive measures to address risks before they happen. This includes making necessary adjustments to procurement strategy and processes. In addition, artificial intelligence (AI) and machine learning (ML) may be used to automate compliance checks in response to a constantly evolving regulatory environment. This helps ensure that procurement operations adhere to international standards and local legislation.

Big data analytics enhances the capabilities of procurement specialists by providing in-depth analysis of the supply chain, allowing for a more detailed comprehension of risk concerns. Organizations may thoroughly understand their supply network by analyzing extensive data from many sources. This will enable them to detect weaknesses and evaluate the possible impact of hazards on their operations. This comprehensive method of assessing risk enables firms to make better-informed decisions, enabling them to prioritize efforts to reduce risk based on the seriousness and probability of various situations. Blockchain technology enhances transparency and security in procurement, namely in risk management and compliance. Blockchain technology establishes an unchangeable record of transactions and interactions between suppliers, ensuring reliable and secure supply chain documentation. This level of openness is essential for confirming the genuineness of products, tracing the origin of components, and guaranteeing adherence to agreements and regulations. Furthermore, the decentralized structure of blockchain diminishes the need for intermediaries, hence decreasing the likelihood of fraud and improving the overall integrity of the procurement process.

Procurement 4.0 greatly enhances organizations' ability to manage risks effectively and ensure compliance in the digital supply chain. The strategic use of AI, ML, big data analytics, and blockchain technology provides a more agile and proactive method for identifying and reducing risks while simplifying compliance procedures. With the evolution of the digital world, these technologies will have a vital impact on developing robust, efficient, and transparent procurement strategies that can effectively handle the problems posed by a quickly changing global market. Organizations must use these digital developments to effectively manage the intricacies of contemporary procurement and stay ahead of the competition in the era of Procurement 4.0.

4.4.4 E-Procurement Platforms

Procurement 4.0, a crucial element of the digital revolution in supply chains, has dramatically enhanced the significance and capabilities of e-procurement platforms. These platforms, powered by cutting-edge technology, are revolutionizing the procurement process, improving efficiency, transparency, and simplicity. E-procurement systems utilize the Internet to carry out all procurement-related tasks, including sourcing, purchasing, tendering, and contract administration. By adopting this digital method, organizations may streamline their procurement procedures, save expenses, and improve communication with suppliers, all inside a centralized online platform. E-procurement solutions are primarily designed to automate and digitize conventional procurement procedures. Automation decreases the need for manual processes and minimizes mistakes, enhancing operational efficiency and allowing procurement specialists to dedicate their attention to strategic duties. These systems can automatically pair buy requests with authorized suppliers, create purchase orders, and handle bills, substantially accelerating the procurement process. In addition, digitization enables data precision and availability previously unachievable, monitoring expenses, adherence to regulations, and evaluating supplier effectiveness in real time.

E-procurement platforms are essential for strategic sourcing since they provide tools and functions that improve decision-making and the selection of suppliers. Procurement teams may utilize advanced analytics skills to assess supplier offers using several factors such as pricing, quality, delivery time, and adherence to industrial ecology and sustainability norms. Moreover, these platforms enable e-auctions and reverse auctions, promoting a competitive bidding process that can result in improved pricing and conditions. The strategic sourcing process is facilitated by extensive supplier databases that offer in-depth information on supplier capabilities, track record, and performance. One notable benefit of e-procurement systems is their managing and ensuring compliance with risks during the procurement process. By centralizing procurement data, these solutions provide exceptional insight into procurement operations throughout the organization. Clear insight is crucial to detect possible hazards early in the procurement process and ensure compliance with business policies and legal obligations. Furthermore, several e-procurement platforms have functionalities that automate validating supplier certificates and ensuring that purchases comply with set budgets and approval protocols.

E-procurement platforms promote collaboration and enhance connections between buyers and suppliers. These systems facilitate a more involved and transparent procurement process by utilizing integrated communication capabilities, document sharing, and collaborative planning elements. Suppliers can control their profiles, upload documents, and provide responses to RFQs and RFPs directly via the platform, guaranteeing that all involved parties have immediate access to identical information. This cooperative setting not only enhances the effectiveness of the procurement procedure but also fosters the establishment of enduring, strategic partnerships with crucial suppliers. Ultimately, e-procurement platforms play a central role in Procurement 4.0, facilitating the digitalization of procurement activities throughout the supply chain. These systems provide a holistic solution for contemporary procurement difficulties by automating and digitizing procurement operations, boosting strategic sourcing, strengthening risk management and compliance, and simplifying cooperation between buyers and suppliers. With the continuous advancement of technology, e-procurement systems will inevitably progress, providing enhanced capabilities and advantages for organizations seeking to optimize their procurement processes in the digital era.

4.4.5 Blockchain for Transparency and Efficiency

Procurement 4.0 utilizes blockchain technology to transform transparency and effectiveness in supply chain operations. Blockchain, also known as a decentralized ledger, offers a secure and visible log of transactions immutable and resistant to tampering, guaranteeing confidence and responsibility in the procurement procedure. Blockchain in procurement provides a significant advantage by improving transparency by creating a shared and unchangeable record of transactions and interactions between parties. This level of transparency allows organizations to monitor the flow of commodities, confirm the genuineness of items, and guarantee adherence to regulatory norms and ethical sourcing methods. In addition, blockchain technology improves the effectiveness of procurement procedures by automating and optimizing transactional activities such as purchase orders, invoicing, and payments. Smart contracts may execute themselves, with the terms of the agreement put straight into code. They allow for the automatic execution of procurement procedures depending on predetermined criteria and triggers. This automated procedure decreases the requirement for human involvement, reduces mistakes, and speeds up the time it takes to execute transactions, thus enhancing overall efficiency and lowering operating expenses.

In addition, blockchain enhances cooperation and confidence among supply chain participants by providing a common platform for safe data sharing and collaboration. Organizations may utilize blockchain technology to exchange information, monitor the flow of products, and cooperate on collaborative endeavors, including product creation, quality assurance, and risk management. This collaboration promotes stronger connections among stakeholders, improves the capacity to track and monitor the supply chain, and facilitates quicker decision-making and problem-solving. In addition, blockchain technology allows organizations to enhance the capacity to track and verify the origin and history of items in their supply chains, guaranteeing that they are obtained and manufactured ethically and sustainably. Using blockchain technology, organizations can track the origins of raw materials, authenticate products, and ensure adherence to environmental and social responsibility norms by documenting all transactions and movements of items. This level of openness not only assists organizations in reducing risks such as the production of counterfeit items, unethical labor practices, and environmental infractions, but it also improves brand reputation and fosters customer confidence.

Procurement 4.0 utilizes blockchain technology to improve transparency and efficiency in supply chain processes. Blockchain facilitates the monitoring of commodities, automates procurement procedures, and promotes collaboration and confidence among supply chain participants by offering a secure and transparent record of transactions. Furthermore, blockchain technology improves the capacity to track and verify the origin and history of products, allowing organizations to guarantee ethical and sustainable sourcing methods and enhance their brand image. However, harnessing blockchain's capabilities in procurement necessitates organizations to allocate resources to the appropriate technological framework, establish alliances with blockchain providers, and cultivate a culture of ingenuity and cooperation throughout the supply chain.

4.4.6 Predictive Analytics for Demand Forecasting

Procurement 4.0 utilizes predictive analytics to transform demand forecasting in supply chain operations. Predictive analytics employs statistical algorithms and machine learning approaches to analyze past data, detect patterns, and predict future demand trends accurately. By integrating predictive analytics into procurement operations, organizations may accurately forecast changes in demand, optimize inventory levels, and enhance overall supply chain efficiency. Predictive analytics in demand forecasting offers a significant benefit in its capacity to analyze extensive amounts of data from many sources, including sales data, market trends, social media, and weather patterns, resulting in more precise and dependable projections. Furthermore, predictive analytics empowers organizations to improve the precision and detail of demand projections by considering other aspects that might impact demand, including seasonality, promotions, and external events. Organizations can utilize advanced forecasting techniques, such as time series analysis, regression analysis, and machine learning algorithms, to accurately capture intricate demand patterns and make more precise predictions about future demand. Organizations may enhance their ability to anticipate future outcomes by utilizing predictive analytics. This can lead to improved accuracy in forecasting, a decrease in mistakes while making forecasts, and a reduction in the likelihood of insufficient or excessive inventory. As a result, inventory management can be optimized, and costs can be minimized.

Predictive analytics also enables demand sensing, continuous monitoring, and analysis of demand signals to detect developing trends and patterns. Organizations may utilize data from point-of-sale systems, online transactions, and other sources to promptly identify shifts in customer behavior and market dynamics. This allows them to make necessary adjustments to their production and procurement strategies. By adopting a proactive approach to demand sensing, organizations may promptly adapt to shifts in demand, mitigate supply chain interruptions, and improve customer satisfaction. Moreover, predictive analytics empowers organizations to enhance procurement decisions by offering valuable insights about forthcoming demand trends and supplier performance. By examining past data about supplier lead times, quality indicators, and delivery performance, organizations may detect possible hazards and advantages within their supplier network and formulate tactics to minimize risks and enhance supplier relationships. Furthermore, predictive analytics may assist organizations in identifying potential areas for reducing expenses and improving operational efficiency by pinpointing inefficiencies and bottlenecks in the procurement process.

Procurement 4.0 utilizes predictive analytics to transform demand forecasting and procurement procedures in the supply chain. Through sophisticated analytics methods, businesses may provide more precise and dependable predictions of demand, optimize the amount of inventory, and improve the overall efficiency of the supply chain. Furthermore, predictive analytics empowers organizations to enhance their ability to sense demand, optimize their decisions on procurement, and uncover chances for reducing costs and improving processes. However, harnessing the possibilities of predictive analytics in procurement requires organizations to allocate resources towards the appropriate technological framework, data analytics competencies, and people cultivation to achieve long-term success in today's ever-changing business landscape.

4.4.7 Customization and Personalization

Procurement 4.0 signifies the merging of technical progress with procurement procedures, resulting in a notable transition towards tailored and individualized methods in the digital supply chain. The transformation is primarily propelled by the capabilities of big data, artificial intelligence (AI), machine learning (ML), and the Internet of Things (IoT). These technologies collectively offer a previously unachievable degree of agility and responsiveness. Customization, in this sense, pertains to the procurement systems' capacity to adjust to different organizations' distinct demands and preferences. On the other hand, personalization entails adapting the procurement experience to users' unique requirements and behaviors. The aspects of Procurement 4.0 are revolutionizing the procurement process by shifting it from a rigid and uniform approach to a more adaptable, flexible, and customer-focused model.

Procurement 4.0 enables organizations to tailor procurement software and platforms to match their unique operational workflows, regulations, and strategic objectives. This is especially crucial in intricate supplier chains when conventional procurement methods must sufficiently tackle distinct industry issues or legal prerequisites. Advanced e-procurement systems have modular architectures and flexible APIs (Application Programming Interfaces), allowing organizations to connect customized modules or third-party apps that improve functionality and compliance. Procurement systems may adapt and develop alongside the business by allowing for a high degree of customization, eliminating the need for frequent and expensive overhauls. Personalization, in contrast, improves the user experience by utilizing artificial intelligence and machine learning algorithms to examine user behavior, preferences, and previous purchasing behaviors. By comprehending these patterns, procurement systems can offer tailored suggestions, automate repetitive operations, and notify users of prospective possibilities or hazards. For example, a system may recommend favored vendors based on previous engagements or emphasize volume-based discounts based on regular order amounts. This enhances productivity, user contentment, and procurement choices by guaranteeing pertinent facts and discernments influence them.

The significance of data analytics in customization and personalization cannot be exaggerated. Organizations may obtain more profound insights into supplier performance, market trends, and internal procurement habits by utilizing the potential of big data. This knowledge is crucial for customizing procurement plans adaptable to present market conditions and anticipating future changes. Data analytics provide a more refined categorization of suppliers and goods, empowering procurement experts to tailor criteria and policies for selection and assessment. This ensures that procurement operations are tightly synchronized with business objectives.

Furthermore, IoT technology enhances the level of customization in Procurement 4.0 by allowing immediate monitoring and control of assets across the supply chain. Internet of Things (IoT) devices can observe and track the state and whereabouts of commodities throughout transportation, offering tailored notifications and information to individuals involved. This feature guarantees that purchase choices are made using the most precise and up-to-date information, improving procurement processes' responsiveness and efficiency.

Procurement 4.0 is revolutionizing procurement processes by introducing advanced customization and personalization features. These features are raising the bar for agility, efficiency, and user involvement in procurement. Organizations may develop procurement ecosystems customized to their operational requirements and adaptable to users' distinct tastes and behaviors by utilizing advanced technology and data-driven insights. The ongoing development of these technologies offers immense promise for advancing customization and personalization in procurement, leading to significant improvements in strategic value and competitive advantage.

4.4.8 Industrial Ecology and Ethical Sourcing

Procurement 4.0 represents a significant shift in the development of procurement, as it combines cutting-edge digital technology with conventional procurement methods. This emerging model prioritizes sustainability (industrial ecology) and ethical sourcing, demonstrating an increasing acknowledgment of firms' social and environmental obligations. Procurement 4.0 utilizes advanced technologies such as big data analytics, blockchain, and artificial intelligence (AI) to enhance supply chain transparency, efficiency, and accountability. This promotes organizations' adoption of sustainable and ethical procurement practices. Procurement 4.0 enhances sustainable and ethical sourcing by improving supply chain visibility. Advanced analytics and artificial intelligence provide procurement experts with profound insights into suppliers' environmental and social behaviors. This level of visibility enables organizations to evaluate and oversee their supply chain's sustainability performance and ethical standards. This simplifies recognizing and addressing issues associated with environmental harm, labor infractions, and unethical business conduct. Organizations may encourage adopting superior ecological stewardship and social responsibility practices across their supply chains by including industrial ecology factors in selecting and evaluating suppliers.

Blockchain technology enhances sustainable and ethical procurement by providing an unchangeable and easily understood record of transactions and contacts with suppliers. This feature is handy for confirming the genuineness of assertions concerning ethical sourcing and sustainability (industrial ecology) certifications. Blockchain technology may be used to track the origin of raw materials, verifying that they come from vendors that follow sustainable harvesting procedures and fair labor conditions. Enhancing traceability at this level reinforces adherence to legal mandates and industry benchmarks and fosters confidence among customers progressively seeking company openness and responsibility. Procurement 4.0 incorporates technical progress and adopts the ideas of circular economy and sustainable development. Organizations can minimize their environmental impact and contribute to the shift towards a sustainable and regenerative economic model by prioritizing suppliers that employ eco-friendly materials and production methods and by incorporating sustainability into their procurement strategy. This strategy involves incentivizing suppliers to implement recycling, reusing, and minimizing waste and emissions, fostering environmental sustainability across the supply chain. In addition, Procurement 4.0 enables organizations to establish more cooperative partnerships with their suppliers, promoting advancements in sustainability and ethical standards. By leveraging digital platforms and technologies, businesses may collaborate directly with suppliers to create innovative solutions that mitigate environmental effects and enhance social consequences. This cooperative method not only improves the sustainability performance of individual enterprises but also stimulates more comprehensive industry-wide enhancements, contributing to the attainment of global sustainability objectives.

Procurement 4.0 offers organizations a substantial chance to strengthen their dedication to sustainability and ethical sourcing. Through digital technology, businesses may enhance the transparency of their supply chain, guarantee the ability to track and trace products and promote cooperation. This enables them to address the increasing expectations for industrial ecology and ethical practices. As businesses become more aware of the significance of sustainable and ethical practices for their brand image and financial performance, Procurement 4.0 will have a vital role in influencing the future of sustainable procurement.

4.4.9 Integration with ERP Systems

Integrating procurement technology with Enterprise Resource Planning (ERP) systems is essential for obtaining a comprehensive perspective on organizational data. Procurement 4.0 refers to the digitalization of procurement procedures, which significantly improves the efficiency and efficacy of supply chain management. An essential element of this change is incorporating procurement systems into Enterprise Resource Planning (ERP) systems. This integration is crucial for providing a smooth and uninterrupted transfer of information across various corporate divisions, facilitating more strategic procurement choices and operational efficiency. ERP systems support financial and operational procedures, including finance, HR, manufacturing, and supply chain management. Integrating procurement activities into the ERP system gives organizations a consolidated perspective of their operations, resulting in greater data precision, reduced procedures, and improved decision-making abilities. The ERP system integrates procurement data, including supplier information, purchase orders, and contract details, with financial records, inventory levels, and production plans. This holistic approach guarantees that procurement decisions are made using thorough and current information, enabling improved inventory management, demand planning, and budget control.

The integration of Procurement 4.0 with ERP systems is strengthened by modern technologies such as artificial intelligence (AI) and machine learning (ML). These technologies can analyze large quantities of procurement data stored in the ERP system to detect patterns, forecast future demand, and suggest the most advantageous purchase options. Artificial intelligence (AI) can automate repetitive procurement operations, such as invoice processing and supplier performance tracking. This automation reduces the need for human work and decreases the likelihood of mistakes. By using automation and intelligence, procurement teams can shift their attention towards strategic operations, such as managing supplier relationships and conducting negotiations, instead of being occupied with administrative duties. Integrating blockchain technology with ERP systems provides supplementary advantages in Procurement 4.0. Blockchain technology offers a reliable and easily accessible system for documenting transactions along the supply chain. This improves confidence and cooperation among suppliers and customers by ensuring security and transparency. Integrating blockchain technology with an ERP system enables immediate monitoring of merchandise and financial transactions, guaranteeing the accuracy of procurement information and adherence to contractual obligations. This degree of openness and security is especially crucial in businesses where the origin and genuineness of the supply chain are of utmost importance.

Ultimately, incorporating Procurement 4.0 into ERP systems is a crucial facilitator for organizations aiming to enhance the efficiency of their procurement and supply chain activities. This integration facilitates better decision-making, efficiency, and collaboration throughout the supply chain. The rapid growth of digital technology presents significant opportunities for future innovation in procurement and ERP integration. Organizations that embrace these innovations should expect to reap even more substantial benefits. The future of procurement relies on harnessing digital transformation to establish procurement activities that are more adaptable, responsive, and strategic. Integrating with ERP systems is a crucial milestone in this process.

4.4.10 Challenges and Opportunities

Procurement 4.0 is a notable advancement in the procurement function, brought about by the digital revolution and the incorporation of cutting-edge technologies like artificial intelligence (AI), the Internet of Things (IoT), blockchain, and big data analytics. This change allows organizations to improve efficiency, transparency, and strategic decision-making in their procurement processes. Nevertheless, adopting Procurement 4.0 also brings about a series of obstacles that organizations must overcome to capitalize on their potential advantages fully. Procurement 4.0 offers a significant opportunity to utilize data for making strategic decisions. Advanced analytics and artificial intelligence (AI) can efficiently analyze large volumes of data in real-time, offering procurement experts valuable insights into market trends, supplier performance, and risk factors. This facilitates enhanced decision-making, enabling organizations to optimize their choice of suppliers, negotiate more favorable terms, and uncover possibilities to save costs. Moreover, predictive analytics can anticipate forthcoming market situations and demand trends, empowering procurement to adopt a more proactive and adaptable approach to addressing fluctuations.

Nevertheless, the shift towards Procurement 4.0 presents notable data management and integration obstacles. A significant challenge several organizations face is the presence of isolated data systems and the absence of the capacity to combine and analyze data from various sources efficiently. Ensuring the quality and consistency of data becomes a top priority, as the effectiveness of AI and analytics in making decisions relies only on the data provided. Incorporating new digital tools and platforms into the current IT infrastructure necessitates a substantial commitment of time and money. It also demands meticulous planning to prevent any delays in procurement processes. Procurement 4.0 offers a notable chance to improve supplier relationships and foster collaboration. Digital platforms can enhance transparency and efficiency in interactions between buyers and suppliers by enabling improved communication, real-time information exchange, and collaborative problem-solving. Blockchain technology offers an unchangeable record of transactions and certificates, which enhances confidence and minimizes conflicts. The enhanced collaboration has the potential to foster innovation, improve product quality, and promote more sustainable procurement methods. However, digitizing procurement needs to change the organization's capabilities and culture. Procurement professionals must acquire additional skills in data analysis, utilization of digital tools, and strategic thinking. Organizations may encounter resistance to change, as staff may exhibit caution about new technology or uncertainty over the effects on their jobs and duties. It is essential to overcome cultural and skill-related obstacles to execute Procurement 4.0 efforts effectively.

Procurement 4.0 offers several opportunities for organizations to transform their procurement methods. These opportunities include making data-based decisions, collaborating more effectively with suppliers, and improving agility and creativity. However, to gain these advantages, one must overcome substantial obstacles such as data integration problems, technology, cultural adjustment, and skills enhancement. Organizations that effectively overcome these hurdles may strategically utilize Procurement 4.0 to fully exploit its potential fully, gaining a competitive edge in a progressively digitized and intricate global economy.

4.5 Smart Warehousing—Logistics

Smart warehousing is a significant advancement in logistics within the digital supply chain since it combines modern technology to bring forth disruptive possibilities. This complete strategy simplifies processes and dramatically improves efficiency, precision, and customer satisfaction. The emergence of technologies such as the Internet of Things (IoT), artificial intelligence (AI), robots, and big data analytics has triggered a transformation towards smarter, automated, and adaptable warehouse operations. The Internet of Things (IoT) is essential to smart warehousing since it allows for a linked environment where devices and equipment may interact effortlessly with management systems. The Internet of Things (IoT) sensors and devices can monitor inventory in real-time, observe environmental conditions, and offer essential data to enhance warehouse operations. This connectivity guarantees decision-makers receive real-time information regarding stock levels, product locations, and possible concerns, enabling prompt and well-informed judgments.

Artificial intelligence and machine learning algorithms boost smart warehousing by analyzing large quantities of data from Internet of Things (IoT) sensors. Artificial intelligence can forecast sudden increases in demand, optimize the amount of stock, and even automatically generate orders for restocking, guaranteeing that the warehouse is consistently ready to fulfill client requirements. In addition, robots and trucks powered by artificial intelligence (AI) can carry out monotonous duties like selecting, packaging, and moving items, hence minimizing human mistakes and enhancing the efficiency of operations. Robotics and automation are fundamental components of intelligent warehousing, revolutionizing the physical transportation of items within warehouse facilities. Automated guided vehicles (AGVs), robotic arms, and drones can efficiently transport inventory, decreasing the need for manual labor and lowering the likelihood of workplace mishaps. This automation not only expedites the warehousing process but also enables uninterrupted operations, so significantly augmenting production. Big data analytics is essential in smart warehousing since it offers valuable information on operational efficiency, consumer habits, and market patterns. Through the analysis of this data, warehouses may optimize their physical arrangements, better their inventory management, and improve the overall effectiveness of the supply chain. Predictive analytics can anticipate possible interruptions and recommend solutions to minimize their impact, ensuring that warehouse operations are flexible and responsive to changes in the supply chain environment.

Smart warehouse integration in the digital supply chain also prioritizes sustainability and energy efficiency. Intelligent systems can oversee and regulate lighting, heating, and cooling systems to reduce energy usage. In addition, optimizing space and inventory levels may decrease waste and improve the overall sustainability of warehouse operations. Nevertheless, the shift towards intelligent storage poses obstacles, such as the requirement for substantial investments in technology and infrastructure. Integrating new systems with old legacy systems can be intricate, necessitating proficient staff to oversee the transfer and run the latest technology.

Furthermore, as warehouses become increasingly interconnected and dependent on digital technologies, there is a growing worry regarding data security and privacy. Notwithstanding these difficulties, the advantages of intelligent storage are persuasive. The obvious benefits of implementing this technology include improved precision, faster execution of tasks, and lower expenses. Additionally, the capacity to adjust to market fluctuations and meet consumer expectations provides a competitive edge. Innovative warehousing facilitates the customization of client experiences by efficiently managing various items and catering to individual consumer preferences. The future of intelligent warehousing in the digital supply chain is optimistic since continuous technological improvements present fresh prospects for optimization and innovation. For example, blockchain technology might significantly improve transparency and security in warehouse operations by creating an unalterable record of inventory movements and transactions.

Smart storage is essential to contemporary logistics since it enhances operational effectiveness and fosters innovation in the digital supply chain. Warehouses are transforming by utilizing IoT, AI, robots, and big data analytics. This transformation allows warehouses to become dynamic and responsive components of the supply chain, making a substantial contribution to the success of organizations. As technology advances, the capabilities and advantages of intelligent warehousing will also improve, strengthening its crucial role in addressing the needs of a swiftly changing global market. Embarking to achieve fully optimized smart storage involves overcoming obstacles, but the potential benefits make the investment and effort worthwhile. This endeavor holds the promise of a future where logistics operations are characterized by increased intelligence, efficiency, and sustainability.

4.5.1 Robotic Transport

Robotic transport in the digital supply chain is an advanced development in logistics that significantly alters the movement of items within and between warehouses, industrial sites, and distribution centers. The significant progress is driven by developments in robotics, artificial intelligence (AI), and automation technologies, which allow for unparalleled levels of efficiency, precision, and safety. Robotic transport systems are becoming crucial elements in contemporary logistics strategies as firms face the growing need to fulfill the rising requirements for speed and dependability brought about by the e-commerce surge and globalized supply chains. The core of robotic transportation uses automated guided vehicles (AGVs) and autonomous mobile robots (AMRs). Automated Vehicles (AGVs), directed using wires, magnetic strips, or markers, have been employed for several decades to efficiently transfer large and heavy goods inside plant premises and warehouses. Nevertheless, the introduction of AMRs signifies notable progress since these robots maneuver through intricate algorithms, sensors, and maps, enabling them to move more flexibly throughout facilities without requiring physical guides. The adaptability of AMRs allows them to adjust to changes in their surroundings, dynamically navigate around barriers, and optimize routes for maximum efficiency.

AI integration with robotic transport systems amplifies their capabilities. Artificial intelligence algorithms utilize data obtained from robot sensors and the surrounding environment to enhance the process of planning routes, avoiding obstacles, and managing loads. Over time, these systems acquire knowledge from their operations, leading to ongoing enhancement in performance and efficiency. Artificial intelligence also enables predictive maintenance, detecting possible robot problems in advance to prevent downtime and maintain a continuous operating flow. Cobots, short for collaborative robots, are a new and exciting development in robotic transportation. Cobots are specifically designed to operate near human workers without causing harm. They are equipped with advanced sensors and artificial intelligence technology to detect and prevent collisions with humans and other obstacles. This partnership between people and robots synergizes the accuracy and resilience of robotic systems with human workers' cognitive talents and adaptability, enhancing the overall efficiency of logistical operations. Robotic transport is not limited to warehouses and manufacturing sites but is also used in the wider supply chain. Companies worldwide are already investigating and testing drone delivery systems as a potential answer to the difficulties in delivering goods over short distances, especially in metropolitan areas or isolated regions. Uncrewed aerial vehicles, or drones, can significantly decrease deliveries' duration and expenses while mitigating traffic congestion and pollutants typically linked to conventional delivery vehicles.

Another creative use of robotic transportation in the digital supply chain is the advancement of self-driving trucks and delivery vehicles. These vehicles, outfitted with sophisticated sensors, cameras, and artificial intelligence, aim to revolutionize long-distance transportation by enhancing safety, decreasing fuel usage, and optimizing the efficiency of freight transportation. Despite the existing legislative and technological obstacles, the potential impact of autonomous vehicles on logistics is substantial. The increasing adoption of automated transit in logistics is also causing alterations in supply chain strategy and infrastructure. Warehouses and distribution centers are being modified to incorporate robotic technologies, including more extensive aisles, automatic loading docks, and sophisticated charging stations for electric-powered robots. The advancement of infrastructure is crucial to facilitate the smooth functioning of autonomous transport systems and optimize their advantages. Implementing robotic transport systems highlights the importance of data security and privacy. Given the heavy reliance on data and connection, it is crucial to prioritize safeguarding against cyber attacks and guaranteeing data integrity in these systems. Cybersecurity measures and robust data management procedures are essential to protect operations and establish stakeholder confidence. The advantages of sustainability make robotic transport an appealing option. Robotic transport systems make supply chains more environmentally friendly by improving routes, minimizing energy usage, and facilitating the use of electric cars and drones. The connection with environmental sustainability aims is becoming increasingly crucial for businesses, consumers, and regulators.

Nevertheless, implementing robotic transportation in logistics is challenging. The initial capital outlay for technology and infrastructure can be significant, and there is a requirement for proficient staff to implement, oversee, and sustain these systems. Furthermore, there are apprehensions regarding the influence of automation on employment since it alters the character of labor in logistics and other fields.

Robotic transport is poised to revolutionize the logistics industry in the digital supply chain, providing significant benefits in terms of efficiency, adaptability, and environmental friendliness. With the ongoing progress of technology, incorporating robotic systems into logistics strategies is expected to become increasingly common. This trend will stimulate innovation and provide a competitive edge in the era of digitalization. Successfully harnessing robotic transport requires overcoming technological, legislative, and workforce-related obstacles. However, the advantages can potentially revolutionize the future of supply chain management.

4.6 Information and Communication

Information and communication are crucial components of the digital supply chain, providing the foundation for efficient coordination, collaboration, and decision-making throughout the whole supply chain ecosystem. Organizations in the digital era depend on sophisticated information technology and communication platforms to swiftly exchange data, disseminate knowledge, and synchronize operations with suppliers, partners, and customers in real-time. An essential element of information and communication within the digital supply chain is utilizing cloud-based collaboration platforms and enterprise resource planning (ERP) systems. These systems let organizations combine and optimize their supply chain activities, encompassing procurement, manufacturing, warehousing, and distribution. This enhances the ability to observe, understand, and manage supply chain operations with clarity, openness, and authority.

Furthermore, the utilization of information and communication technology empowers organizations to improve the visibility and traceability of their supply chain, facilitating more effective tracking and monitoring of commodities at every stage of the supply chain. Technological tools such as RFID tags, barcode scanners, and GPS tracking devices enable organizations to monitor the movement of items in real-time, starting at the place of origin and ending at the final destination. By using these technologies, organizations may enhance inventory precision, diminish cycle durations, and promptly adapt to fluctuations in demand or disturbances in the supply chain. Organizations may improve their decision-making, optimize inventory levels, and increase supply chain efficiency by having the capacity to see and track information in real time. Moreover, information and communication technology facilitates enhanced collaboration between organizations and their suppliers, partners, and consumers, promoting increased transparency and trust within the supply chain ecosystem. Organizations can utilize web-based portals and digital platforms to disseminate information, facilitate document sharing, and engage in collaborative efforts, such as product creation, quality assurance, and risk management. Organizations may enhance communication and cooperation by using a centralized platform. This will help simplify communication channels, minimize mistakes, and improve decision-making throughout the supply chain.

Furthermore, information and communication technology empowers organizations to improve consumer engagement and happiness by providing instantaneous access to order status, delivery updates, and product information. Organizations may utilize customer relationship management (CRM) systems and e-commerce platforms to offer tailored shopping experiences, monitor order status, and promptly address consumer queries. Organizations may use these technologies to enhance consumer happiness, cultivate brand loyalty, and gain a competitive advantage in the marketplace. Furthermore, information and communication technology empowers organizations to improve demand forecasting and inventory management by employing data analytics and predictive modeling. By examining past sales data, market trends, and consumer behavior, organizations may improve their ability to forecast future demand, optimize inventory levels, and minimize stock-outs or surplus inventory instances. Organizations may utilize machine learning algorithms to analyze extensive quantities of data and detect trends or anomalies that could indicate demand shifts or supply chain interruptions. By using predictive analytics, organizations may optimize inventory planning, save expenses, and boost customer happiness.

Moreover, information and communication technology empowers organizations to bolster supply chain resilience and risk management through improved visibility and control over supply chain processes. Organizations may utilize risk management software and supply chain visibility systems to detect hazards such as supplier interruptions, geopolitical concerns, or natural calamities. Organizations may manage risks, minimize the effect of interruptions, and maintain business continuity by monitoring key risk indicators and implementing contingency plans. Organizations may promptly address emerging risks and develop successful mitigation plans by engaging in real-time communication and cooperation with suppliers and partners.

Information and communication technologies are essential in the digital supply chain since they allow organizations to enhance visibility, collaboration, and decision-making throughout the whole supply chain ecosystem. Organizations may strengthen their supply chain operations, improve customer happiness, and gain a competitive advantage in the marketplace by utilizing cloud-based collaboration platforms, real-time tracking, traceability systems, and advanced analytics tools. However, to fully harness the capabilities of information and communication technologies in the digital supply chain, organizations must invest in the appropriate technology infrastructure, data analytics capabilities, and talent development. These investments are necessary for long-term success in today's ever-changing business environment.

4.7 Production 4.0

Production 4.0, also known as the fourth industrial revolution, is a significant change in manufacturing that incorporates advanced technologies like the Internet of Things (IoT), artificial intelligence (AI), robotics, and additive manufacturing into the digital supply chain. These technologies facilitate the development of intelligent factories that are networked, automated, and highly adaptable, resulting in substantial enhancements in production, quality, and efficiency. An essential element of Production 4.0 is incorporating Internet of Things (IoT) devices and sensors into the production process. These devices are used to monitor the functioning of equipment, gather real-time data, and enhance the efficiency of production workflows. For instance, sensors integrated into machinery can monitor the machine's condition, identify malfunctions, and activate maintenance notifications to avoid unexpected periods of inactivity, thereby decreasing expenses and enhancing the overall efficiency of the equipment.

Additionally, artificial intelligence (AI) and machine learning algorithms are essential in Production 4.0 since they provide predictive maintenance, quality control, and process optimization. AI algorithms can examine production data and detect trends, abnormalities, and correlations. This allows manufacturers to anticipate equipment breakdowns, enhance production schedules, and enhance the quality of their products. AI-driven quality control systems can examine goods in real-time, detect flaws, and autonomously modify production settings to uphold quality requirements. Moreover, AI-powered predictive analytics may enhance production planning and scheduling by accurately estimating demand, projecting material needs, and finding bottlenecks or inefficiencies in the manufacturing process. In addition, robotics and automation technologies play a crucial role in Production 4.0, transforming production processes by substituting manual labor with self-governing systems and collaborating robots (cobots). Robotic arms and automated guided vehicles (AGVs) can carry out repetitive operations like material handling, assembly, and packing more quickly, accurately, and consistently than human workers. This results in increased productivity and reduced production expenses. Collaborative robots, outfitted with sensors and AI algorithms, may operate with human operators to execute intricate jobs safely and efficiently, facilitating versatile and adaptable industrial procedures. Companies like BMW and Siemens have used collaborative robots to improve efficiency and adaptability in their manufacturing facilities.

Furthermore, additive manufacturing, sometimes called 3D printing, is a revolutionary technology in fabrication 4.0. It allows the fabricating of intricate parts and components in a customized and on-demand manner. Additive manufacturing systems construct products by adding layers based on digital designs, providing more design flexibility, faster production times, and decreased waste compared to conventional manufacturing techniques. This technique enables firms to efficiently make prototypes, replacement parts, and small quantities of personalized items at a lower cost, eliminating the requirement for costly tooling or molds. Companies such as General Electric and Airbus utilize additive manufacturing to fabricate airplane components, resulting in reduced weight, enhanced fuel economy, and streamlined supply chain logistics.

Furthermore, digital twins are becoming increasingly valuable in the context of Production 4.0 since they allow manufacturers to generate virtual duplicates of tangible assets, procedures, and systems. Digital twins combine real-time data from IoT sensors, industrial equipment, and supply chain systems to form a digital model of the physical world. This allows businesses to simulate, analyze, and optimize production processes virtually. Automotive manufacturers utilize digital twins to mimic production lines, detect bottlenecks, and enhance assembly processes before implementation. This results in reduced time-to-market and improved operational efficiency.

In addition, Production 4.0 enables the incorporation of supply chain partners and stakeholders by utilizing digital platforms and collaborative networks. Manufacturers may use cloud-based platforms and blockchain technology to exchange manufacturing data, cooperate on product development, and monitor the origin of raw materials and components across the supply chain. The openness and visibility provided by this system allow for improved coordination, traceability, and compliance during the production process. This ultimately reduces risks and strengthens confidence amongst supply chain parties. IBM and Maersk offer blockchain technology to enhance supply chain transparency and facilitate cooperation. Production 4.0 is a significant change in manufacturing driven by modern technologies like IoT, AI, robots, additive manufacturing, and digital twins. It brings about a transformation in the digital supply chain. These technologies facilitate the development of intelligent factories that are networked, automated, and highly adaptable, resulting in enhancements in production, quality, and efficiency. Additionally, manufacturing 4.0 enhances cooperation and integration across supply chain collaborators via digital platforms and networks, improving coordination, traceability, and adherence throughout manufacturing. Businesses that use Production 4.0 may obtain a competitive advantage, stimulate innovation, and attain success in the current high-speed and ever-changing business landscape.

4.7.1 3D Printer

3D printing, or additive manufacturing, is a revolutionary technique in the digital supply chain that allows for the creation of intricate parts and components quickly, flexibly, and with high levels of customization. An essential feature of 3D printing in the digital supply chain is its capacity to create items as needed, reducing the requirement for conventional manufacturing methods like machining or casting. By using this, organizations may effectively decrease the time it takes to complete a process, lower the expenses associated with storing excess inventory, and promptly adapt to fluctuations in customer demand or modifications in design specifications. Companies such as GE Aviation and Boeing utilize 3D printing technology to manufacture airplane parts as needed, resulting in shorter manufacturing periods and enhanced flexibility in the supply chain. Furthermore, 3D printing allows for the creation of intricate shapes and personalized patterns that are impossible with conventional manufacturing techniques. The ability to have design flexibility enables organizations to enhance the performance of parts, minimize material wastage, and develop groundbreaking products that fulfill precise client demands. Medicinal device makers utilize 3D printing technology to fabricate customized implants and prostheses designed to fit particular patients' unique anatomical structures and medicinal requirements. By utilizing the capabilities of 3D printing technology, these organizations have the potential to better patient outcomes and optimize consumer pleasure.

In addition, 3D printing facilitates decentralized manufacturing and distributed supply chains, hence decreasing dependence on centralized production facilities and long-distance transportation. Companies can utilize 3D printers to manufacture components in nearby locations closer to end-users or assembly factories. This practice helps minimize shipping expenses and carbon emissions linked to transportation. Using this decentralized manufacturing approach boosts the ability of the supply chain to withstand and recover from disruptions, decreases the likelihood of encountering hazards within the supply chain, and promotes the advancement of sustainable practices. Companies such as Adidas and Nike utilize 3D printing technology to manufacture personalized shoes directly at their facilities, resulting in shorter production times and less waste. In addition, 3D printing facilitates rapid prototyping and iteration, enabling enterprises to expedite product development cycles and provide new items to the market more quickly. Automotive manufacturers utilize 3D printing technology to create prototypes of car components, allowing engineers to test and refine ideas before large-scale manufacturing efficiently. This efficient prototyping capacity decreases the amount of time it takes to bring a product to market, reduces the expenses associated with development, and enhances the overall quality of the product.

Furthermore, 3D printing allows enterprises to tailor items to niche markets or specific client groups, generating novel prospects for innovation and distinctiveness. In addition, 3D printing enhances the ability to withstand disruptions in the supply chain by decreasing reliance on conventional suppliers and supply chain networks. Companies may utilize 3D printers to manufacture spare components as needed, removing the necessity for extensive inventory or dependence on suppliers from other countries. This results in decreased lead times, mitigating supply chain interruptions and enhancing overall resilience. In addition, 3D printing enables enterprises to manufacture components with intricate shapes or in small quantities at a reasonable cost, minimizing the likelihood of becoming outdated and having surplus inventory.

Moreover, 3D printing facilitates sustainable production by minimizing material wastage and energy usage in contrast to conventional manufacturing techniques. 3D printers utilize the necessary material to manufacture a component, minimizing waste and decreasing environmental impact. In addition, enterprises can use recycled or bio-based materials as raw materials for 3D printing, which can further reduce the environmental impact of production. In addition, 3D printing allows enterprises to enhance part designs to achieve lightweight and material efficiency, resulting in improved fuel economy and reduced emissions in the transportation and aerospace sectors. In addition, 3D printing enhances supply chain innovation by allowing for the development of novel business models and providing value-added services. For instance, corporations can provide clients with 3D printing services, enabling them to personalize items or create replacement components as needed. This generates additional sources of income and improves client contentment by providing customized solutions. In addition, enterprises may utilize 3D printing technology to fabricate intricate assemblies or multi-material components that are challenging or unfeasible to produce using conventional methods. This enables them to explore novel prospects for product innovation and differentiation.

Moreover, 3D printing facilitates supply chain flexibility and agility by allowing organizations to adjust promptly to fluctuations in demand or market conditions. Companies may utilize 3D printing technology to manufacture small quantities of personalized items efficiently, decreasing the requirement for extensive production runs and mitigating inventory-related uncertainties. Furthermore, organizations can modify the settings of 3D printers to manufacture various components or goods quickly, allowing them to adapt promptly to changing client preferences or disruptions in the supply chain.

Ultimately, 3D printing transforms the digital supply chain by facilitating manufacturing, customization, and innovation as needed. Through 3D printing technology, firms may decrease the time it takes to complete a task, raise the overall excellence of their products, and strengthen the ability of their supply chain to withstand and recover from disruptions. In addition, 3D printing facilitates sustainable production methods by minimizing material waste and energy use. In addition, 3D printing enables the development of new business models and value-added services, which creates chances for increasing revenue and engaging with customers. The ongoing development and advancement of 3D printing technology are anticipated to significantly influence the digital supply chain, leading to additional advancements and changes in many sectors.

4.7.2 Wearable Technology

Wearable technology is now more common in the digital supply chain and provides creative solutions to improve productivity, safety, and efficiency in many sectors. An essential element of wearable technology in the digital supply chain is its use in labor management and logistical operations. Wearable gadgets like smart glasses, wristbands, and intelligent helmets are coupled with sensors and augmented reality (AR) capabilities. These devices allow workers in warehouses, factories, and distribution centers to receive real-time information, instructions, and assistance while completing activities. These gadgets offer the convenience of hands-free operation, enhancing ergonomics and enabling workers to concentrate on activities more efficiently. In addition, wearable technology allows for continuous monitoring of worker health and safety throughout the digital supply chain. Wearable sensors can monitor worker health and avoid accidents or injuries caused by weariness by tracking physiological data such as heart rate, body temperature, and physical effort levels. In addition, integrating position tracking and proximity sensors into wearable devices may notify workers of possible dangers or collisions in the workplace, enhancing safety and minimizing the likelihood of accidents. Boeing and DHL utilize wearable technologies to improve worker safety and ergonomics.

Moreover, wearable gadgets enhance managing inventories and fulfill orders in the digital supply chain. For instance, using wristbands or gloves equipped with RFID technology enables workers to instantly scan and monitor inventory items, minimizing mistakes and enhancing the precision of inventory records. Wearable barcode scanners and ring scanners allow for scanning items and packages without using hands, which improves efficiency and decreases the time it takes to complete selecting, packing, and shipping tasks. Furthermore, wearable gadgets integrated with location tracking and indoor positioning systems (IPS) allow for accurate monitoring of items at every stage of the supply chain, starting from the warehouse and ending at the customer's doorstep. In addition, wearable technology improves the visibility and transparency of the supply chain by offering real-time tracking and monitoring of assets, commodities, and shipments. Wearable GPS trackers and RFID tags integrated into containers, pallets, or vehicles allow organizations to monitor the whereabouts and condition of products during transportation, facilitating improved coordination and optimization of transportation routes. In transit, wearable devices with environmental sensors can track and measure temperature, humidity, and other environmental factors. This ensures that delicate or perishable commodities maintain their quality and remain intact. UPS and FedEx utilize wearable technologies to enhance visibility and traceability in their supply chain operations.

In addition, wearable technology facilitates the ability to anticipate and address maintenance needs and manage assets within the digital supply chain. For example, wearable sensors connected to equipment and machinery can track performance parameters like temperature, vibration, and energy usage to identify defects or breakdowns before they happen. Implementing this proactive maintenance strategy aids in avoiding expensive periods of inactivity, prolongs the lifespan of assets, and enhances overall equipment effectiveness (OEE). In addition, using augmented reality (AR) glasses and headsets supplied with maintenance manuals and troubleshooting instructions allows personnel to carry out maintenance operations more efficiently and precisely. This results in reduced service times and improved asset uptime.

Moreover, wearable technology enhances the ability of supply chain stakeholders to collaborate and communicate, regardless of their geographical position or closeness. Wearable devices with speech recognition and communication capabilities allow workers to connect with colleagues, supervisors, or specialists in real time, making it easier to provide remote support and fix problems. This enhances cooperation, the exchange of knowledge, and the decision-making process across geographically dispersed teams, resulting in quicker resolution of issues and enhanced operational efficiency. Walmart and Amazon utilize wearable technologies to improve employee communication and collaboration in warehouses and fulfillment centers.

Furthermore, wearable technology facilitates customized and situationally aware services inside the digital supply chain. Wearable gadgets with biometric sensors and AI algorithms may analyze user behavior and preferences. This analysis allows them to offer personalized recommendations, warnings, and notifications tailored to individual requirements. This optimizes user experience, promotes efficiency, and cultivates increased employee involvement and contentment. Moreover, wearable gadgets with language translation capabilities facilitate multilingual communication and training in various worldwide supply chain contexts, thereby overcoming language obstacles and enhancing accessibility.

Furthermore, wearable technology enables the gathering and examination of data in the digital supply chain, offering practical observations on operational efficiency, worker conduct, and process enhancement. Wearable devices produce extensive data about worker activities, movement patterns, and ambient circumstances. This data may be examined using sophisticated analytics techniques to detect trends, patterns, and areas for enhancement. By utilizing a data-driven strategy, firms can make well-informed decisions, streamline workflows, and achieve ongoing improvements in their supply chain operations.

Ultimately, wearable technology transforms the digital supply chain by offering inventive solutions to improve labor management, safety, productivity, and visibility. Wearable gadgets, such as augmented reality (AR) glasses, intelligent helmets, wristbands, and sensors, have diverse uses in many sectors. They enhance operational efficiency and facilitate improved decision-making. The ongoing development and maturation of wearable technology are anticipated to have an increasing influence on the digital supply chain, leading to more advancements and changes in supply chain operations.

4.8 Supply Chain Stages

The digital supply chain fundamentally changes how goods and services are transported from suppliers to customers. It leverages digital technology to improve efficiency, visibility, and responsiveness at every process step. This transition involves many crucial stages: planning, procurement, production, distribution, and product returns. Every step has been dramatically influenced by digital technologies, including the Internet of Things (IoT), artificial intelligence (AI), blockchain, and big data analytics. Here is a detailed analysis of each step in the digital supply chain, supported with examples to demonstrate the effects of digitalization.

4.8.1 Planning

A vital component of supply chain management, planning ensures the effective coordination of activities, processes, and resources to satisfy customer demand while minimizing expenses and optimizing productivity. Planning is vital for each phase of the digital supply chain to ensure the uninterrupted movement of products and services from suppliers to customers. Demand planning is the initial phase of digital supply chain planning, during which organizations predict consumer interest and purchase of their products or services. To accurately forecast future demand, sophisticated forecasting techniques, including predictive analytics and machine learning algorithms, are applied to historical sales data, market trends, and customer behavior. As an illustration, corporations such as Amazon and Walmart employ advanced demand forecasting models to predict consumer demand and optimize inventory levels, thereby mitigating the occurrence of stockouts and surplus inventory. Organizations determine the optimal production schedule to meet anticipated demand while balancing resource availability, capacity constraints, and production costs during the production planning phase two. Advanced planning and scheduling (APS) software and manufacturing execution systems (MES) are digital technologies that assist organizations in optimizing production processes, sequencing tasks, and allocating resources effectively. Automotive manufacturers, for instance, optimize production schedules, minimize transition times, and maximize throughput in their assembly facilities through APS software.

In inventory planning, the third stage of the planning process, organizations ascertain the most favorable inventory levels to satisfy customer demand while minimizing transporting costs and the occurrence of stockouts. To balance the cost of stockouts against the cost of inventory storage, economic order quantity (EOQ) and just-in-time (JIT) inventory management are implemented as inventory optimization techniques. Furthermore, digital supply chain inventory planning employs cutting-edge technologies to enhance inventory visibility and precision, including barcode scanning, RFID, and real-time inventory tracking systems. As an illustration, retailers utilize RFID tags and inventory management systems to optimize inventory replenishment processes and monitor the movement of products in real time. Distribution planning is the fourth stage of the planning process, during which businesses maximize the passage of products from manufacturing facilities to distribution centers and, ultimately, to consumers. Transportation management systems (TMS) and route optimization software are examples of digital technologies that assist businesses in optimizing shipping routes, consolidating cargo, and reducing transportation costs. TMS software is utilized by logistics companies, for instance, to maximize distribution routes, decrease petroleum consumption, and enhance delivery precision. Supplier planning, the fifth stage of the planning process, entails collaborative efforts between organizations and suppliers to guarantee a consistent and effective provision of materials and components. Digital supply chain platforms enable businesses to exchange demand projections, inventory levels, and production schedules by facilitating collaboration and real-time communication with suppliers. As an illustration, vendors-managed inventory (VMI) systems and supplier portals are utilized by manufacturers to coordinate production schedules, share information with suppliers, and reduce lead times and inventory costs.

In capacity planning, the sixth stage of the planning process, organizations ascertain the most advantageous capacity levels for their transportation resources, production facilities, and warehouses to satisfy consumer demand. Digital simulation and modeling tools assist organizations in examining various scenarios, assessing capacity limitations, and detecting potential obstacles within the supply chain. As an illustration, manufacturers employ discrete event simulation software to optimize production layouts, efficiently allocate resources, and enhance the overall performance of the supply chain. Risk management planning constitutes the seventh stage of the planning process. During this phase, organizations identify, evaluate, and mitigate potential threats to supply chain operations. Predictive analytics and digital risk management tools assist organizations in identifying potential hazards, including but not limited to supplier disruptions, geopolitical uncertainties, and natural disasters. To monitor key risk indicators, assess the impact of disruptions, and develop contingency plans to mitigate risks and ensure business continuity, for instance, organizations implement supply chain risk management software. Sustainability planning constitutes the eighth phase of the planning process, during which organizations integrate ethical, social, and environmental factors into their supply chain strategies. Digital supply chain platforms facilitate the monitoring and assessing sustainability indicators throughout the supply chain, including carbon emissions, water consumption, and labor practices. As an illustration, corporations employ blockchain technology to ascertain adherence to sustainability standards and trace the origin of raw materials, thereby ensuring accountability and transparency in their supply chain operations.

Financial planning constitutes the ninth stage of the planning process, during which organizations allocate resources, create budgets, and manage costs throughout the supply chain. Digital finance and accounting systems enable organizations to optimize expenditures, reduce expenses, and increase profits by providing real-time visibility into financial performance. For instance, enterprises implement enterprise resource planning (ERP) software to facilitate financial data integration with supply chain operations, enhancing financial planning and decision-making. In the eleventh phase of the planning process, known as technology planning, organizations evaluate digital innovations and emergent technologies to improve supply chain capabilities and bolster competitiveness. By implementing digital transformation strategies, including but not limited to cloud computing, big data analytics, and artificial intelligence, businesses can automate operations, enhance decision-making throughout the supply chain, and capitalize on data-driven insights. For instance, organizations employ sophisticated analytics and machine learning algorithms to streamline supply chain operations, forecast customer demand, and detect prospects for financial savings and increased revenue. In contingency planning, the eleventh stage of the planning process, organizations formulate alternative strategies and fallback plans to mitigate the effects of disruptions and unanticipated events. Digital supply chain platforms facilitate the simulation of various scenarios, evaluating contingency plan efficacy, and developing response strategies to mitigate disruptions. Organizations employ scenario planning and simulation tools to develop strategies to reduce risks and ensure business continuity in supply chain disruptions such as natural disasters, labor strikes, or transportation delays.

Planning is pivotal in supply chain management during the digital age. It comprises many phases, including but not limited to supplier planning, capacity planning, risk management planning, sustainability planning, financial planning, technology planning, and contingency planning. In today's dynamic and interconnected business environment, digital technologies such as blockchain, artificial intelligence, and advanced analytics are reshaping how organizations strategize and oversee their supply chains. This transformation empowers these entities with enhanced visibility, agility, and resilience.

4.8.2 Sourcing

Sourcing is an essential component of the supply chain phases, comprising tasks such as supplier identification, evaluation, and selection to obtain the vital products and services required for production and distribution. Sourcing has experienced substantial changes in the digital age of supply chains, as it has become increasingly dependent on data-driven approaches and sophisticated technologies to optimize supplier relationships, minimize expenses, and avert potential hazards. Implementing e-procurement platforms and online marketplaces is a critical element of digital sourcing. These tools empower organizations to optimize the sourcing procedure, consolidate supplier information, and automate procurement workflows. As an illustration, e-procurement solutions offered by Alibaba and Ariba enable businesses to procure goods and services from a worldwide network of vendors, thereby facilitating judicious sourcing choices.

In addition, digital procurement facilitates enhanced transparency and visibility across the entire supply chain, granting organizations instantaneous access to data regarding suppliers, performance indicators, and market developments. For example, supplier portals and supplier relationship management (SRM) systems enable businesses to track supplier compliance, identify cost savings or process enhancement opportunities, and monitor supplier performance. Organizations can strategically utilize digital tools and analytics to negotiate advantageous contracts, streamline supply chain risks, and make data-driven sourcing decisions.

In addition, digital sourcing promotes innovation and strategic collaboration with suppliers, thereby cultivating enduring partnerships that generate mutual value. For instance, collaborative procurement platforms facilitate supplier engagement during the initial stages of product development, allowing organizations to harness their specialized knowledge to create inventive solutions and enhance product excellence jointly. Moreover, supplier innovation programs and crowdsourcing platforms incentivize suppliers to contribute technologies, ideas, and expertise, fostering supply chain-wide innovation and continuous improvement. Procter and Gamble and Toyota, among others, have adopted digital procurement as a strategic approach to foster supplier collaboration and stimulate innovation within their specific sectors. Moreover, by granting increased visibility into supplier operations and supply chain networks, digital sourcing empowers organizations to adopt sustainable and ethical sourcing practices. An illustration of this is how blockchain technology enables businesses to ascertain the origin of their products and raw materials, thereby assuring adherence to environmental and social regulations. Organizations can enhance brand repute, mitigate the risk of counterfeit products, and verify the authenticity and sustainability of their supply chains by utilizing blockchain-based traceability solutions. Organizations such as Nestlé and Unilever implement blockchain technology to enhance supply chain transparency and encourage responsible procurement.

In addition, by utilizing data analytics and predictive modeling, digital procurement enables businesses to optimize supplier qualification and selection processes. As an illustration, supplier risk management platforms evaluate supplier performance, financial stability, and regulatory compliance by utilizing sophisticated algorithms. Organizations can enhance the efficiency of their supply chains by identifying high-risk suppliers and devising mitigation strategies by examining historical data and external risk factors. Moreover, machine learning algorithms and artificial intelligence (AI) can identify patterns and trends in enormous amounts of data, allowing businesses to optimize supplier portfolios and make informed procurement decisions. In addition, digital procurement empowers organizations to negotiate more favorable terms and conditions with suppliers by utilizing real-time market intelligence and competitive insights. As an illustration, market analytics tools and dynamic pricing algorithms grant organizations insight into supplier capacities, demand–supply dynamics, and pricing trends, thereby empowering them to negotiate contract terms and prices that are most favorable. Moreover, digital sourcing platforms enable organizations to solicit competitive offers from numerous suppliers and select the most cost-effective alternatives by facilitating reverse auctions and competitive tendering processes. Prominent organizations such as SAP Ariba and Amazon Business offer digital sourcing solutions that streamline procurement procedures and generate financial benefits.

Additionally, through expedited supplier onboarding and integration, digital sourcing empowers organizations to improve supply chain responsiveness and agility. As an illustration, cloud-based supplier management platforms enable organizations to efficiently manage supplier registration and qualification procedures, facilitating the swift integration of fresh suppliers and the expansion of supplier networks as required. Moreover, digital procurement platforms allow organizations to promptly respond to fluctuations in demand or disruptions in the supply chain by providing real-time visibility into the performance and availability of suppliers. This agility enables businesses to meet customer demands and ensure business continuity amidst volatile market conditions. As an additional benefit, digital sourcing enables adherence to industry standards and regulatory requirements by providing centralized documentation and audit traces. As an illustration, supplier compliance management systems facilitate monitoring supplier licenses, certifications, and regulatory approvals by businesses, guaranteeing compliance with ethical and legal principles. In addition, digital procurement platforms offer the advantage of automated workflows and documentation management functionalities, effectively minimizing the risk of non-compliance and streamlining audit procedures. Pharmaceutical and aerospace companies, among others, utilize digital procurement solutions to guarantee adherence to stringent regulatory mandates and quality criteria.

In summary, implementing digital sourcing in the digital supply chain has revolutionized sourcing procedures, allowing businesses to maximize supplier connections, minimize expenses, and efficiently address risks. Using e-procurement platforms, data analytics, and collaborative sourcing tools, organizations can optimize the sourcing procedure, augment the visibility of the supply chain, and foster strategic collaboration and innovation with their suppliers. In addition, digital sourcing facilitates the implementation of ethical and sustainable sourcing practices, optimizing supplier qualification and selection procedures and negotiating more favorable terms and conditions with suppliers. Organizations that persistently adopt digital procurement technologies have the potential to attain success in the ever-evolving business landscape of the present while also attaining a competitive edge.

4.8.3 Manufacturing

Manufacturing is a critical phase within the digital supply chain, during which unprocessed materials are refined into finished goods by implementing sophisticated technologies and automated procedures. Manufacturing operations have been transformed by digitization, which has allowed businesses to increase flexibility, quality, and productivity while decreasing costs and lead times. An imperative element of the digital supply chain manufacturing process entails implementing intelligent factories outfitted with interconnected sensors, Internet of Things (IoT) apparatus, and automata. These technologies facilitate real-time monitoring and control of production processes, enabling organizations to maximize output, reduce outages, and enhance equipment efficacy (OEE). For instance, IoT-enabled manufacturing systems are utilized by Siemens and Bosch to collect and analyze data from production equipment, thereby facilitating initiatives such as predictive maintenance and continuous improvement.

In addition, sophisticated production techniques, including computer numerical control (CNC) machining and additive manufacturing (3D printing), can be implemented via digital manufacturing. By eliminating the need for expensive tooling or molds and reducing lead times, additive manufacturing enables businesses to produce prototypes and complex parts with incomparable design freedom and adaptability. In addition, CNC machining ensures production consistency and quality by facilitating the precise and efficient machining of plastic and metal components. For example, additive manufacturing is utilized by General Electric and Boeing to manufacture aircraft components, thereby reducing weight, increasing fuel efficiency, and streamlining supply chain logistics. In addition, digital manufacturing enables the incorporation of artificial intelligence (AI) and data analytics into the production process, facilitating process optimization, quality control, and predictive analytics. By identifying patterns, anomalies, and opportunities for advancement in immense quantities of production data, AI algorithms enable businesses to forecast equipment failures, optimize production schedules, and enhance product quality. Moreover, machine learning algorithms can optimize production parameters dynamically, reducing waste and guaranteeing consistent quality. Automobile manufacturers, for instance, employ quality control systems propelled by artificial intelligence to inspect and validate components during production, thereby reducing defects and increasing customer satisfaction.

Moreover, digital manufacturing facilitates enhanced personalization and customization of products to cater to customers' specific demands and preferences. By utilizing flexible manufacturing processes and digital design tools, businesses can customize products to meet the particular needs of each consumer while maintaining operational efficiency and scalability. As an illustration, prominent corporations such as Nike and Adidas provide customizable apparel and footwear via sophisticated manufacturing processes and digital design platforms. This enables patrons to personalize their purchases electronically and obtain them relatively quickly. Additionally, digital manufacturing promotes agility and resilience in supply chains through the ability to swiftly adjust production processes in light of fluctuations in market conditions or disruptions. Utilizing cloud-based manufacturing systems and digital twins empowers organizations to simulate and enhance production operations virtually, facilitating prompt adjustments in response to emerging demands or unanticipated circumstances. In addition, collaboration platforms and digital supply chain networks enable organizations to share production resources and expertise with partners, facilitating the implementation of agile and collaborative manufacturing strategies. Prominent corporations such as BMW and Toyota utilize digital manufacturing technologies to enhance the adaptability and promptness of their production processes, thereby sustaining a competitive advantage in ever-changing markets.

In summary, digital manufacturing is revolutionizing the manufacturing phase of the supply chain by allowing organizations to improve efficiency, flexibility, and productivity while decreasing expenses and lead times. Businesses can enhance product quality, optimize production processes, and respond swiftly to shifting market demands by utilizing AI, intelligent factories, additive manufacturing, and data analytics. In addition, digital manufacturing facilitates increased product customization and personalization, improves the resilience and agility of supply chains, and encourages innovation and collaboration throughout supply chain networks. It is anticipated that the ongoing development and maturation of digital manufacturing technologies will have an increasing influence on the supply chain, stimulating additional advancements and alterations in manufacturing processes.

4.8.4 Delivery

Logistics, or the delivery phase, encompasses the movement of products from producers to distributors, retailers, or even directly to consumers. As an illustration, UPS utilizes its ORION (On-Road Integrated Optimisation and Navigation) system, an artificial intelligence-powered instrument designed to optimize delivery routes by integrating parcel information, real-time traffic data, and weather conditions. By utilizing this technology, UPS can reduce delivery times and conserve millions of liters of gasoline annually. Delivery signifies the ultimate phase within the digital supply chain, during which completed goods are conveyed from production facilities to ultimate consumers or distribution centers. The delivery process has been significantly transformed by digitization, which has empowered organizations to improve customer satisfaction, functionality, and efficiency by utilizing cutting-edge logistics technologies and data-centric approaches. An essential element of digital supply chain delivery is the implementation of real-time monitoring and tracking systems, including telematics solutions and GPS-enabled tracking devices. These technological advancements furnish organizations with up-to-date information regarding the whereabouts and condition of cargo, thereby facilitating the proactive administration of delivery routes, timetables, and possible interruptions. As an illustration, corporations such as Amazon and UPS employ real-time tracking systems to give customers precise delivery notifications and estimates, augmenting the delivery process's transparency and dependability.

Furthermore, the advent of mobile applications and digital delivery platforms has revolutionized how clients engage with delivery services, facilitating enhanced levels of convenience, adaptability, and personalization. Mobile applications enable customers to monitor the progress of their shipments, engage in conversations with delivery carriers, and specify delivery preferences, including time windows and instructions. In addition, digital delivery platforms provide functionalities, including geofencing, dynamic routing, and route optimization, which enhance delivery operations' efficacy. As an illustration, prominent companies such as DoorDash and Uber employ digital delivery platforms to instantaneously pair drivers with delivery orders, optimize delivery routes, and ensure customers have a seamless delivery experience. In addition, digital delivery facilitates businesses' implementation of novel delivery models, including autonomous delivery vehicles and crowdsourced delivery. By connecting companies with local couriers or independent contractors who can fulfill delivery orders on demand, crowdsourced delivery platforms enable businesses to implement flexible and cost-effective delivery strategies. Furthermore, autonomous delivery vehicles, such as robotics and drones, can fundamentally transform the last-mile delivery sector by providing expedient, eco-friendly, and effective delivery alternatives. For last-mile delivery, businesses like FedEx and Domino's Pizza are investigating using drones and robotics to reduce delivery costs and increase customer delivery speed.

Moreover, digital delivery enables delivery operations to undergo continuous refinement and data-driven decision-making. Through acquiring and examining delivery performance data, organizations can discern patterns, obstacles, and opportunities for enhancement within the delivery procedure. Utilizing machine learning algorithms and predictive analytics empowers organizations to anticipate delivery demand, optimize delivery routes, and more efficiently allocate resources. Furthermore, digital delivery platforms enable organizations to collect feedback from customers and delivery carriers to improve the quality and dependability of delivery services, enabling them to identify problems and implement corrective measures. In summary, digital delivery is revolutionizing the supply chain's delivery phase by empowering businesses to improve customer satisfaction, visibility, and operational effectiveness via data-centric approaches and sophisticated logistics technologies. Innovative delivery models, including crowdsourced and autonomous delivery vehicles, real-time tracking systems, and digital delivery platforms, transform how products are transported to consumers. In the contemporary, dynamic business landscape, organizations can enhance delivery performance, optimize delivery operations, and maintain competitiveness by harnessing the potential of digital delivery technologies.

4.8.5 Returns

The returns process includes managing items that have been returned due to faults, consumer discontent, or other causes. Digital technologies enhance this process by offering increased transparency and management. For instance, Zara uses RFID tags to handle inventory and streamline returns effectively. Utilizing these tags enables the organization to categorize efficiently returned products and then replenish their stock or divert them as required, resulting in enhanced customer satisfaction and operational effectiveness. Returns management is critical in the digital supply chain when customers return items for flaws, damages, or changing tastes. The digitalization trend has wholly transformed the returns process, enabling organizations to make it more efficient, improve customer satisfaction, and optimize the activities related to reverse logistics. An essential element of returns management in the digital supply chain is the incorporation of digital returns portals and self-service platforms. These technologies help clients initiate returns, monitor return shipments, and request refunds or swaps online. This offers ease and flexibility while decreasing the administrative workload on customer support personnel. Companies like Amazon and Walmart provide digital returns portals, enabling customers to efficiently and effortlessly conduct returns, enhancing happiness, and fostering loyalty.

Furthermore, utilizing digital analytics and data-driven insights allows organizations to clearly understand return patterns, underlying reasons for returns, and potential areas for enhancement. By analyzing return data, organizations may discern patterns, identify reoccurring difficulties, and adopt remedial measures to diminish return rates and enhance product quality. Predictive analytics and machine learning algorithms can predict the number of returns, anticipate when products will be returned, and optimize methods for managing and replenishing inventories. In addition, sentiment analysis technologies allow organizations to comprehend consumer feedback and sentiment-related return experiences, enabling them to customize their return procedures and policies to align with customer expectations. In addition, digital returns management allows for efficient and cost-effective reverse logistics activities, which help organizations expedite the processing, refurbishing, and disposal of returned items. Automated returns processing systems utilize barcode scanning, RFID technology, and automated sorting equipment to accelerate the processing of returned products, decreasing cycle times and saving labor costs. In addition, digital returns systems allow organizations to optimize their decision-making when disposing of returned items. This includes restocking, refurbishing, recycling, or liquidating the products based on their condition and worth. Companies such as Best Buy and Apple utilize digital returns management systems to effectively handle returned gadgets and devices, optimizing the value of recovery and minimizing waste.

In addition, digital returns management allows organizations to improve customer communication and transparency during the returns process, minimizing ambiguity and consumer dissatisfaction. Through real-time updates and alerts, customers are kept informed about the progress of their returns, including return shipments, reimbursements, and exchanges. This helps to enhance satisfaction and loyalty. Additionally, digital returns systems allow consumers to access and understand return rules, methods, and schedules, enabling them to make well-informed choices and handle their returns more efficiently. Organizations may enhance customer support and help during returns by utilizing digital communication channels like email, SMS, and chatbots. This enables them to address concerns promptly and effectively. Ultimately, digital returns management is transforming the returns phase of the supply chain by allowing organizations to simplify returns procedures, improve customer satisfaction, and optimize reverse logistics operations. Organizations may enhance their returns management by utilizing digital returns portals, analytics, and automation technologies. These systems improve visibility, efficiency, and transparency, reducing costs and increasing customer satisfaction. Organizations can convert returns management from a cost center into a strategic asset by utilizing digital technology and employing data-driven strategies. This transformation may increase customer loyalty and a competitive edge in the digital economy.

4.8.6 Integration

The integration stage focuses on establishing efficient communication and smooth data transfer throughout various phases of the supply chain. SAP's ERP solutions exemplify the integration of different components of the supply chain into a cohesive system. This connection offers immediate and up-to-date insight throughout the supply chain, enabling improved decision-making and streamlined operations. Integration is crucial at the many phases of the supply chain, as it ensures smooth coordination and collaboration across numerous stakeholders, processes, and systems. In the current era of digital supply chains, the integration process has gained even more importance as organizations want to utilize technology and data to improve their supply chain's visibility, flexibility, and effectiveness. An essential element of integrating the digital supply chain is connecting different systems and data sources via application programming interfaces (APIs), middleware, and integration platforms. These technologies provide the connection and synchronization of data between various systems, including enterprise resource planning (ERP), customer relationship management (CRM), and supply chain management (SCM) systems. This ensures that there is only one reliable source of information and enables immediate decision-making. Companies like SAP and Oracle provide integration platforms, allowing organizations to connect data and operations seamlessly throughout their complete supply chain ecosystem, from suppliers to customers.

Furthermore, integration facilitates cooperation and the exchange of information across partners in the supply chain, promoting stronger connections and allowing for coordinated planning and implementation. CPFR efforts facilitate the sharing of demand projections, inventory levels, and production plans among trade partners, promoting improved coordination and synchronization of supply chain operations. In addition, collaborative platforms and digital marketplaces facilitate information interchange, enable product design and innovation collaboration, and facilitate real-time production and distribution coordination among suppliers, manufacturers, distributors, and retailers. Companies like Alibaba and Walmart utilize digital platforms to establish connections between suppliers and customers, facilitating smooth cooperation and communication across the supply chain. Moreover, integration facilitates complete visibility and traceability across the supply chain, providing organizations with immediate access to inventory levels, order status, and shipment tracking information. Advanced visibility tools and control towers utilize data from several sources, including IoT sensors, RFID tags, and GPS tracking devices, to oversee the transportation of commodities from suppliers to customers. By integrating data from both internal and external sources, organizations can detect areas of congestion, streamline logistical procedures, and effectively anticipate and handle potential hazards and interruptions in the supply chain. Companies like Amazon and UPS utilize control tower systems to oversee and govern their worldwide supply chain operations, facilitating proactive decision-making and improving customer service.

Furthermore, integration allows organizations to utilize cutting-edge technologies like artificial intelligence (AI), machine learning (ML), and blockchain to improve their supply chain's visibility, efficiency, and resilience. Artificial intelligence (AI) and machine learning (ML) algorithms can analyze large quantities of supply chain data to detect patterns, trends, and anomalies. This analysis allows for predictive analytics, demand forecasting, and optimization of supply chain operations. Blockchain technology facilitates the safe and transparent exchange of data and transaction processing, reducing instances of fraud, improving the capacity to track and trace items, and enhancing confidence among parties in the supply chain. Companies like IBM and Maersk utilize blockchain technology to digitize and optimize global trade operations, facilitating safe and transparent transactions while minimizing paperwork and delays. Integration is crucial for maximizing supply chain efficiency and staying competitive in the digital age. Organizations may improve cooperation, visibility, and agility by integrating systems, processes, and data throughout the supply chain. This allows them to promptly adapt to shifting market dynamics and client requirements. Furthermore, integration will enable organizations to utilize new technologies such as artificial intelligence (AI), machine learning (ML), and blockchain to improve their supply chain's efficiency, robustness, and innovation. As organizations increasingly adopt digital transformation, integrating systems and processes will continue to be crucial in achieving success in the digital supply chain.

4.8.7 Collaboration

Effective collaboration among suppliers, manufacturers, and distributors is essential for ensuring a supply chain that can quickly adapt and respond to changes. TradeLens, a platform created by Maersk and IBM utilizing blockchain technology, facilitates the safe and transparent exchange of information between parties in the supply chain. This kind of coordination enhances shipment efficiency and mitigates delays. Collaboration is essential to supply chain management, facilitating tight cooperation between organizations, partners, suppliers, and consumers to achieve objectives such as boosting efficiency, minimizing expenses, and improving customer happiness. Collaboration has undergone considerable changes in the era of digital supply chain, primarily due to technological improvements and the emergence of data-sharing platforms. An essential element of cooperation in the digital supply chain is the implementation of collaborative planning, forecasting, and restocking (CPFR) efforts. CPFR facilitates real-time information exchange between trade partners, including demand projections, inventory levels, and production plans. This promotes improved coordination of supply chain operations. Companies like Walmart and Procter and Gamble utilize Collaborative Planning, Forecasting, and Replenishment (CPFR) to cooperate in demand planning and inventory management. This approach helps to minimize stockouts and enhance inventory turnover.

Furthermore, utilizing digital collaboration platforms and marketplaces facilitates enhanced cooperation between organizations and their suppliers, manufacturers, distributors, and customers. These platforms serve as a centralized center for exchanging information, communicating, and coordinating operations within the supply chain ecosystem. Companies like Alibaba and Shopify provide digital marketplaces that link buyers and sellers, facilitating smooth cooperation and transaction processing. Moreover, collaborative platforms enable the joint development of products, fostering innovation and customization, enabling organizations to adapt to evolving market needs and client preferences promptly. Moreover, digital collaboration allows organizations to utilize cutting-edge technology like artificial intelligence (AI), machine learning (ML), and blockchain to improve cooperation and decision-making across the supply chain. Artificial intelligence (AI) and machine learning (ML) algorithms can analyze large quantities of supply chain data to detect patterns, trends, and anomalies. This analysis allows for predictive analytics, demand forecasting, and optimization of supply chain operations. Blockchain technology facilitates the safe and transparent exchange of data and transaction processing, reducing fraudulent activities, improving the capacity to track and trace information, and enhancing confidence among parties in a supply chain. Companies like IBM and Walmart are now investigating utilizing blockchain technology to improve transparency and traceability in food supply chains. This would allow for quicker and more precise recalls in the event of contamination or quality problems.

In addition, digital cooperation promotes a culture of trust, openness, and creativity among supply chain partners, leading to ongoing enhancement and value generation. Organizations can discern potential avenues for streamlining processes, diminishing expenses, and augmenting income by exchanging information and pooling resources. Furthermore, digital cooperation allows organizations to promptly adapt to market fluctuations and disturbances, minimizing risks and capitalizing on emerging prospects. Companies like Zara and Uniqlo utilize digital collaboration tools to cooperate with suppliers and manufacturing partners. This allows them to promptly modify production schedules and adapt to evolving fashion trends and consumer preferences. Ultimately, digital cooperation is crucial for maximizing supply chain efficiency and maintaining a competitive edge in the digital age. By utilizing digital tools and platforms, organizations may improve collaboration, communication, and decision-making throughout the supply chain ecosystem. In addition, the digital partnership allows organizations to use emerging technologies such as artificial intelligence (AI), machine learning (ML), and blockchain to improve transparency, traceability, and trust among partners in the supply chain. As organizations increasingly adopt digital transformation, cooperation will continue to be crucial in achieving success in the digital supply chain.

4.8.8 Customization

Customization in the supply chain phases of the digital era pertains to the capacity of organizations to adapt products, services, and experiences to fulfill the distinct requirements and preferences of particular consumers or market groups. Utilizing technology like data analytics, artificial intelligence, and digital manufacturing has allowed for unparalleled customization in the digitalization process. An essential element of customization in the digital supply chain is leveraging customer data and analytics to get insights into preferences, behaviors, and trends. Organizations may get valuable insights into client preferences and predict demand for tailored products or services by analyzing data from many sources, including social media, online interactions, and purchase history. Companies like Nike and Adidas utilize consumer data and analytics to create and manufacture customized footwear and gear, allowing customers to personalize colors, styles, and features based on their preferences.

Furthermore, digital technologies empower organizations to adopt adaptable and nimble manufacturing processes that can create personalized goods as needed. Additive manufacturing, often known as 3D printing, enables organizations to create highly personalized and intricate components and products while minimizing tooling and setup expenses. Organizations may provide customized items more efficiently and at a reduced cost by utilizing digital design tools and 3D printing technology. Companies such as BMW and Airbus use 3D printing technology to manufacture personalized components for automobiles and airplanes, allowing for enhanced design adaptability and customization. Moreover, digital customization encompasses more than just product design and manufacture. It also encompasses personalized services, experiences, and interactions along the consumer journey. Organizations may utilize digital technologies like artificial intelligence and machine learning to provide customized suggestions, offers, and messages to specific clients, considering their interests and behaviors. For instance, e-commerce platforms like Amazon and Netflix utilize AI algorithms to scrutinize consumer data and offer tailored product recommendations and content ideas, improving the customer experience and stimulating user involvement.

In addition, digital customization allows organizations to collaborate with consumers, resulting in value co-creation through interactive and participatory processes. Crowdsourcing systems and online communities enable organizations to actively include customers in collecting feedback, generating ideas, and obtaining input about product creation and innovation. By engaging consumers in the customization process, organizations may develop goods and services that more effectively cater to their requirements and preferences, enhancing customer happiness and fostering loyalty. Companies such as Lego and Starbucks use crowdsourcing and consumer feedback to engage customers in product creation and innovation. This allows them to design goods and create experiences that appeal to their intended audience. Customization is essential in the digital supply chain, allowing organizations to fulfill clients' varied and changing requirements in the dynamic marketplace. Through digital technology and data analytics, organizations may acquire valuable knowledge about consumer preferences, create and manufacture tailored goods as needed, and provide individualized experiences and interactions at every stage of the customer journey.

Furthermore, the use of digital customization allows organizations to engage in collaborative efforts with consumers, resulting in the creation of value and the promotion of innovation. This, in turn, leads to better connections and increased loyalty. As organizations increasingly adopt digital transformation, the ability to customize will continue to be a crucial factor in maintaining competitiveness and achieving success in the digital supply chain.

4.8.9 Sustainability

The growing recognition of environmental, social, and ethical concerns, legal mandates, and customer preferences has made sustainability a crucial factor at every supply chain step. Organizations are utilizing technology and data to improve sustainability practices at every level of the supply chain in the digital era. An essential element of sustainability in the digital supply chain is employing data analytics and digital technology to enhance resource efficiency, diminish waste, and mitigate environmental harm. Organizations can utilize data analysis from several sources, including industrial processes, transportation routes, and energy usage, to discover areas for improvement and execute policies aimed at reducing carbon emissions, energy consumption, and waste generation. Companies like Walmart and Unilever utilize data analytics and digital supply chain platforms to monitor and enhance their carbon footprint, reduce greenhouse gas emissions, and attain sustainability goals.

Furthermore, digital technologies empower organizations to improve transparency and traceability in the supply chain, allowing them to observe and confirm adherence to sustainability requirements and laws closely. Blockchain technology has become a potent tool for improving transparency and traceability. It achieves this by providing an unchangeable record of transactions and the movement of products across the supply chain. Through blockchain technology, organizations can trace the source of raw materials, authenticate ethical sourcing methods, and guarantee adherence to sustainability certifications such as Fair Trade or organic. Companies such as Nestle and Bumble Bee Foods utilize blockchain technology to track the production process, starting from the farm and ending at the consumer's plate. This allows consumers to make well-informed decisions and encourages sustainable and ethical practices.

Moreover, digital supply chain platforms facilitate the collaboration between organizations and their suppliers, partners, and other stakeholders to promote sustainability efforts and exchange best practices. Collaboration platforms and digital marketplaces encourage the exchange of information and ideas and cooperation among partners in the supply chain. This allows them to work together to develop and implement programs and initiatives focused on sustainability. Companies like Patagonia and IKEA collaborate with suppliers and partners to effectively implement sustainable procurement methods, minimize waste, and advocate for circular economy ideas. In addition, digital collaboration platforms facilitate organizations' interaction with consumers and stakeholders, allowing them to increase awareness of sustainability challenges and promote beneficial transformations. Social media platforms, online forums, and interactive websites offer avenues for organizations to effectively communicate their sustainability initiatives and actively involve customers in promoting sustainable lifestyles and consumption habits.

Finally, digital technologies empower organizations to create and advance novel sustainable goods, products, and business models that reduce environmental harm and foster social accountability. Additive manufacturing, often known as 3D printing, allows organizations to create items with little material waste and energy usage. This technology provides higher customization and design freedom. Moreover, digital platforms facilitate sharing and circular economy adoption, which enhance resource efficiency and minimize waste production. Companies such as Airbnb and Rent the Runway utilize digital platforms to facilitate the sharing and reusing resources like lodgings and apparel. This helps to decrease the need for new production and consumption. In essence, sustainability in the digital supply chain encompasses mitigating adverse effects and promoting innovation and value creation for all parties involved. Ultimately, organizations at every level of the supply chain are now placing significant importance on sustainability, and the advent of the digital era offers fresh possibilities for fostering substantial transformation and ingenuity. Organizations may maximize resource usage, improve transparency, and promote sustainability in the supply chain by utilizing data analytics, digital technology, and collaborative platforms. In addition, digital technology empowers organizations to create novel goods, procedures, and economic frameworks that reduce environmental harm and foster social accountability. As companies increasingly adopt digital transformation, they will continue prioritizing sustainability as a critical area of attention. This emphasis on sustainability will contribute to value development and a competitive advantage in the digital supply chain.

4.8.10 Traceability

Traceability in the digital supply chain pertains to the capacity to monitor and trace the progression of goods, components, and materials over the whole supply chain, from acquiring raw materials for the final client. Digitalization has fundamentally transformed how traceability is carried out, allowing organizations to use advanced technologies like RFID (Radio-Frequency Identification), barcoding, GPS (Global Positioning System), and blockchain to improve visibility, transparency, and accountability. Traceability in the digital supply chain relies on implementing RFID and barcoding technologies to accurately and distinctively identify and monitor individual goods or batches as they progress through the supply chain. RFID tags and barcodes are affixed to items or packaging, enabling organizations to acquire and document data on the precise whereabouts, current state, and overall condition of commodities immediately and continuously. Companies such as Walmart and Target utilize RFID technology to monitor the flow of items throughout their warehouses and shops, facilitating precise inventory control and minimizing instances of stockouts and overstocking.

Furthermore, digital traceability allows organizations to confirm the genuineness, source, and adherence to items' regulations at every stage of the supply chain. Blockchain technology has become a potent tool for improving traceability by establishing an unchangeable and easily understood record of transactions and goods movements. Through blockchain technology, organizations can track the product development process, from acquiring raw materials to manufacturing, distributing, and final selling. This ensures that all regulatory obligations and sustainability criteria are met. Companies such as Maersk and IBM utilize blockchain technology to monitor and trace the transportation of products in worldwide supply chains, facilitating increased transparency and confidence among supply chain collaborators and customers.

Moreover, digital traceability enhances proactive risk management and mitigation by allowing organizations to identify and address supply chain interruptions and difficulties promptly. Organizations can detect and address possible risks, such as delays, quality faults, or compliance breaches, to mitigate their effects by capturing and analyzing real-time data about supply chain activities. Nestle and McDonald's utilize digital traceability systems to oversee and control food safety and quality across their supply chains. This allows them to track the origin of ingredients and promptly address recalls or instances of contamination. Digital traceability is crucial for improving visibility, openness, and accountability in the digital supply chain. It allows organizations to optimize operations, reduce risks, and establish confidence with consumers and stakeholders.

4.8.11 Risk Management

Effective risk management is essential to supply chain management, particularly in the modern era of intricate and interrelated supply networks. Digitalization has presented both advantageous prospects and formidable obstacles to risk management in the supply chain. Technological advancements have enhanced visibility and control and introduced novel vulnerabilities and uncertainties. An essential element of risk management in the digital supply chain is employing data analytics and predictive modeling to detect and evaluate possible risks and disruptions. By examining past data, up-to-date information, and external influences such as market patterns and geopolitical occurrences, organizations may predict and alleviate potential dangers such as interruptions in the supply chain, changes in demand, and cybersecurity hazards. Companies like IBM and Cisco utilize sophisticated analytics and artificial intelligence algorithms to forecast and mitigate supply chain interruptions, allowing them to oversee risks and maintain uninterrupted company operations actively.

Furthermore, digital technologies empower organizations to improve the robustness and flexibility of their supply chains, allowing them to promptly and efficiently address unexpected disruptions and uncertainties. Cloud computing allows organizations to store and access essential data and applications remotely, decreasing reliance on physical infrastructure and enhancing resilience to shocks like natural disasters or pandemics. Moreover, digital supply chain platforms and collaboration tools improve cooperation between organizations and their suppliers, partners, and other stakeholders. This enables real-time communication and decision-making to manage new risks and difficulties effectively. Companies like Flex and Schneider Electric utilize digital collaboration systems to effectively coordinate and communicate with their extensive network of suppliers and partners worldwide. This enables them to address supply chain interruptions and mitigate their consequences promptly.

In addition, blockchain and IoT (Internet of Things) technologies have novel prospects for improving the supply chain's transparency, traceability, and security, diminishing the likelihood of fraud, counterfeiting, and unauthorized entry. Blockchain technology generates an unchangeable and easily understood log of transactions and product movements, allowing organizations to confirm the genuineness and reliability of items and transactions across the supply chain. Through blockchain and IoT, organizations may enhance their ability to observe and analyze supply chain processes, monitor and trace the flow of items in real time, and guarantee adherence to regulatory mandates and industry benchmarks. Companies such as Walmart and Carrefour utilize blockchain and IoT technologies to improve transparency and traceability in their supply chains. This allows them to effectively detect and address problems associated with food safety, product quality, and ethical sourcing.

Ultimately, risk management is crucial for guaranteeing supply chains' durability, adaptability, and long-term viability in the digital age. Organizations may improve visibility, transparency, and control in their supply chains by utilizing digital technologies like data analytics, cloud computing, blockchain, and IoT. This allows them to detect, evaluate, and reduce risks more effectively. In addition, digital technologies empower organizations to enhance cooperation and communication with suppliers, partners, and other stakeholders, promoting prompt decision-making and responsiveness to new risks and difficulties. As organizations increasingly adopt digital transformation, managing risks will remain a crucial area of attention. This is necessary to ensure the success and competitiveness of supply chains in the current dynamic and linked business environment.

4.8.12 Customer Experience

Customer experience is vital in the digital supply chain, as it significantly impacts gaining a competitive edge and achieving business success. Digitalization has revolutionized how organizations engage with customers across various supply chain stages, facilitating tailored, smooth, and multi-channel experiences. Understanding customer preferences, behaviors, and expectations is crucial in enhancing the customer experience within the digital supply chain. This can be achieved by effectively utilizing data analytics and AI-driven insights. Organizations can acquire valuable insights into customer needs and preferences by analyzing data from touchpoints like online interactions, social media, and purchase history. This allows them to customize products, services, and experiences to meet individual requirements. For instance, companies like Amazon and Netflix utilize AI algorithms to analyze customer data and offer personalized recommendations, offers, and content suggestions. This helps to improve the overall customer experience, driving engagement and fostering loyalty.

In addition, digital technologies empower organizations to provide smooth and integrated customer experiences, allowing customers to engage with brands effortlessly through various channels and touchpoints. Modern technology has revolutionized how customers engage with brands, offering them convenient and flexible options to browse, shop, and interact. E-commerce platforms, mobile apps, social media, and chatbots have enabled customers to connect with brands anytime and anywhere. In addition, digital supply chain platforms and logistics solutions provide organizations with the ability to offer fast and reliable delivery options, real-time order tracking, and easy returns, ultimately improving convenience and customer satisfaction. Companies like Zara and H&M provide omnichannel shopping experiences, enabling customers to shop online or through a mobile app. They offer a range of delivery and pickup options, making the process more flexible and convenient.

In addition, digital technologies empower organizations to improve transparency and communication throughout the customer journey, fostering customer trust and loyalty. Digital supply chain platforms allow organizations to keep customers informed and engaged throughout purchasing by providing real-time updates and notifications about order status, delivery schedules, and product availability. In addition, digital communication channels like email, SMS, and social media allow organizations to connect actively with customers, respond to inquiries and concerns quickly, and offer personalized assistance and support. Companies such as Sephora and Nordstrom utilize digital communication channels to connect actively with customers. They provide beauty tips, recommendations, and customized assistance, all of which contribute to an enhanced customer experience and the development of long-term relationships.

Ultimately, customer experience plays a vital role in the digital supply chain, allowing organizations to cultivate satisfaction, loyalty, and advocacy among their customers. Organizations can understand customer needs deeply by utilizing data analytics, AI-driven insights, and digital technologies. This enables them to provide personalized and seamless experiences across various channels, improving transparency and communication throughout the customer journey. In addition, customer experience goes beyond meeting expectations. It involves surpassing them, pleasing customers, and establishing lasting relationships that foster business growth and success in the digital age.

Chapter 5
Digital Supply Chain Reshaping Industrial Ecology: A Study from the Pharma Industry

The relationship between digital supply chain, inventory management, and industrial ecology is multifaceted and increasingly significant in pursuing sustainable development. Integrating digital technologies like blockchain and artificial intelligence (AI) into supply chains can enhance transparency, traceability, and accountability. This leads to more efficient resource use, better coordination, and reduced environmental impacts. Effective inventory management can optimize stock levels, reduce waste, and support circular business models when combined with digital tools. This is crucial for minimizing the ecological footprint of industrial activities. Maximizing the use of resources and reducing environmental impacts through designing industrial systems that mimic natural ecosystems. The digital supply chain and smart inventory management are vital to achieving these goals. This chapter details a digital supply chain tool a manufacturing company can use to stabilize their schedules and improve inventory management. A pharmaceutical company in Pakistan is studied based on exploratory research. The study suggests using data analytics tools as it helps all stakeholders communicate clearly about the current status of the whole supply chain and also lets the organization find and fix delays that affect meeting deadlines and the overall efficiencies and improve the processes.

5.1 History of the Pharmaceutical Industry

The earliest pharmacies date back to the Middle Ages. Isaac Abu Quraysh founded the first pharmacy (drugstore) in Baghdad in 774 AD. Soon after, pharmacies began to operate throughout the medieval Islamic world and, eventually, in medieval Europe. By the nineteenth century, many drug businesses in Europe and North America had progressively grown into giant pharmaceutical corporations (Kumar, 2019).

1910, Alembic Chemicals opened in Baroda after Bengal Chemical and Pharmaceutical Works in Calcutta began producing pharmaceutical goods. As time went on,

A. Shah et al., *Industrial Ecology*, Industrial Ecology,
https://doi.org/10.1007/978-981-97-3619-5_5

additional institutions, such as the King Institute of Preventive Medicine in Chennai, the Central Drug Research Institute in Kasauli, Himachal Pradesh, and the Pasteur Institute of India in Coonoor, Nilgiris, Tamil Nadu, also emerged. New medications replaced the old ones in antimicrobials and cardiovascular pharmaceuticals due to new therapeutic advancements in the Western world; medicines that regulate hormones and antimalarials are only a couple. The locally produced pharmaceuticals that used local raw materials were replaced, and it is now unavoidable to import large quantities of drugs (Kumar, 2019).

With more than $1228.45 billion in sales worldwide in 2020, the pharmaceutical industry (PI) is one of the fastest-growing parts of the economy. The market for medicines has multiplied since 2017, about 5.8 percent per year. This business made 1143 billion USD in sales in the year 2017. That number will increase to 1462 billion USD in 2021 (Crawley, 2013). International Trade Administration claims that (ITA) "The research, manufacture, development, and marketing of medicines and biological for human or veterinary use" are the foundation of the sector. Digitalization uses data from systems integration, connected devices, and other sources to understand and predict customers' wants and improve the supply chain (Hole et al., 2021).

5.2 Manufacturing and Supply Chain in Pharma Company

Making the drugs, vaccines, and consumer healthcare products and ensuring they are harmless and of the highest quality is a huge accountability. It drives the Manufacturing Company to revolutionize across every stage of the procedure. Its supply chain supports each of its business units.

Pharmaceutical Supply Chain of the company understudy consists of 02 leading sub-supply chains: Biopharmaceuticals and Vaccines Operations and Consumer Health Supply Chain. Biopharmaceuticals are more focused on biological pharmaceutics, whereas the same division of the global supply chain also includes vaccine operations at different sites that share both. In contrast, some are only dedicated to either one of them. The Consumer Health Supply Chain solely focuses on consumer health products and biopharmaceutical manufacturing at some manufacturing sites. For data privacy, the name and location of the pharma is not disclosed. The establishment's name is "company throughout the chapter."

5.2.1 *Biopharmaceuticals and Vaccines Operations*

The Biopharmaceutical operations usually include the pharmacological and toxicological drugs on a large scale and clinical drug manufacturing of the product. Pharmacological and toxicological manufacturing usually includes drugs, generally well-established products with a known product history. However, the drug regulator always has a very detailed eye on the commercial scale manufacturing of the products since it can potentially affect the patients directly, which might sometimes be on a smaller scale and sometimes on a much larger scale. The operations in the biopharmaceuticals and vaccines include the sourcing or procurement of the raw materials of active pharmaceutical drugs, raw materials of excipients of the product, as well as the packaging components for the products commercially manufactured at sites, which are done from a mixture of local and foreign suppliers in the company.

Inventory management is crucial for both the raw materials and the finished goods to be shipped to the distribution network and then to the patients; at the company, inventory management is monitored by the utilization of the enterprise resource planning systems, which are all connected to the central servers whereas the warehousing is handled both at the manufacturing sites and in the distribution networks. The information technology team is responsible for the management and introduction of updated technological solutions at the manufacturing sites and the distribution network inside the company, where all the new technologies are introduced in the pharmaceutical supply chain.

The pharmaceutical industry is regulated; hence, the price of the products cannot be increased based on the organization's decision. Therefore, to meet the price of the different raw and packaging materials, the company develops the suppliers and works as a team to reduce the cost per pack and ensure business continuity. The transportation of biopharmaceutical drugs is done within specific environmental conditions. In contrast, the vaccines are transported in a highly controlled environment with transportation logistics services to avoid any threat to product stability.

5.2.2 *Consumer Health Supply Chain*

The consumer health supply chain is less regulated than biopharmaceuticals and vaccine supply chain operations. The consumer health supply chain consists of the supply chain of the consumer health products that are available to maintain the daily health and life of the consumers. The company takes pride in being the market leader for consumer health products. The consumer health supply chain consists of the procurement/buying of raw and packaging materials from local sources, either local, available in Pakistan, or suppliers in other countries around the world. The company works with the suppliers to develop them to meet their quality standards.

The Inventory management and warehousing of the consumer products are handled via SAP software at the Pharma Manufacturing Company's manufacturing sites and warehouses of the company's distributors. Demand and supply targets are assessed and planned by the supply chain team, which is adhered to by the production teams at manufacturing sites. The pharmaceutical industry is regulated.

Hence, the price of the products cannot be increased based on the organization's decision. Therefore, to meet the price of the different raw and packaging materials, the company develops the suppliers and works with them as a team to reduce the cost per pack to ensure business continuity. Transportation of consumer health products is done within specific environmental conditions, whereas transportation logistics services are used to avoid any threat to product stability.

5.2.3 Digitization and Industrial Ecology

In 2018, Büyüközkan and Göçer revealed that supply chains have certain levels of impact on the environment. Digitalization in the supply chain can significantly reduce the usage of paper, which substantially affects the industrial ecological systems regarding waste reduction. As per the article published in the Environmental News and articles from Pakistan, up to 70% of organizational waste includes paper waste. Therefore, digitizing different paper-based processes in industries can reduce the significant amount of waste generated, eventually reducing carbon footprints; this would also contribute to the United Nations' sustainable development goals agenda for 2030.

Lange et al. (2020) stated that industrial ecology depends on efficient and effective stakeholder communication. Moreover, it has also been discussed that digitization can reduce carbon footprints and natural resource consumption across various industrial sectors. Digitization can help achieve environmentally sustainable solutions, such as paperless systems, to manage supply chain operations across different supply chains.

5.2.4 Digitization in Pharmaceutical

Future contributions from the pharmaceutical industry will go toward intelligent automation technologies and may enable augmented production, including additive manufacturing, tailored medication, localized 3D printing of therapies, etc. Following COVID-19, digitalized technology is more vital than ever in allowing businesses to perform better through increased industrial productivity, precise planning and forecasting, enhanced competitiveness, and financial sustainability (Faraj et al., 2021). Together with this, Patent expirations, escalating competition, rising customer demand, and price pressures that are getting worse are all problems for the traditional product-focused business model. The sectors are changing because digital services outside the product are being added to the range of services offered. This is part of the digital revolution. Thanks to digitization, the face of healthcare is changing, and "connected health" has the potential to be advantageous to all stakeholders by achieving the "triple aim" of delivering a better patient experience, increasing health outcomes, and lowering per capita costs (Hole et al., 2021).

The pharmaceutical industry is characterized by large profit margins, strict regulations, high risk, lengthy and investment-intensive research and development periods, and extensive marketing. Data analytics and digitization in the PI can support lowering the high levels of downtime that pharmaceutical facilities are prone to (Anthony Jnr & Abbas Petersen, 2021). Automated remedial actions, seamless processes, and machine-to-machine communication are made possible by artificial intelligence and machine learning, using the Internet of Things, and preventative maintenance (Ngamvichaikit, 2021). There are a lot of rules and regulations in the pharmaceutical industry, so even small mistakes can have terrible results for patients and the company's business, legal situation, and reputation (Sehlstedt et al., 2016). For instance, a big drug company recalled almost 500,000 tablets a few years ago because of factory problems with human monitoring and packaging.

Companies will already reduce such errors in the future thanks to digitalization and automation, reducing financial and reputational harm (Kitson et al., 2018). Some pharmaceutical companies have invested in high-availability computing systems and implemented digital sensors and robotics to eliminate concerns about data transfer between units. Due to this, a fully automated production line has been created, making it much simpler to uphold cleanroom regulations, record and manage electronic batch records, and more, as well as process performance analysis (using root-cause analysis) to identify and make adjustments. Upstream and downstream supply chain digital information integration also significantly improves demand–supply management. Pharmaceutical firms are still experimental when providing digital services outside traditional products because the pharmaceutical industry is undergoing substantial changes (Parida et al., 2019).

Digitalization in the PI can benefit small and large businesses (Lakshmi & Patel, 2020). For instance, adopting digitalization to create counterfeit-proof drugs with serial numbers that can be tracked throughout the supply chain should ensure quality and satisfy upcoming serialization rules (Anderson, 2018). Rosenbaum et al. (2017) elaborated that pharmaceutical firms might use digitization to anticipate an upsurge in demand from international markets. Utilizing cloud-based information exchanges, they might utilize digitalization to follow the rules, find cost-saving manufacturing efficiencies, and communicate more quickly with distributors and suppliers. It can also track medication compliance and forecast accurate demand across regions using health applications, smart sensors, automation, and social media, enabling in-the-moment manufacturing (Van Veldhoven & Vanthienen, 2019).

Unlike other industries, healthcare is experiencing difficulty delivering digital tools to customers (Chilukuri et al., 2014). As we moved closer to the digital revolution, we saw that it had also altered the business models in several sectors related to health (Lakshmi & Patel, 2020). However, the pharmaceutical industry has a notoriously sluggish adoption of new technologies and implementation of digital solutions. Consequently, implementing digital services within the pharmaceutical industry has evolved cautiously (Lee et al., 2019). Despite this, digitization is one of the most effective methods for ensuring all individuals have access to secure medications in light of the ongoing threat and significant COVID-19 pandemic issues (Ayati et al., 2020).

There is no question that PI needs to use making digital tools. Digitalization is essential to ensure the continued provision of medical supplies in light of the developing demands of an ever-changing world and population. While numerous sectors have effectively adopted digitalization, the PI industry has lagged significantly behind. Consequently, this study aims to investigate the concepts of successful digitization that apply to PI. The paper will emphasize Contract Manufacturing and Development Organizations (CDMOs). Since GMP is a crucial component of industrial digitalization deployment, general GMP aspects are also covered and explored in PI (Hole et al., 2021).

The data regarding the Conformance to Supply (CtS)/Conformance to Plan (CtP) and Conformance to Release Plan (CTRP) is handled manually which increases the risk of delay in delivering the product to the patient and misinformation, which leads to an impact on the CtS (conformance to schedule)/CtP (conformance to plan) and CTRP (conformance to release plan) of the site eventually impacting the performance of supply chain. This research study aims to identify the following objectives. First, to identify and understand the role of digital supply chain tools in achieving schedule stability. Second, the importance of digital supply chain tools in optimizing inventories should be studied.

5.3 Data Analytics

The terms Business Analytics, Data Analytics, Big Data, and Supply Chain Analytics are alternatively used to define the same concept in the recent literature on supply chain management (Wang et al., 2016). Data Analytics is a process of evaluating and managing data with the help of some software to drive supply chain competitiveness (Fosso Wamba et al., 2015; Stedman, 2020). Data Analytics is an evolving area of interest for all supply chain professionals; it is the competence of an organization in how well they collect, analyze, and integrate the big data across the supply chain and install data analytics capability with other essential resources (Akter et al., 2016). Data Analytics capability strengthens firms' ability to forecast supply chain information correctly (Tiwari et al., 2018).

Data analytics can accelerate data processing capability and allow firms to capture and evaluate data from inside the organization or upstream and downstream through various sources such as RFID, internet, and mobile records and locations (Srinivasan & Swink, 2018).

The concept of supply chain data analytics has now become deep-rooted. Three types of analytics can be incorporated across supply chain management, i.e., Predictive, Descriptive, as well as Prescriptive (Wang et al., 2016). Analytics ability provides firms the opportunity to make the decision based on data rather than experience and intuition (Dremel et al., 2017; Wang et al., 2020).

Successful businesses nowadays have trusted data analytics capabilities for decision-making (Dubey et al., 2019), but little knowledge of technology hinders the transformation (Mikalef et al., 2019). Ferraris et al. (2018), also highlighted that people are reluctant to use data analytics because of its useability and reliability. It has been proven that to gain performance competitiveness, and organizations should positively incorporate data analytics capability (Mikalef et al., 2019).

An organization is more competitive with robust data analytics capability (Ferraris et al., 2018). Decisions extracted from data are more accurate in supply chain management; it improves forecasts, reduces delays, and makes the supply chain more effective.

5.3.1 Benefits of Data Analytics

Literature has given an obvious concept of Data Analytics and its importance in transforming a supply chain into a digital supply chain (Büyüközkan & Göçer, 2018; Cagliano et al., 2021). The following are the benefits of Data analytics:

Inventory Optimization: Customer demands are dynamic and change rapidly. Data Analytics enabled the supply chain to monitor inventory in real time and keep a sufficient amount of inventory on hand by analyzing past trends (Büyüközkan & Göçer, 2018). It eases warehouse management and reduces inventory levels because analytics capability enhances visibility (Agrawal & Narain, 2018). Data analytics investigate data promptly to predict consumer behavior, demand planning, and inventory management (Cagliano et al., 2021).

Transparency and Accuracy: The supply chain should be transparent in reacting as per requirement; otherwise, it will be disrupted. Data Analytics empowers companies to act wisely in case of any disruption because it anticipates data and predicts accordingly (Büyüközkan & Göçer, 2018), hence enabling supply chain transparency.

Productivity and Efficiency: A supply chain's digital transformation enables the company to access relevant documents and data, and when needed, this advantage increases employees' efficiency and raises firms' productivity (O'Connor, 2022).

Schedule Stability: Lack of data and poorly anticipated data will ultimately hurt the supply chain and cause delays; without scheduling data visibility, the supply chain will only perform poorly. Data analytics capability equipped companies to handle data efficiently, which increases supply chain responsiveness and enables them to track shipment schedules (Chen et al., 2015).

5.4 Research Aims and Objectives

The research aims to assist the company in achieving release schedule stability by establishing a robust process to handle the critical performance indicators, which are conformance to plan and conformance to release plan, which would eventually help in the achievement of the schedule stability, leading to improvement of the inventory management of the organization.

The study is about improving performance and how digital supply chain and data analytics factors or tools impact the release schedule stability and inventory targets. The idea behind the study is to assist in how data analytics reduce the risk of delay in the supply of products to patients.

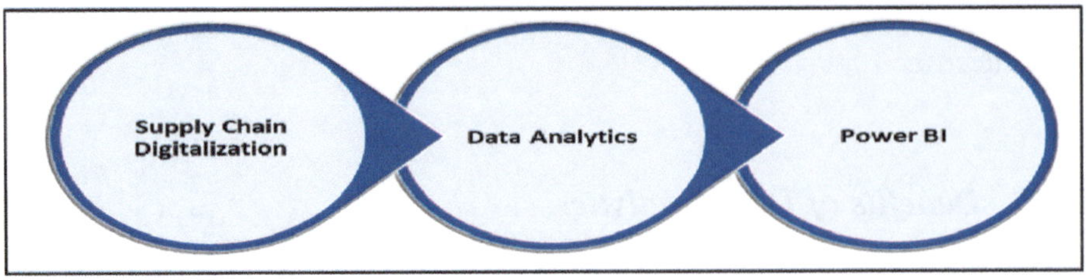

Fig. 5.1 Framework of selection of data analytics tool

This research strategy is based on an exploratory study, which generates an in-depth, multi-faceted understanding and acceptance of a compound issue in its real-life surroundings and context. This research emphasizes supply chain digitalization. Figure 5.1, which illustrates the framework for how organizations can digitalize the supply chain using data analytics tools (Power BI). Microsoft Power BI is used as a data analytics tool for the following reasons:

1. Power BI has a user interface different from other business analytics software (Town & Thabtah, 2019).
2. The Power BI dashboard does not require hardcore programming or coding skills.
3. Power BI licenses are available on the company's IT portal.
4. Power BI can easily integrate with the SAP ECC 6.0 to fetch the data from the SAP Supply chain.

5.4.1 Research Approach

Due to the idea of the exploration, the study utilizes a qualitative methodology. Qualitative research includes ideas, implications, similitude, definitions, images, and portrayals. Qualitative research shows that this study has every one of the fundamental apparatuses that can be utilized to tackle the issue (Berg & Howard, 2012). Qualitative data instruments such as observation, in-depth interviews, open-ended questions, and field notes are utilized to collect data from respondents in their natural settings. The Research approach of this study is inductive in nature. The techniques associated with information assortment give interested respondents a portrayal of the significant exploration. Subjective examination approaches provide a more extensive comprehension of conduct and extract more information about genuine circumstances (Leedy et al., 2014). The qualitative data was collected through interviews with the company supply chain manager, supervisors, and staff and through field observation.

Further data was collected from literature reviews to learn the benchmark digital supply chain being followed around the world in the pharmaceutical industry. Interviews were unstructured, and meetings were informal. Data was also collected through observation at the site. The research participants who contributed directly to this research were the company's supply chain and production department representatives.

5.4.2 Research Data

The data utilized in this study are primary sources. Primary data refers to information gathered in real-time by the researchers. This process is highly involved and time-consuming for both researchers and sources. The primary data was collected through observations and personal interviews with employees of the Pharma Manufacturing Company. The research findings were reported based on evidence taken from interviews and general observation. The interview consisted of open-ended questions covering general information about supply chain digitalization through data analytics. A detailed literature review made us ask relevant questions; the study aims to discover data analytics' importance in supply chain digitalization and the challenges in implementation.

This study formulated the following questions to answer through interview findings, inspired by the work of Järvinen (2016) and vetted by a Faculty member of a private university in Pakistan.

- To what extent can pharmaceutical companies use data analytics to overcome schedule stability?
- What are the opportunities and limitations of data analytics perceived by pharmaceutical industries?

5.5 Discussion and Findings

The primary purpose of this case study is to understand the company's current process, how data analytics capabilities help the company in supply chain digitalization, and its effectiveness. Table 5.1 summarizes the theme of the discussion, which is based on interviews and general observations.

Table 5.1 Case theme

S. No.	Title	Theme explanation
1.	Process flow	The first part of the chapter discusses the department's process, from raw material demand to dispatch to the customer
2.	Interviews' outcomes	The second part of the chapter discusses the findings of the interviews, which focus on the importance of data analytics in supply change digitalization and its challenges
3.	The dashboard	The last part elaborates on implementing a Power BI dashboard and how it benefits the company supply chain

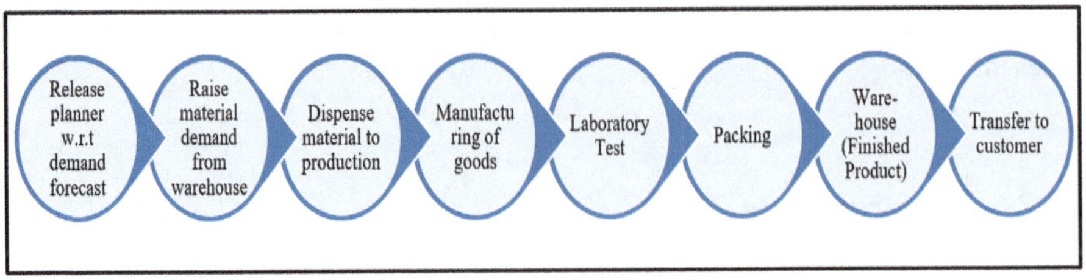

Fig. 5.2 Process flow

5.5.1 *Flow of the Process Under Scope*

The scope of the study is from material dispensing to material packing and outbound to distributors. The organization's process is identified through observation and interviews during the factory visit. The process starts with the planner; the planning department issues a planner at the start of the month based on the demand forecast. The production department then raises the raw material requirement to the warehouse, and the warehouse dispenses the required material in the form of a batch. The batch undergoes production; the production batch will not be transferred to packing unless a lab test is done. After passing the lab test, the batch is transferred to the packing department. Packed material was sent to the warehouse after all the documentation was completed. The material is finally outbound to the customer as per their demand. The general Flow of the process is sketched in (Fig. 5.2).

5.5.2 *Interviews' Outcome*

The research findings were reported based on evidence taken from interviews and general observation. The interview consisted of open-ended questions covering general information about supply chain digitalization through data analytics. A detailed literature review made us ask relevant questions; the study aims to discover data analytics' importance in supply chain digitalization and the challenges in implementation.

The data collected during the interview was interpreted to focus on the relevant information. After being reviewed several times, the findings were presented in Table 5.2.

The interview findings suggested that data analytics capabilities improved digital supply chain efficiency. However, data analytics alone cannot resolve any firm's challenges. Data Analytics capabilities enhanced data visibility and communicated the information to the relevant stages without delay. On-time communication enables the managers to make timely decisions, reducing schedule instability and optimizing inventory levels. On the other hand, manual data handling causes delays and limits the decision-making process. Data feeding to the system should be reliable and valid

Table 5.2 Interview outcome

Importance of data analytics	Opportunities	Challenges/limitation
Reduce schedule instability	It can be used for better demand forecasting	Unreliable or wrong data feeding
Information visibility at the right stage at the right time	Can help in better communication within the department	Employee reluctance to use new technology
Optimize and reduce excessive inventory on the shop floor	On-time data communication can affect manufacturing decision	Manual data handling

to reap all the benefits of data analytics. For the correct result, you must have the correct data to feed. People often fail to implement technology because the data feeding to the system is invalid, giving inappropriate analysis.

Findings suggest that to implement any data analytics tool; you must secure or mature the process for obtaining valid data. To get valid data, you must connect all process steps with the help of a digital tool, i.e., machine learning (machine-to-machine communication). If they connect all machines, they can quickly sort out the causes of time loss or any delays, so the process's efficiency will be increased with swift action. All steps are connected, and all stakeholders can easily view those steps that cause delays on the screen. Reliable and concrete data analysis gives valuable insights from which they enjoy the above-stated benefits and enable the digital supply chain to prosper.

Resistance to adopting new technologies is one of the challenges faced by the organization. Power BI opted to implement data analytics capability in the Pharma Manufacturing Company. In the next part, this study will discuss the process of Power BI implementation and its benefits for the company.

5.6 Digital Supply Chain Tools Evolution in the Company

The production planning used to perform on the customized software (Business et al.–BPCS) by the company was old, un-friendly to users, and, most importantly, the database of the software was not maintained in alignment with the central servers of the company instead it was maintained locally by the Information Technology (IT Tech) team of the company in Pakistan. Pharma Manufacturing Company, being a multinational organization, has multiple key performance indicators governed centrally in which conformance to schedule (CTS)/conformance to plan (CTP) and conformance to release plan (CTRP) are reported to the central team on a weekly and monthly basis manually. Figure 5.3 is the BPCS user interface, which was in an MS-DOS format and was not very user-friendly.

The company strategically decided to implement a one-stop central solution for the enterprise resource plan (ERP), which would be implemented at all company

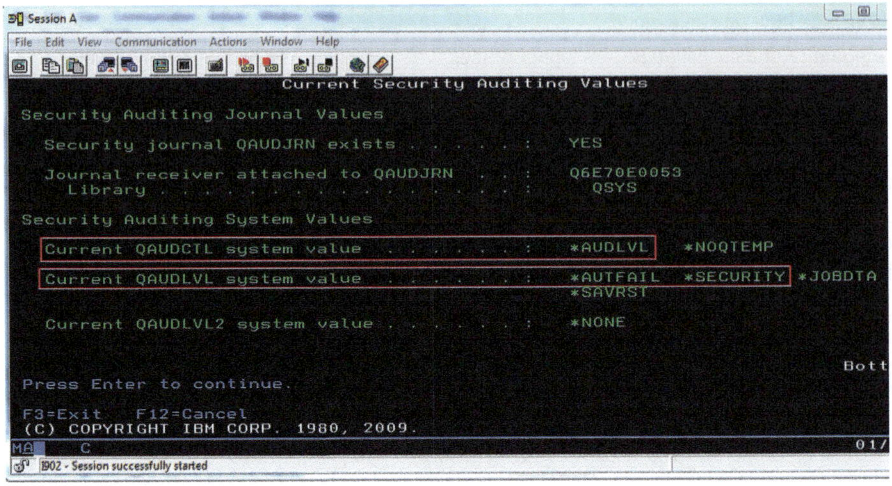

Fig. 5.3 BPCS user interface

sites worldwide. All the data would reside in one central database, which will be maintained and handled centrally for all the company sites. Therefore, the company implemented SAP GCC version 6.0, domestic manufacturing enterprise resource planning (d-EMRP) software. After implementing the d-MERP at the company sites, all the enterprise resource planning data was migrated to the company's MERP central servers. All the data from the previous BPCS was cleaned and structured during the implementation of the d-MERP to align with the company's central database for similar materials and suppliers used at multiple sites worldwide. Figure 5.4 describes the SAP Interface for supply chain planning.

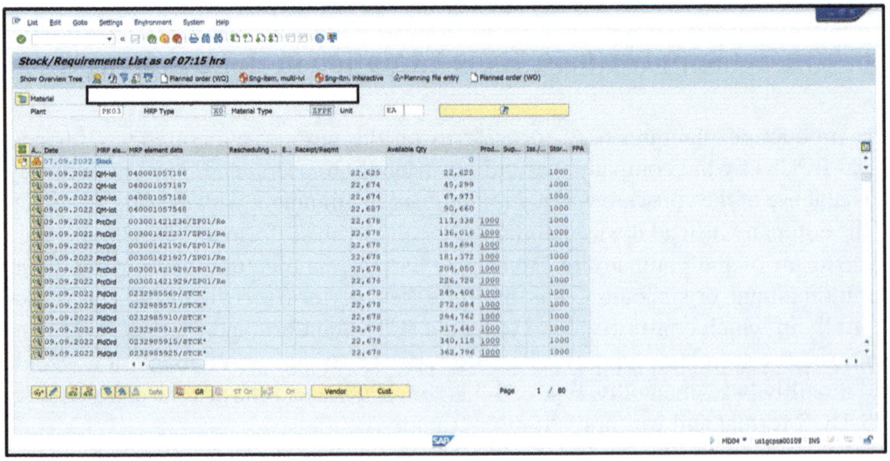

Fig. 5.4 SAP user interface

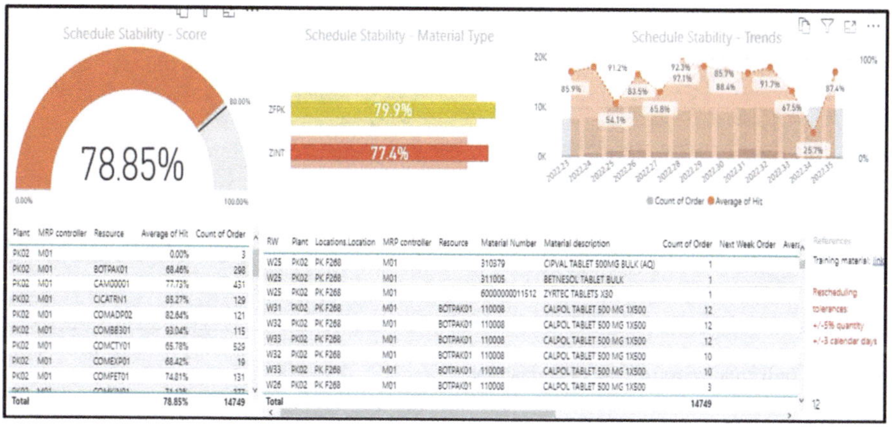

Fig. 5.5 Schedule stability report

The successful implementation of the enterprise resource planning system allowed a platform to have all the data for resource planning in a one-stop solution d-MERP, which can further used to make quick decisions to maintain the KPIs of conformance to schedule (CTS)/conformance to plan (CTP) and conformance to release plan (CTRP) which would eventually improve the release schedule stability of the company as well as the inventory targets would automatically be maintained if the release schedule is maintained following the targets.

5.6.1 *Manufacturing Enterprise Resource Planning (MERP)*

All MERP is being managed in the domestic enterprise resource planning system (d-MERP); the data inside the d-MERP was connected to a digitalization tool, Microsoft Power-BI, to visualize the entire planning of the manufacturing resources along with a central Dashboard for a site which would be updated for each batch/lot of each product at each workstation which would be visible to all the stakeholders involving production department, supply chain department, and quality department along with the status of each batch/lot of the product as well as the standing of site against the targets. Figures 5.5 and 5.6 represent the Schedule Stability and Conformance to Release.

5.6.2 *Central Dashboard for CTS/CTP and CTRP*

The dashboard captured the status of the conformance to schedule (CTS)/ conformance to plan (CTP) and conformance to release plan (CTRP) against the

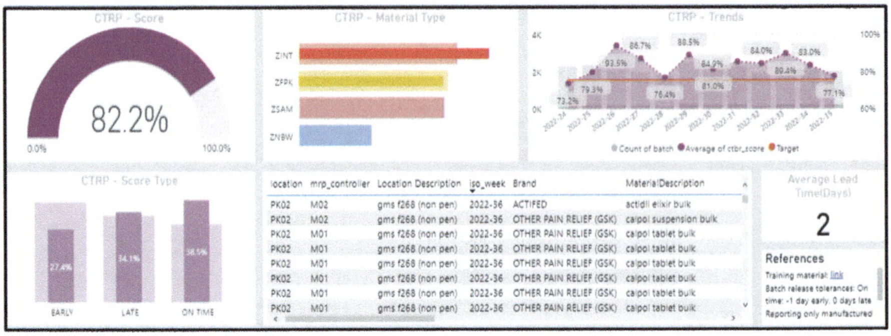

Fig. 5.6 Conformance to release plan dashboard

targeted values. Therefore, if the product batch/lot is not released from one workstation to the next, the status would be visible to all the stakeholders. Consequently, it will be immediately investigated and expedited to the next workstation. The utilization of the dashboard could achieve the following benefits:

- Measuring, monitoring, and reporting the conformance status to schedule/ conformance to plan for every stakeholder.
- Quick decision-making is needed to resolve the hurdles.
- Achievement of conformance to release plan enabling the achievement of the supply plain release schedule stability.
- Achievement of the release supply stability would lead to effective and efficient management of the organization's inventory, enabling the effective use of the organization's resources for inventory management.

5.7 Conclusion and Limitations

The pharmaceutical business deals with the health of living things; it requires special treatment for raw materials and finished products, so an intelligent (digital) supply chain must be followed by its end-to-end process.

This deep study observed that due to the non-visibility of target data concerning present data, delays occurred in the key performance indicators, i.e., conformance to schedule, conformance to plan, and conformance to release plan, which are directly hurting the lead time and so customers are surviving with low stock and the company revenue is not meeting the forecasted numbers. This study analyzed data to determine the causes of a percentage of crucial performance indicators hurting the planning and scheduling team performance.

While conducting the interviews, the authors found some new insights into the causes, as the live data was not displayed on the dashboard that was shared with upper management, highlighting the Production Planning department for all the delays. However, Production Planning refused to take responsibility for the causes of the delay. This happens in almost all meetings, and no one is ready to accept their team's mistake. So, the authors studied all likely cases and found that with the help of digital supply chain tools, these kinds of problems can be quickly resolved. To resolve the issue, the company should introduce a tool that connects data from all stakeholders to a digital form and can be shared anywhere at any time with live status. So that one can easily mark their department's performance concerning KPIs and all hurdles of blaming each other could be resolved. Power BI is one of the best tools that give you leverage to select your KPIs from the data and convert them into a form dashboard where you can see where you are standing right now from the targeted KPIs. In this way, you can easily find out which stakeholder machine is causing a delay in communicating with another stakeholder. In this way, you can easily find the gap in your supply chain process with the help of a digital supply chain tool.

The limitations confronted in this study were as follows:

- The research was based on the company, which shares limited data due to confidentiality issues.
- Due to the packed schedules of the company's Production Planning team, unable to visit the site weekly. But tried best to align online meetings to clear the problems regarding research.

References

Agrawal, P., & Narain, R. (2018). Digital supply chain management: An overview. In *IOP conference series: materials science and engineering*. Institute of Physics Publishing. https://doi.org/10.1088/1757-899X/455/1/012074

Akter, S., et al. (2016). How to improve firm performance using big data analytics capability and business strategy alignment? *International Journal of Production Economics, 182*, 113–131. https://doi.org/10.1016/j.ijpe.2016.08.018

Anderson, S. (2018). The digitization of the pharmaceutical historian archive. *Pharmacy in History, 48*(2).

Anthony Jnr, B., & Abbas Petersen, S. (2021). Examining the digitalization of virtual enterprises amidst the COVID-19 pandemic: A systematic and meta-analysis. *Enterprise Information Systems, 15*(5), 617–650.

Ayati, N., Saiyarsarai, P., & Nikfar, S. (2020). Short and long term impacts of COVID-19 on the pharmaceutical sector. *DARU Journal of Pharmaceutical Sciences, 28*, 799–805.

Berg and Howard. (2012). *Qualitative research methods for the social sciences* (8th edn).

Büyüközkan, G., & Göçer, F. (2018). Digital supply chain: Literature review and a proposed framework for future research. *Computers in Industry, 97*, 157–177. https://doi.org/10.1016/j.compind.2018.02.010

Cagliano, A. C., Mangano, G., & Rafele, C. (2021). Determinants of digital technology adoption in supply chain. An exploratory analysis. *Supply Chain Forum, 22*(2), 100–114. https://doi.org/10.1080/16258312.2021.1875789

Chen, D. Q., Preston, D. S., & Swink, M. (2015). How the use of big data analytics affects value creation in supply chain management. *Journal of Management Information Systems, 32*(4), 4–39.

Chilukuri, S., Rosenberg, R., & Van Kuiken, S. (2014). *A digital prescription for pharma companies*. McKinsey & Company (pp. 1–6).

Crawley, M. J. (2013). *The R book. Chichester* (edn. 2). Wiley.
Dremel, C., et al. (2017). *How AUDI AG established big data analytics in its digital transformation.*
Dubey, R., et al. (2019). Empirical investigation of data analytics capability and organizational flexibility as complements to supply chain resilience. *International Journal of Production Research, 59*(1), 110–128. https://doi.org/10.1080/00207543.2019.1582820
Faraj, S., Renno, W., & Bhardwaj, A. (2021). Unto the breach: What the COVID-19 pandemic exposes about digitalization. *Information and Organization, 31*(1), 100337.
Ferraris, A., et al. (2018). Big data analytics capabilities and knowledge management: impact on firm performance. *Management Decision, 57*(8), 1923–1936. https://doi.org/10.1108/MD-07-2018-0825
Fosso Wamba, S., et al. (2015). How "big data" can make big impact: Findings from a systematic review and a longitudinal case study. *International Journal of Production Economics, 165*, 234–246. https://doi.org/10.1016/j.ijpe.2014.12.031
Hole, G., Hole, A. S., & McFalone-Shaw, I. (2021). Digitalization in pharmaceutical industry: What to focus on under the digital implementation process? *International Journal of Pharmaceutics X, 3*, 100095. https://doi.org/10.1016/J.IJPX.2021.100095
Järvinen, J. (2016). *The use of digital analytics for measuring and optimizing digital marketing performance.*
Kitson, P. J., et al. (2018). Digitization of multistep organic synthesis in reactionware for on-demand pharmaceuticals. *Science, 359*(6373), 314–319.
Kumar, U. S. (2019). Case study: Digitization and digitalization in pharmaceutical industry: Opportunities and challenges in India. *Advances in Management, 12*(4), 23–29.
Lakshmi, B., & Patel, S. (2020). Digital marketing in pharmaceutical industry—An overview and assessment. *International Journal of Pharmaceutical Science & Innovation, 1*(1).
Lange, S., Pohl, J., & Santarius, T. (2020). Digitalization and energy consumption. Does ICT reduce energy demand? *Ecological Economics, 176*. https://doi.org/10.1016/j.ecolecon.2020.106760
Leedy, P. D., Ormrod, J. E., & Johnson, L. R. (2014). *Practical research: Planning and design.* Pearson Education.
Lee, J., Cameron, I., & Hassall, M. (2019). Improving process safety: What roles for Digitalization and Industry 4.0? *Process Safety and Environmental Protection, 132*, 325–339.
Mikalef, P., et al. (2019). Big data analytics and firm performance: Findings from a mixed-method approach. *Journal of Business Research, 98*(July 2018), 261–276. https://doi.org/10.1016/j.jbusres.2019.01.044
Ngamvichaikit, A. (2021). Leveraging design thinking for pharmaceutical digital marketing. *Asian Journal of Business Research, 11*(1), 120–140.
O'Connor, E. (2022). *Why are data and analytics key to digital transformation.* EPC Group. https://www.epcgroup.net/why-are-data-and-analytics-key-to-digital-transformation/
Parida, V., Sjödin, D., & Reim, W. (2019). *Reviewing literature on digitalization, business model innovation, and sustainable industry: Past achievements and future promises.* Sustainability. MDPI (p. 391).
Rosenbaum, M. S., et al. (2017). The digitization of health care retailing. *Journal of Research in Interactive Marketing, 11*(4), 432–446.
Sehlstedt, U., et al. (2016). Embracing digital health in the pharmaceutical industry. *International Journal of Healthcare Management, 9*(3), 145–148.
Srinivasan, R., & Swink, M. (2018). An investigation of visibility and flexibility as complements to supply chain analytics: An organizational information processing theory perspective. *Production & Operations Management, 1*, 42–49. https://doi.org/10.1111/poms.12746
Stedman, C. (2020). *Data analytics (DA), Tec target.* https://www.techtarget.com/searchdatamanagement/definition/data-analytics
Tiwari, S., Wee, H. M., & Daryanto, Y. (2018). Big data analytics in supply chain management between 2010 and 2016: Insights to industries. *Computers and Industrial Engineering, 115*, 319–330. https://doi.org/10.1016/j.cie.2017.11.017

Town, P., & Thabtah, F. (2019). Data analytics tools: A user perspective. *Journal of Information and Knowledge Management, 18*(1), 1–16. https://doi.org/10.1142/S0219649219500023

Van Veldhoven, Z., & Vanthienen, J. (2019). *Designing a comprehensive understanding of digital transformation and its impact.*

Wang, C., Zhang, Q., & Zhang, W. (2020). Corporate social responsibility, Green supply chain management and firm performance: The moderating role of big-data analytics capability. *Research in Transportation Business and Management, 37*(September), 100557. https://doi.org/10.1016/j.rtbm.2020.100557

Wang, G., et al. (2016). Big data analytics in logistics and supply chain management: Certain investigations for research and applications. *International Journal of Production Economics, 176*, 98–110. https://doi.org/10.1016/j.ijpe.2016.03.014

Chapter 6
IR 4.0 and Health Care Industry: A Case Study on Technology Transformation to Implement Industrial Ecology

IR 4.0 promising solution would make the world a better place. Industry Revolution 4.0 focuses on operations, creating higher visibility of the interconnection between all processes, networking, and cyberspace/security with low or no human interaction. IR 4.0 has shown excellent results in developed countries. However, in developing countries, the change level is not primarily focused. With time, data and information would play a significant role in making the organization achieve its high excellence and bottom line performance.

Similarly, diagnostics operations in developed countries are getting automated with time. However, in developing countries, organizations focus on data and utilize resources to make themselves capable of optimizing operations. This chapter gives an in-depth look at the Industrial Revolution (I.R.) and its place in history. It looks at healthcare markets worldwide and Pakistan, focusing on diagnostic market trends.

6.1 History and Timeline of the Industrial Revolution

The first industrial revolution came between 1760 and 1830 (Era of Mechanical Production). In the eighteenth Century, the first revolution appeared after the steam engine's invention. In Fig. 6.1, the use of steam engines was shown in the textile and agriculture industries. Further, the steam engine was used in trains and ships, which changed the transportation aspect entirely.

The second industrial revolution occurred between 1850 and 1914 (Era of Mass Production). It started after the invention of the internal combustion engine to replace steam engines. At this time, the most important invention was electricity because it supported mass production. As shown in Table 7.1, communication improved through radio and telephone during this period. The invention of transportation made traveling to considerable distances feasible.

A. Shah et al., *Industrial Ecology*, Industrial Ecology,
https://doi.org/10.1007/978-981-97-3619-5_6

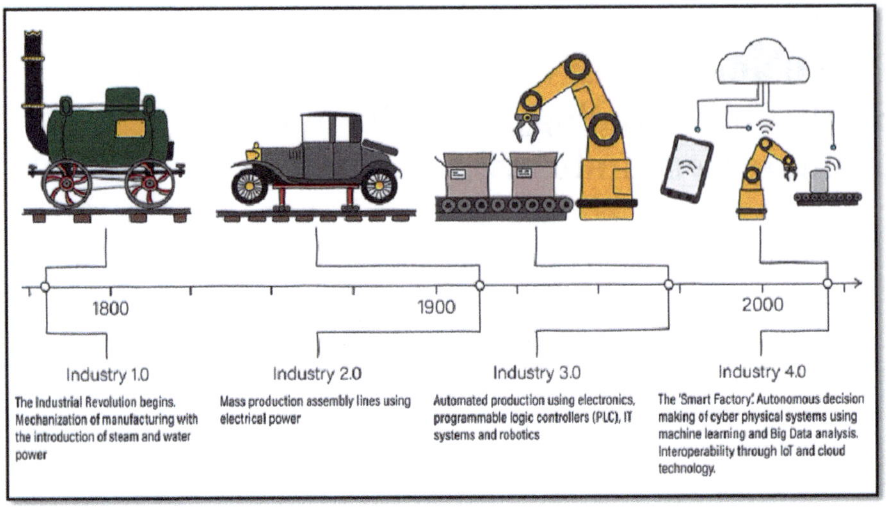

Fig. 6.1 Timeline of the industry revolution

The third industrial revolution is considered present. This era started with the invention of computers. Computers played a significant role when the USA and Russia competed in space. The usage of computers increased, and that is where the internet was born. The smartphone was introduced, allowing humans to know anything with a click away. The problem solution gets easier to explore through the fascinating discovery of the internet. Combining computers or smartphones with the internet allows humans to search, learn, and study online. The further application introduced is presented in Table 6.1.

Table 6.1 Industrial revolution invention (Duignan, 2023)

	First revolution	Second revolution	Third revolution	Fourth revolution
Years	Mids 1700–1830	Mids 1800s to Mid-1900s	1960s to 2000s	Now
Geographical	Mostly in Britain	Europe, North America, and Japan	Everywhere	Everywhere
Inventions	Spinning Jenny, Water Wheel, Steam engine, Cotton gin, concrete, modern roads, batteries	Assembly-line manufacturing, electric rails, plastic, lightbulbs, telephone, telegraph, automobiles, skyscrapers, typewriters, and motion pictures	Computer programming, machine automation, internet, 3D printers, biotechnology, robotics, transistors, space research	Machine Learning, Deep Learning, Big Data, Artificial Intelligence, Cloud Computing, Predictive Programming

The fourth industrial revolution introduced computational power, and the need for IOT, Big Data, Blockchain, Machine Learning, and Artificial Intelligence will become necessary. As presented in Fig. 6.1, intelligent factories are considered part of Industry 4.0, which involves the prediction; using sensors in any field significantly impacts the operation, strategic, and bottom-line performance.

6.2 Healthcare Market—Global and Pakistan

The healthcare industry was worth about $8.45 trillion in the world market in 2018. It is expected that it could reach $10 trillion by 2022. Many people think IoT can cut operational costs and waste in healthcare by $100 billion a year. Most people worldwide use health devices, about 28% of the population. Many businesses are still getting over the effects of COVID-19 and changing how they work. This is because of the strict containment measures that were put in place, such as social isolation, working from home, and closing down commercial activities, which made things more challenging to run. At an annual growth rate of 8.4%, the market should reach $10,414.36 billion in 2026.

Pakistan has a high demand for healthcare services and medical devices, expected in the private sector. Healthcare providers include government, NGOs, and the private sector. At the end of fiscal June 2021, the Government of Pakistan (GOP) will likely spend about $785 million on healthcare. Even though there is a complex and significant health infrastructure, there are some major problems with health care delivery. These include a growing population, an uneven distribution of health professionals, a lack of workers, insufficient money, and limited access to high-quality healthcare services.

6.2.1 Diagnostic Market and Trends

The market for medical and diagnostic laboratory services globally was anticipated to increase at an 11.5% rate from $152.68 billion in 2021 to $170.17 billion in 2022 (Report Linker, 2022). Primarily as a consequence of businesses reorganizing their operations and recovering from the repercussions of COVID-19, which previously required the cessation of commercial activities, social isolation, and remote work, among other restrictive containment measures that caused operational difficulties, the market is currently experiencing growth. The medical and diagnostic laboratory services market is projected to attain $248.64 billion by 2026, expanding at a CAGR of 9.9%.

The "Market Analysis of Equipment for Clinical Laboratories and Availability of and Need for Formal Financing in Pakistan" report says diagnostic services in Pakistan's private sector were worth between USD 500 million and USD 600 million in 2017 (Karandaaz, 2019). They are also expected to grow 15% yearly for the next five years. However, between 2002 and 2017, data from the Pakistan Bureau of Statistics shows that total consumer spending went up by an average of 9.2% year-over-year. An estimated 418 billion rupees ($3983 million) was spent on healthcare 2017. Diagnostic labs occupy a one-fourth market worth Rs. 114 billion ($1.087 million) annually.

6.3 IR 4.0 Impact on Supply Chain of Developing Countries

Greater Transparency and Accuracy: Digital platforms help companies track the entire supply chain in real time, such as the in-transit location of goods. The IoT supports the solution of inventory for easy tracking with relevant data. This optimizes inventory, reduces associated costs, and improves order accuracy and ETAs (thereby minimizing out-of-stock situations), lot and batch control, and inventory optimization.

Data-Backed Decision Making: Technology algorithms like machine learning can predict results based on data (such as sensors, RFID, trends, patterns, and images). Implementing these predictive analytics techniques has nearly reduced forecasting errors by fifty percent. Accurate demand forecasting enables businesses to optimize their inventory, leading to cost savings.

Improved Warehouse Management: The usage of IR 4.0 significantly impacts the warehouse management capabilities (supply chain stages). For example, they can track lab tests from one place to another, track customer cases, and tell customers which department causes delays. It helps in transparency delivery. RFID technology can forecast the precise whereabouts of a product, including its location within a truck. This accuracy lets managers provide workers with location-based instructions, thereby conserving time. Additionally, labor hours required per order are decreased.

Additionally, machine-to-machine communication reduces transportation expenses by optimizing the number of carriers per shipment. Also optimized per square foot of inventory storage is precise demand forecasting. Thus, plant managers can efficiently regulate the global flow of inventory.

6.4 Background and Research Questions

In Pakistan, the present practice is relevant to IR 3.0. However, big names like Systems Affinity are advancing in machine learning and artificial intelligence. Many firms are working on it in their prototyping phases. Nowadays, companies are working on detecting early cancer stages and detecting causes based on images. These are examples based on healthcare, following the phase of IR 4.0. Although it is in process, Pakistan will move towards IR 4.0 by 2027.

The study aims to develop the IR 4.0 transformation model in the diagnostic laboratory to optimize operations during pandemics like COVID-19. To fulfill the objective, this will try to answer the following research questions:

1. What are the identified factors of IR 4.0 for developing countries?
2. How are diagnostic lab bottom-line performance and operation excellence achieved through IR 4.0?
3. How long would it take for diagnostic labs in developing countries to transform entirely into IR 4.0?
4. How does IR 4.0 impact the diagnostic lab's transparency, visibility, and accountability?

6.5 Exploring Perspectives: Literature Insights

6.5.1 An Introduction to Industrial Revolution 4.0

In the twenty-first Century, the world is experiencing the fourth industrial revolution and digitalization, commonly known as Industry 4.0 (Ghobakhloo, 2019). It is not a hype but a hit. The term "Industry 4.0" instantly grabbed the attention of governments and industrialists worldwide. This idea of IR 4.0 is supported by the growth of communication and information technologies (Nascimento et al., 2019). According to Mabkhot et al. (2021), IR 4.0 is evolving from the introduction of many developments that instantly modify the manufacturing, services, design, and operations linked with production systems.

One foremost part of this change is that technology does not seem to encourage economic prosperity, but it is a crucial element that will make the future more sustainable. As the latest digital technologies are being introduced, companies are creating new technical and social interactions between the virtual and physical elements, which leads to technical, organizational, and human-related changes (Kadir & Broberg, 2020). Schroeder et al. (2019) stated that IR 4.0 includes the digital transformation of the consumer and industrial markets, from manufacturing to the digitalization of delivery networks. Harun-Or-Rashid (2020) states that modern technology is transforming the production industry. In the long term, the equipment and machinery will upgrade the production methods through self-optimization and digitization. In this way, it is viable to constantly improve modern technologies' workflows, integrating factors including augmented reality, IoT, cloud computing, data, simulation, cyber-security, and industrial automation (Barreto et al., 2017). Each of these factors changes the core of Industry 4.0. Each element is related and, when linked together, has potential that has never been possible (Harun-Or-Rashid, 2020). There are big expectations in business communities and research that these modern technologies will pervade the vast range of service sectors and manufacturing chains (Nascimento et al., 2019).

6.5.2 Industrial Revolution 4.0 in Emerging Economies

There are several advantages to implementing IR 4.0 in emerging economies to enable industrial development. All these factors mentioned above lead to a dynamic system where systems and employees can predict and act before challenges and concerns arise rather than reacting to them after they occur (Santos et al., 2017). The developing countries that are middle class, like South Africa, Mexico, Cambodia, Bangladesh, and Vietnam, need process innovation and automation in production plants, manufacturing cost reduction, product design, and profit maximization to compete with the developed countries to maintain their economic prosperity in the future (Larson et al., 2012). As said above, complying with IR 4.0 has many advantages for developing countries, but these advantages can only be attained if the possibilities are available. The countries can support the concerns that might arise during the transformation. Because of the following revolutions, it has caused many progressive changes in production, and the processes have become sustainable and automated (Marr, 2018). According to Acton (2018), automation will improve efficiency and productivity and thus make resource management and energy better. These IR 4.0 technologies will provide better resource consumption and encourage the country's competitive advantage.

6.5.3 Industrial Revolution 4.0 in Healthcare

Healthcare 4.0 refers to something that has emerged and derived recently from IR 4.0. Today, the healthcare sector is more digital than in past years. It is a common term for data-driven digital health technologies, such as mobile health, medical I.T., e-health, digital medicine, wireless health, health informatics, and health information systems. It defines breakthrough innovation in the health sector as leading to valuable networks and recently developed business models. Developments in Healthcare 4.0 are arising in many technologically advanced economies worldwide. The digital health market is anticipated to grow to $223.7 billion by 2023 (Jayaraman et al., 2020).

According to Ahsan and Siddique (2022), healthcare has been moving slowly in terms of digitization and advancement. Thanks to the growth of healthcare technologies, medical equipment, and smartphone devices, which can exchange and connect data. The healthcare sector has come across considerable transformations in the past few years. From making antibiotics and vaccines to establishing patient care and moving on to next-generation genomic medicines. A significant innovation has altered how healthcare providers and patients interact at each stage of this evolution. Healthcare 4.0 refers to adopting the principles and technologies of IR 4.0 in healthcare processes and systems. Healthcare 4.0 allows customization in real-time, which makes a patient-focused atmosphere easier. Healthcare 4.0 is an approach of technology similar to IR 4.0 that needs fundamental organizational transformations concerning social, cultural, and technical aspects (Tortorella et al., 2020).

From a technical standpoint, this Industrial Revolution 4.0 depends mainly on Cyber-Physical Systems and is based on three pillars of technologies: the Internet of Things, Cloud Computing, Smartphones, and big data Analytics (Aceto et al., 2020). These emerging technologies are digital data foundations supporting advancing healthcare 4.0 services and applications (Jayaraman et al., 2020). The evolution of information and communication technology is completely changing all human activities, significantly impacting the healthcare sector as well. Based on this analysis, it can be deduced that these technologies and, above all, the developing mindsets brought forward by healthcare 4.0 are modifying the vision of the healthcare future and the continuous presence of personalized healthcare services (Aceto et al., 2020). According to him, it not only offers advancements for traditional procedures like Cloud health information systems, Advanced monitoring of pathological and physiological signals, and medication activities but also creates the possibility for new unpredictable processes, approaches, and applications, for example, Home based Rehabilitation, Advanced living Environments, and personalized healthcare. Li and Carayon (2021) say that digitally enabled healthcare should focus on people, including patients and healthcare professionals, in theory and practice.

When conceptualizing and implementing intelligent and interconnected healthcare, it is imperative to consider their characteristics, demands, capacities, and constraints. It is essential to ensure that Health Treatment 4.0 is designed to reduce and improve these injustices and provide universal access to high-quality patient safety while addressing gaps and inequities. Additionally, the potential harm that technology could do to socio-technical systems must be addressed. To do this, the public healthcare system may need to make an effort, and the importance of social determinants may need to be recognized.

6.5.4 Diagnostics 4.0 and the Digital Transformation

Diagnostics 4.0 refers to a digitalized healthcare environment, making clinical information accessible in real-time. It will allow easy access to electronic healthcare records and provide the laboratories with new diagnostic opportunities. These days, diagnostics measurement is also known as wet lab laboratory parameters, in which blood through wearable machines is not acceptable in a diagnostic environment. In contrast, diagnostics machines for patient monitoring are being actively developed by nanotechnology (Neumaier, 2019). According to Lippi and da Rin (2019), automation is one of the foremost developments in laboratory diagnostics. In automated laboratories, numerous analysts run tests on diverse samples while being linked as modular systems or production lines.

By integrating the various strengths of diagnostics into a single track, it is possible to enhance standardization, efficiency, safety, and quality of lab testing, giving a valuable investment return for the long term and requalifying staff as well. Zaninotto and Plebani (2010) say that the quality of laboratory performance has improved over the past few years. This improvement seems aligned with the advancement of technology. The criteria for total laboratory automation are robotic laboratories, which allow improved functions and secure human errors. In this way, the ability to automate the majority of manual activities like day-to-day clinical lab activities, for example, manual preparation of specimens, manual testing, and weighing of reagent preparation, facilitates the laboratory experts to generate critically more accurate results and concentrate on higher-value procedures, like the evaluation of results, interpretation, test development, and the conversion of tests into medical practice.

According to Genzen et al. (2018), you must have the support of the laboratory automation vendor to ensure that shutdowns are limited once you introduce automation and digitization. Ongoing conversations between the company and the clinical laboratories are crucial to solve problems. There are several things to consider that affect downtime. The automation vendor you select must have a solid history of providing excellent customer service. Moreover, analyzing the automation performance for the test conducted in your medical laboratory and engaging in real-time discussions with the company are essential steps. Water, electrical supply, and temperature conditions around your laboratory environment can impact the automation system's efficiency. Before going live, the medical laboratory staff needs enough time to train on the system. Under-training employees can result in problems down the road and possible delays. The clinical team must feel confident using the system and utilizing all of its features.

6.5.5 Cloud Computing in Diagnostics 4.0

Wilson et al. (2022) suggest that medicine laboratory has an essential role in the health sector by giving healthcare experts important data to guide the assessment of risks, prevention of diseases, prognosis, diagnosis, monitoring, and treatment of patients, which assures accurate, secured and efficient clinical as well as medical decision making. In the last few years, most of the technological developments in medicine laboratories have enhanced monitoring and medical diagnostics, improving the quality of patient care. Kricka et al. (2015) say that I.T. and cloud computing have come up with many features of laboratory medicine, and there is a massive usage of computers in data interpretation, transmission, manipulation, graphical reporting, and I.T., like cloud computing and remote servers. As per Toby Blackburn (2022), Cloud lab is a technical term similar to cloud computing in information technology. The I.T. experts also struggled to describe what this term means and how it differs from the other technologies.

According to Bhosale (2018), Cloud Computing is always a catalyst. Cloud computing generates a list of clinics nearby to schedule an appointment with a family physician or locate the nearest clinic, including their respective appointment times. This is called cloud computing. It also possesses many apps, including detecting tumors in MRI scans and patterns in patient data. They also make applications for diagnostic centers and medical laboratories, enabling them to become fully mobile, making lab operations faster, and giving them better patient-customer experiences.

According to Aceto et al. (2020), there are several cases in which cloud computing architectures have adapted to simplify and strengthen the development, design, and application of information systems for processing, sharing, and collecting clinical records.

According to McMullin (2014), in the diagnostics laboratory, cloud computing differs in communication, security, power, and cost. Utilizing a cloud-based computerized system may increase a lab's profitability. Doctors can order direct blood tests from labs, and the findings can be safely transferred to the patient's electronic health records. Thanks to the cloud's safe access to consolidated documents, every practitioner can access the latest, up-to-date data. As long as cloud service providers maintain rigorous security requirements, the cloud has a tremendous, bright future in healthcare. Information theft in a well-secured digital cloud is more challenging in a medical practice. In this regard, cloud computing is helping us achieve automation, improved quality, productivity, and speed. It allows the lab experts to work efficiently, deliver good quality, and give a better patient experience.

6.5.6 Smartphones in Diagnostics 4.0

Without any need to send a sample to a distant laboratory, diagnostic tools that may be used with a smartphone may enable medical professionals to diagnose illnesses or check on patients immediately. According to researchers, this could alter how people care for their health by placing a disease-monitoring tool in their hands (Halford, 2015). Scientists use these smartphones as microscopes and medical tools to help fight diseases in distant parts of the world. These devices essentially served as a low-cost, portable laboratory, eliminating the need for a lab technician. It was one example of a growing trend in medical smartphone applications beyond essential functions like detecting heartbeats or sending your doctor pictures of a worrisome tumor. "This technology is significant" (Lauran Neergaard, 2015). Smartphone-based diagnostics is an effective strategy that has the potential to change the way medicine is currently conducted. Integrating these systems into healthcare could enable the decentralization of medical specialties and quick disease data collection and storage in oncology, hematology, and virology.

The implications for epidemiology and public health may be significant. Since the diagnostic test result is attached, precise geographic and demographic information can be kept (Hernández-Neuta et al., 2019). According to Wilson et al. (2022), Smartphones can now be used as read-out platforms for diagnostic tests thanks to developments in molecular analysis, bio-sensing, mathematical algorithms, microfabrication, 3D printing, etc. This allows complex and expensive tests to be moved from the central clinical laboratory to the point of care. Applications of smartphone-based diagnostics include oncology (such as prostate, skin, and colorectal cancer), hematology (such as leukemia anemia), endocrinology (such as hormones), infectious disease (such as malaria), pulmonology (such as cystic fibrosis), microbiology (such as gastroenteritis), virology (such as HIV, hepatitis, and flu), and more. Recently, smartphone-based devices have been developed using paper-based sensors, microfluidic chips, and digital fluid tests for quantitative nucleic acid detection, notably for viral detection, which has recently become more common (Rajendran et al., 2021). The combination of communication technologies and microfluidic devices has facilitated the e-health movement.

6.5.7 Digitalization in Medicine

Medicine 4.0, the fourth stage of medical progress, is almost synonymous with Industry 4.0 in today's digitalized society. The internet and artificial intelligence (A.I.) assist in digitizing advanced medications, which first debuted roughly 150 years ago. The global healthcare industry relies heavily on the overview of A.I. medical systems. Modern medical therapies must be developed to their maximum potential using cutting-edge computational technologies. A.I. systems are primarily changing how the world's healthcare system operates by enabling a fundamental reshaping of the diagnostics in medicine, growth of new pharmaceuticals, testing, and treatment, and advancements in the field of transplant surgery (von Eiff & von Eiff, 2020).

A crucial step toward the digitalization of medicine is the application of finite element analysis (FEA) in computational simulation (Ammarullah et al., 2021). Industrial designers and medical engineers can assess several interrelated perceptions, such as device constancy and durability, with the help of FEA (e.g., predicting the end-of-life of patient-specific implants). The pressures inside a material under altered thermodynamic conditions can be displayed using FEA (Driscoll, 2019). The FEA approach identifies weak regions and allows for design adjustment. AI and digitalization typically lower medical clinics' costs while improving the quality of healthcare services.

6.5.8 On-Demand Healthcare

As per Fox's investigation (2013), As a result of the following factors, consumers are progressively adopting online platforms to obtain medical information:

- Forty-seven percent of customers are willing to learn more about their doctor.
- Thirty-eight percent of consumers want to look at medical and hospital facilities.
- Seventy-seven percent of people prefer making medical appointments online.

Medical professionals can now deliver their services, capabilities, and competency directly to patients in the new digital economy, much like independent experts. Numerous healthcare organizations connect medical professionals directly and offer an online marketplace with the necessary medical facilities. These findings provide patients with the required medical treatments and procedures. As a result, healthcare professionals now account for a portion of the digital healthcare market that offers patient-centered treatments (Reddy, 2022).

6.5.9 Telemedicine and AI

In a way, Telemedicine is similar to a recent concept that gained popularity during the COVID-19 epidemic era. Telemedicine enables the care and assistance of patients not in life-threatening situations. Unwell patients have fewer connections with one another as a result. It is also critical to promote education to combat chronic illnesses and patients (Portnoy et al., 2020). According to Johns Hopkins and his data, there were around 50–70 television viewers each month at the time of the first global lockdown in March 2020. The number has significantly increased to 94,000 through May 2020. Moreover, the monthly tele visits counts continued to hover around 35,000 after healthcare facilities had been generally renewed (Smith, 2021).

Technically, this communication method allowed for transmitting medical information directly in various formats, including X-ray images, results of C.T. scans, ultrasounds, and photos, and videoconferencing between patients in real-time, doctors and medical facilities. In areas where patients cannot take advantage of the opportunity to receive direct assistance from specialized professionals at a medical facility, medical counseling services have been able to be established thanks to telemedicine.

Additionally, in industrialized nations, telemedicine is significant, has considerably decreased incorporation costs along with its associated treatments, has improved diagnostic accuracy, and has made remote patient health monitoring possible. Particularly for elderly patients and those with chronic illnesses, this philosophy is essential. For instance, video conferencing is frequently used to help elderly patients at St. Luke University Health Network in Pennsylvania. They are aware that this social group is less likely to adopt mobile applications and is more accustomed to technologies designed for desktops and laptops. The telemedicine market is expected to increase at an average annual rate of about 19.2 percent, reaching USD 56.2 billion in 2020 and USD 175.5 billion by 2026 (Ugalmugle, 2020). The most significant proportion of the global market for patient telemonitoring is between 32 and 48%. China and the U.S. have created the largest telemedicine technology markets in terms of expenditures (Ugalmugle, 2020).

Given some parameters, the worldwide telemedicine market may be segmented, and these include:

- Remote communication (between a clinic and the Home or another clinic).
- Interaction with technology elements such as communication, monitoring systems and channels, measuring sensors and tools, video conferencing systems, "wearable" technologies, databases, and mobiles.
- The primary reasons for applying are (diagnostics, treatment, medical education, consultation, and monitoring).

Different methods and software solutions are developed using various tools depending on how these solutions are designed. The individual should have sufficient knowledge and abilities in the development fields, including familiarity with cloud technologies in mobile, embedded solutions, and particular medical sector protocols. This is because these components are intricately intertwined. Finally, the following telemedicine services are essential to provide:

- Interaction between medical professionals who work remotely.
- Authentication and identification of specific individuals
- Recording their activities through consultation and remote monitoring of medical patients' health.

6.5.10 Big Data Analytics (BDA)

Big Data Analytics (BDA) enables the improvement of diagnostic practice expertise. Additionally, treatment based on BDA is highly accurate. It relates specifically to cases with difficult cancer, diseases, and tumors. Early diagnosis and precise therapy are the key elements (Kudryavtseva et al., 2020). Using BDA in genetic analysis helps calculate genetic pathology and generate accurate knowledge. Even the fastest computers probably need hours to perform a focused study of data, but GPU computing can make this process go faster (Goel et al., 2020).

It helps decrease medication error rates and predict forthcoming charge rates (Harbeck, 2019). BDA helps increase healthcare's predictability. Healthcare providers can increase facilities' efficiencies and resources for active deployment by analyzing predictive data regarding the strain on their systems and admission loading. Predictive analysis should also be evaluated for its potential to assist businesses in effectively mobilizing human resources by predicting potential outbreaks of the flu and colds that could cause a scarcity of workers.

Pharmaceutical companies are negotiating contracts with technological companies familiar with BDA. This gadget creates unique prospects for considering how pharmaceuticals can function in real life to gain access to patient health data. One recent instance is the Roche announcement in 2018, in which the corporation paid USD 2 billion to buy all the shares of Flatiron Health. This startup collects clinical data for cancer patients. Real-world research information that enables pharmaceutical producers to demonstrate the value and usefulness of their products. The areas of oncology, cardiac disease, and respiratory problems see the most active research in this subject (Harbeck, 2019).

Today's gold standard for medication evaluation is the typical randomized clinical trial, although actual drug use data is gathered outside these trials. Today, automated drug discovery is successfully developed using neural networks. Researchers and medicinal chemists collaborate to pinpoint flaws and develop more effective models for developing new drugs (Xu et al., 2019). It is anticipated that developments in interdisciplinary disciplines that combine computational and genomic technology will open up new possibilities for customized medicine (Kovalev & Kudryavtseva,

2020). The well-known Human Genome Project and the Human Microbiome Project began operations in the United States simultaneously (Chial, 2008). Within the scope of the U.S. National Institutes of Health, a unique Data Analysis and Coordination Center was established during its deployment. In this regard, active research is being done as part of the joint Chinese-European project MetaHit.

6.6 Methodological Approach

To accomplish the aim of this study, a comprehensive description of the approaches employed is discussed in this chapter. It defined the research design, demographic and sample design, research procedures, data collecting techniques, and data analysis methodologies used to carry out the study.

The research philosophy is a Case Study for this research. A case study relies on a transformation model in diagnostic laboratories to optimize operations during real-world diagnoses. The first step in taking this case study is to figure out an issue and view it from the broadest perspective. This results in targeted research that addresses the problem by providing a deeper understanding. Finally, research outcomes often result in policy suggestions, novel environmental initiatives, or shifts in societal norms.

This case study research has developed a transformation model based on the Industry Revolution—IR 4.0—concerning diagnostic labs. The unique designs may include philosophical assumptions and theoretical frameworks. The essential aspect of this type of investigation is the AS-IS process of diagnostic laboratories concerning stakeholders and customer order fulfillment with the logistics theme of IR 4.0.

6.6.1 Health in Focus: Exploring the Expertise of a Medical Testing Company

The medical testing company, an NGO (Governmental Organization) headquartered in Karachi, works throughout Pakistan for the betterment of its citizens through its affiliated departments and non-governmental organizations (NGOs) and sub-offices with local and international affiliation see Table 6.2. Thalassemia and Hemophilia Care Center are the divisions under the welfare, where everything is free, and resources are given to their benefit. The revenue generation comes under the division of Laboratory and Blood Bank. The Blood Bank division has 21 collection points. The sample is collected and delivered at a centralized location at the specified time. From where all the processes occur for the test, consumers can get the report online or through social media (WhatsApp). There are a total of 40 locations under the medical testing company. Twenty-one are collection labs (Diagnostic Labs), and 19 are blood banks. The company has managed to collect 200,000 units of blood bags.

Table 6.2 Affiliations of the medical testing company

National affiliations	International affiliations
• National Blood Transfusion Program • Sindh Blood Transfusion Authority • Thalassemia Federation of Pakistan • Sindh NGO's Networks of HIV/AIDS • Sindh Board of Technical Education • Affiliation with Baqai University • Affiliation with Pakistan Medical and Research Council • Karachi University for Training • Dow University of Health Sciences • Hepatitis Advisory Board Pakistan • Pakistan Pharmacists Association	• American Association of Blood Banking (AABB) • External Quality Assurance System (EQAS) • College of American Pathologists (CAP) • International Society of Blood Transfusion (ISBT) • Thalassemia International Federation (TIF) • World Foundation of Hemophilia (WFH)

It is the leading blood bank in Pakistan. The company has maintained the quality all over Sindh, which is FDA-approved. There are some listed on their website, the certification and accreditation.

The medical testing company has five divisions, which;

- Thalassemia Care Center
- Hemophilia Care Center
- Laboratory
- Institutes of Hematology
- Blood Bank.

Under the Blood Bank, they extract three components from one blood bag.

- Plasma (Hemophilia Patients)
- Platelets (Dengue, Cancer Patients)
- Red Cells (Thalassemia Patients).

6.6.2 Data Collection

The process of gathering data information that the researcher needs to answer specific objectives and research questions is known as data collection. Close-ended questionnaires, unstructured interviews, and observations through websites and customer reviews are the data collection methods used in our research study. A questionnaire is a research tool that consists of several questions to collect data from the respondents. It enables the researcher to spend less time gathering massive amounts of data quickly. An unstructured interview is considered an in-depth, open-ended, or long interview and has been a favorable method in research. Observation is regarded as the primary method in all the methods of research. Qualitative research is conducted by observing human activities and physical settings. Quantitative research is conducted in settings that are typically the natural place of the activity.

6.6.3 Sampling Technique

The purposive sampling technique is used in our research Case Study. The sampling technique refers to its subtype of Expert sampling that depicts only field experts and helps us to make generalized findings. Expert sampling is a form of purposive sampling employed when the research requires insights from individuals possessing specialized knowledge. During the exploratory phase of qualitative research, this knowledge may be necessary to identify potential new areas of interest or to facilitate access for other participants. The sampling methods chosen for this study are purposive and expert, which refers to taking information from a specific industry, such as the diagnostic laboratory or the medical testing company. It is bound to such people who have the desired information, or maybe they reach specific criteria for the researcher. Snowball sampling involves selecting the participants and identifying the critical factors from the field experts, creating issues in the success of IR 4.0 in diagnostic labs (Rai & Thapa, 2015).

6.6.4 Data Analysis

Data analysis involves transforming, analyzing, and modeling data gathered in a research. This study employs the Transformation model of Industry Revolution 4.0, grounded in expert opinion and a review of the relevant literature. A literature review will ascertain a few factors for developing the questionnaire. Based on the questionnaire, an interview session will be conducted with the field experts. After that, the collected information will be analyzed and transformed into a model leading us to our industry revolution 4.0.

Below is the questionnaire to understand the usage of technology in healthcare and diagnostic labs:

Cloud Computing

1. In your opinion, is cloud computing a good option in the healthcare industry?
2. Can you describe the importance that it will bring to the healthcare industry?
3. Do you support it because it would benefit hospital performance, better response time, and efficient customer service?
4. What are the challenges if an organization transforms entirely on the cloud? How long would it take to transform from entirely traditional practice to cloud?
5. Based on diagnostic labs, does cloud computing bring any advantages to patients?
6. In the traditional market, it is difficult for an individual to know about lab test rates, doctors' availability for consultation, nearest lab locations, monitor the report, and be notified of updates. Do you believe that cloud computing would solve this problem? How?
7. Does cloud computing improve diagnostic labs' customer experience?

8. From a customer and supplier perspective, Please give your opinion on how cloud computing will enhance transparency, accountability, and visibility in diagnostic labs.

Smartphones Based Application

1. What is your opinion regarding smartphone involvement in IR 4.0?
2. Do you think smartphone-based diagnostics is a powerful approach?
3. Can Virtual personal assistance tools like Siri and Alexa be synced with diagnostic solutions?
4. Separate applications can be developed to connect patients with clinicians and healthcare professionals.
5. Can Smartphones evolve into powerful portable gadgets with physical sensors and connectivity? What would you say about this?
6. Should smartphones be altered permanently by adding biosensors or mechanical attachments to turn smartphones into robust read-out devices for diagnostics?
7. Smartphone-based readers have also been developed to detect cholesterol, blood pressure, steroid hormones, and protein and calorie measurements. Should the results be considered authentic?
8. What Impact do Smartphone companies receive from this revolution?
9. After these smartphone advancements in sensors and rays, could they harm human bodies?
10. In your opinion, This will also increase the cost of smartphones.

Big Data Analytics

1. Could you please describe what big data is in the healthcare sector?
2. How can big data analytics benefit businesses? In terms of cost, response time, and bottom-line performance?
3. What are your opinions on big data?
4. What are some of the challenges of a big data project?
5. What are the different approaches to dealing with big data?
6. How does big data assist in business Decision making?
7. Please let us know if BDA in healthcare can improve diagnostics practices.
8. Do you think BDA helps to decrease the rate of medication errors and to be more predictive in future concerns?
9. How long would it take to implement BDA practices?
10. Will it help the companies to create visibility, accountability, and transparency if BDA is implemented?

Digitalization (A.I., Telemedicine and On-Demand Healthcare)

1. How does the A.I. system change the healthcare system of diagnostics in medicine?
2. Can you please explain what telemedicine in Medicine 4.0 is? What are your opinions about it?
3. How are diagnostic lab bottom-line performance and operational excellence achieved through artificial intelligence?

4. How does telemedicine affect medical diagnostics and treatment?
5. Does the digital healthcare market offer patient-centered treatment? How?
6. How does AI lower costs and improve the quality of healthcare services?
7. What are the challenges that can come with implementing on-demand healthcare?
8. How do telemedicine and AI impact transparency, visibility, and accountability in the diagnostics lab?
9. In what ways is this method of communication, i.e., telemedicine, beneficial for patients?
10. What new challenges does AI bring to the doctors and workers in the healthcare sector?
11. What medical facilities offer on-demand healthcare practices to patients?

6.7 Discussion

6.7.1 Big Data Analysis

Globally, the term "big data" has become increasingly well-known in recent years. Nearly all research fields, including those associated with business and academia, generate and analyze massive amounts of data for various motives. The most challenging undertaking is the management of this enormous, both structured and disorganized mountain of data. Massive data management is unfeasible with traditional software; therefore, technologically advanced software and applications are necessary, capable of efficiently and cost-effectively utilizing high-end computational resources. Utilizing machine learning (ML) techniques, such as neural networks and other A.I. approaches, to automate decision-making would constitute a significant achievement. To conduct practical analysis and generate actionable insights, they must develop more intelligent web applications and improved techniques for managing this "endless sea" of data.

A complex system known as healthcare was established with the sole intention of preventing, diagnosing, and treating human health issues. The primary constituents of a healthcare system are health professionals (physicians or nurses), healthcare facilities (including hospitals and clinics that administer medications and other diagnostic and therapeutic technologies), and a funding institution that provides financial support to the entities above.

Concerning data utilization in the healthcare industry, the appropriate application of Big Data is a subject that is frequently debated. The adoption of electronic medical records, the vast quantity of data transmitted by various sensor types, and the generation of data by patients via social media all contribute to the unceasing expansion of data streams in the healthcare industry. If the data are utilized appropriately, healthcare organizations can contribute to disease surveillance, public health management, and clinical decision-making. Due to the volume of data and the complexity of its processing, clinical data processing is challenging.

The principal dilemma of big data is how to effectively manage and utilize such an enormous volume of information to inform data-driven decision-making across multiple domains. The implementation of ample data storage, analysis, presentation of analytical outcomes, and clinical inference represents an additional noteworthy challenge within healthcare data. Healthcare organizations are adopting data analytics tools that classify, integrate, and appropriately present complex data to enhance comprehension. Implementing this approach would improve the efficiency of collecting, organizing, analyzing, and presenting substantial volumes of healthcare data.

6.7.2 *Smart-Based Application Analysis*

People groups working on the web are 90% procuring through a cell phone. Indeed, even financial specialists and non-specialist individuals who continue their work utilize cell phones. The advanced cell world is obliterating the present business world. Looking at the medical services area, the PDA annihilation needs additional time in Pakistan; however, numerous new companies and innovations are coming in, yet their adaption is time-consuming. i.e., the application is functioning admirably, yet it is not the same as Bykea. Clients are slowly moving towards cell phone innovation in the well-being sector. Yes, it ought to be finished, and organizations are doing it well, yet each person is not utilizing this associate. Be that as it may, the menial helper generally enters our lives and makes clients often like Google Assistant. Yes, this will function as an Aptech and various new companies are dealing with it. However, there is a significant hole in this stage.

This is a need of the market nowadays. Such as late at night, a click can provide us with specialized refreshers. This is the world's fate; however, it is not right now in Pakistan. They got applications in the cell phones but not a successive client. They have seldom been utilized by wellbeing cognizant people groups, such as contraptions and apps. It does not need to be added as an outer device but should be associated with and adjusted inside cell phone equipment and programming technology. Yes, most certainly, it has previously been drilled and will be considered as a legitimate peruser. The vast majority of the specialists and people groups consider the outcome right and doubt that 1% of individuals get misgivings on a survey. They will get extraordinarily extraordinary and explicit items to buy, and they can get you out all. NIC startup was there as a sugar decrease referenced on their Website. The sugar test was directed by cell phones, so assuming medical care associations utilize them, they are not hurtful to human bodies. It will not have a tremendous effect on the cost of Android and Siri since if any organization imagines some innovation connected with well-being promotion, they will get them out, not reexamine them. They are also revealed with their framework's Android and Siri. It will be peanuts for them to purchase, yet their deals and use are expanding due to this adjustment of innovation much of the time.

6.7.3 Cloud Computing Analysis

In the literature review of the chapter, our study develops the concept of a cloud lab, which determines the need for a better customer experience. Considering this aspect of diagnostic labs, cloud labs would play a significant role in helping a patient find the nearest via a few taps through smartphones or get the reports at Home. Furthermore, consult with doctors directly or get opinions from doctors while at home. Cloud computing works as a catalyst that enhances the doctor-patient experience.

According to our interviewee, convenient access is vital in the healthcare industry. This is true for patients and internal organizations; operations, documentation, and online tools would help the organization process smoothly. Our interviewee guessed that a pure cloud computing process would take at least five years to migrate for developing countries because of concerns about transparency, accountability, and visibility on either the supplier's side or the customer's. The solution is cloud computing, which will resolve these problems through a centralized portal.

6.7.4 Digitalization (AI, Telemedicine, On-Demand Healthcare) Analysis

Developing A.I. medical systems will substantially impact the worldwide healthcare sector. In the next five years, no physicians will be required due to AI-driven modifications to the healthcare system. Our cloud-based doctors can make any diagnosis that a human cannot for clients. AI and digitization will often help medical clinics cut expenses while enhancing the standard of care. The telecare and digital telemedicine systems will be linked to the patient's vitals. Every individual has a medical line. Computers will monitor people, and telemedicine 4.0 will undoubtedly rule the world. Healthcare counseling services have been established thanks to telemedicine in locations where patients cannot make use of the chance to obtain direct support from specialized professionals at a medical facility.

Several healthcare organizations offer an online marketplace with the required medical facilities and direct connections between medical specialists. These discoveries provide patients with essential medical procedures and therapies. In the modern world, when technology systems, communication apps, internet accessibility, and smartphones are present, there is no problem with on-demand health care. The only issue that might emerge in the future is with laws and regulations. Physical doctors will not exist shortly. Compared to human doctors, these computerized doctors will diagnose patients more accurately using patient information. They will also communicate with them. These digital doctors will rule the globe in the not-too-distant future.

6.8 Implications

Based on the above analysis, some implications developing countries must focus on and keep up with if they want to participate in IR 4.0.

- **Continuous R&D**: A common factor that makes any country successful is continuous R&D in its respective field. This suggests diagnostic labs should keep up-to-date on technology and new global trends. Besides these four factors, IOT, Cyber Security, Robotics, and many more are also part of IR 4.0. However, the study shows four primary factors that firms must support to become part of IR 4.0.
- **Support by Top Management**: A concept of leadership is shared: Every company moves upward only if the top management supports new ideas and innovation. Without top management, the company would become stagnant and unable to survive in the fast-moving technology world.
- **Resist to Migrate on Technology**: A critical practice that every diagnostic lab has to focus on in the healthcare industry. The customer, at this time, becomes aware and intelligent. Everything is on the Internet, and you have access to the information. If any diagnostic lab does not adopt the technology, the customer will go to that lab, which offers services online.
- **Data Sharing Concerns**: Based on our research, the diagnostic labs are concerned about sharing the data. The problem of sharing data makes their business less profitable. This does not seem right, as a normal diagnostic lab thought. However, to maximize the profit or bottom line performance, they must play with their data to better understand their logistics, suppliers, and customers' data.
- **Cost of Awareness**: The second step is to spread awareness once an organization adapts the technology. It is a costly step but a fruitful one. The problematic move for developing countries is to produce high costs to make a good conversion of customers. However, this would be resolved if better awareness is provided to the customers. Plus, allowing the supplier to negotiate and communicate through the platform will make the supplier more effective and efficient.
- **Industrial Ecology and Sustainability**: Adapting technologies in the healthcare sector will eventually lead to organizations achieving sustainability and implementing industrial ecology. IR4.0 introduces various technologies such as digitization, artificial intelligence, the Internet of Things (IoT), machine learning, big data analytics, and augmented reality into health. These technologies enable proactive interventions for early detection and treatment of diseases, enhancing patient care and comfort. Industrial ecology emphasizes the sustainable use of resources and the reduction of environmental impact. This translates to more energy-efficient medical procedures and equipment, reduced waste through better recycling and disposal systems, and minimized carbon footprint in healthcare.

6.9 Summary

Smartphone-based diagnostics is an incredibly potent method that has the potential to profoundly alter the current state of medicine, presumably and hopefully in the same way that smartphones have transformed the communications industry. Considering the present existence where life has become speedy, diagnostics are additionally expected in a hurry. Cell phones and demonstrative applications can assist clients in overcoming any barrier between specialists and users. Yes, they can be. With artificial intelligence quickly filling all fields, Alex and Siri can be utilized as private help devices to help and match up information. However, devoted clinical help elements can be created and empowered with smartphones. BDA in healthcare has made it possible for doctors to battle dreadful illnesses like AIDS and cancer. The field of health has been dramatically impacted by data science.

Healthcare data science can resolve medical problems, save lives, and provide ample time to take safeguards. It will save a ton of money and your most valuable time. Given our interviewee, advantageous access plays a significant part in the medical services industry. The advancements made by A.I. to the medical services framework will eliminate the need for specialists over the next five years. Our cloud-based specialists will have the capacity to make any conclusion that a human cannot for the clients. Computer-based intelligence and digitization will frequently assist clinical centers in cutting costs while improving the care norm. The telecare and computerized telemedicine frameworks will be connected to the patient's vitals. Each individual has a clinical line. PCS will observe individuals, and telemedicine 4.0 will, without a doubt, control the world.

Adopting Cyber-physical Systems (CPS) leads to the development of intelligent healthcare facilities that can optimize decision-making and improve service delivery. This includes real-time patient data monitoring, cost-effective treatment options, and time-saving processes. While IR4.0 offers numerous benefits, it also presents challenges such as data fragmentation, privacy concerns, and the complexity of integrating new technologies into existing healthcare frameworks. Addressing these challenges is crucial for the successful implementation of IR4.0 in healthcare. As the healthcare sector adapts to the technology, it prepares for the next phase, Industry 5.0. Lastly, IR4.0 in healthcare, viewed through the lens of industrial ecology, holds the promise of creating a more sustainable, efficient, patient-centered medical environment. However, careful consideration of digital transformation's environmental impacts and challenges is also required.

References

Aceto, G., Persico, V., & Pescapé, A. (2020). Industry 4.0 and health: Internet of Things, Big Data, and cloud computing for healthcare 4.0. *Journal of Industrial Information Integration, 18*. https://doi.org/10.1016/J.JII.2020.100129

Acton. A. (2018). *Industry 4.0–the opportunities behind the challenge.*

Ahsan, M. M., & Siddique, Z. (2022). Industry 4.0 in healthcare: A systematic review. *International Journal of Information Management Data Insights, 2*(1), 100079. https://doi.org/10.1016/J.JJIMEI.2022.100079

Ammarullah, M. I., Afif, I. Y., Maula, M. I., Winarni, T. I., Tauviqirrahman, M., Akbar, I., Basri, H., van der Heide, E., & Jamari, J. (2021). Tresca stress simulation of metal-on-metal total hip arthroplasty during regular walking activity. *Materials, 14*(24), 7554.

Barreto, L., Amaral, A., & Pereira, T. (2017). Industry 4.0 implications in logistics: An overview. *Procedia Manufacturing, 13*, 1245–1252. https://doi.org/10.1016/J.PROMFG.2017.09.045

Bhosale, A. (2018, February). *Cloud—The future of laboratory automation.* Elets eHealth.

Chial, H. (2008). DNA sequencing technologies key to the human genome project. *Nature Education, 1*(1).

Driscoll, M. (2019). The impact of the finite element method on medical device design. *Journal of Medical and Biological Engineering, 39*(2), 171–172.

Duignan, B. (2023). Inventors and inventions of the industrial revolution. In *Encyclopedia Britannica.*

Fox, et al. (2013). *Health online 2013.* Pew Research Center. https://www.pewresearch.org/internet/2013/01/15/health-online-2013/

Genzen, J. R., Burnham, C. A. D., Felder, R. A., Hawker, C. D., Lippi, G., & Peck Palmer, O. M. (2018). Challenges and opportunities in implementing total laboratory automation. *Clinical Chemistry, 64*(2), 259–264. https://doi.org/10.1373/CLINCHEM.2017.274068

Ghobakhloo, M. (2019). Industry 4.0, digitization, and opportunities for sustainability. *Journal of Cleaner Production, 252.* https://doi.org/10.1016/J.JCLEPRO.2019.119869

Goel, S., Knaggs, M., Goel, G., Zhou, X. W., Upadhyaya, H. M., Thakur, V. K., Kumar, V., Bizarri, G., Tiwari, A., & Murphy, A. (2020). Horizons of modern molecular dynamics simulation in digitalized solid freeform fabrication with advanced materials. *Materials Today Chemistry, 18*, 100356.

Halford, B. (2015, May 25). *Smartphones put medical diagnostics in your hands.* https://cen.acs.org/articles/93/i21/Smartphones-Put-Medical-Diagnostics-Hands.html

Harbeck, N. (2019). Digitalization in medicine: It is our chance and responsibility now to shape the digital future of breast cancer management. *Breast Care, 14*(3), 128.

Harun-Or-Rashid, M. (2020). Industry 4.0 implementation in emerging countries: Challenges, opportunities and strategies.

Hernández-Neuta, I., Neumann, F., Brightmeyer, J., Ba Tis, T., Madaboosi, N., Wei, Q., Ozcan, A., & Nilsson, M. (2019). Smartphone-based clinical diagnostics: Towards democratization of evidence-based health care. *Journal of Internal Medicine, 285*(1), 19–39. https://doi.org/10.1111/JOIM.12820

Jayaraman, P. P., Forkan, A. R. M., Morshed, A., Haghighi, P. D., & Kang, Y. Bin. (2020). Healthcare 4.0: A review of frontiers in digital health. *Wiley Interdisciplinary Reviews: Data Mining and Knowledge Discovery, 10*(2). https://doi.org/10.1002/WIDM.1350

Kadir, B. A., & Broberg, O. (2020). Human well-being and system performance in the transition to industry 4.0. *International Journal of Industrial Ergonomics, 76*, 102936. https://doi.org/10.1016/J.ERGON.2020.102936

Karandaaz (2019) Karandaaz study estimates total value of private diagnostic services industry between USD 500 to 600 Million—Karandaaz Pakistan. https://karandaaz.com.pk/media-center/news-events/karandaaz-study-estimates-total-value-private-diagnostic-services-industry-usd-500-600-million/. Accessed March 25, 2024.

Kovalev, K. V., & Kudryavtseva, K. E. (2020). Molecular genetic deviations and obstetric pathology. *Akusherstvo Ginekol, 2020*, 26–32.

Kricka, L. J., Polsky, T. G., Park, J. Y., & Fortina, P. (2015). The future of laboratory medicine—A 2014 perspective. *Clinica Chimica Acta, 438*, 284–303. https://doi.org/10.1016/j.cca.2014.09.005

Kudryavtseva, E., Popov, V., Muller-Kamskii, G., Zakurinova, E., & Kovalev, V. (2020). Advantages of 3D printing for gynecology and obstetrics: brief review of applications, technologies, and prospects. In *2020 IEEE 10th international conference nanomaterials: Applications & properties (NAP), 02SAMA09-1.*

Larson, N., Oshiro, R., & Camargo, O. (2012). 3M Brazil ergonomics: A success story in the making. *Work, 41*(SUPPL.1), 5977–5980. https://doi.org/10.3233/WOR-2012-0997-5977

Li, J., & Carayon, P. (2021). Health care 4.0: A vision for smart and connected health care. *IISE Transactions on Healthcare Systems Engineering, 11*(3), 171. https://doi.org/10.1080/24725579.2021.1884627

Lippi, G., & Da Rin, G. (2019). Advantages and limitations of total laboratory automation: A personal overview. *Clinical Chemistry and Laboratory Medicine, 57*(6), 802–811. https://doi.org/10.1515/CCLM-2018-1323

Mabkhot, M. M., Ferreira, P., Maffei, A., Podržaj, P., Mądziel, M., Antonelli, D., Lanzetta, M., Barata, J., Boffa, E., Finžgar, M., Paśko, Ł., Minetola, P., Chelli, R., Nikghadam-Hojjati, S., Wang, X. V., Priarone, P. C., Litwin, P., Stadnicka, D., Lohse, N., & Lupi, F. (2021). Mapping industry 4.0 enabling technologies into United Nations sustainability development goals. *Sustainability, 13*, 2560. https://doi.org/10.3390/SU13052560

Marr, B. (2018, September 2). *What is industry 4.0? Here's a super easy explanation for anyone.* https://www.forbes.com/sites/bernardmarr/2018/09/02/what-is-industry-4-0-heres-a-super-easy-explanation-for-anyone/?sh=79fe45d89788

McMullin, S. H. (2014, August 18). *What does "the cloud" mean for healthcare and clinical labs?* SelectScience.

Nascimento, D. L. M., Alencastro, V., Quelhas, O. L. G., Caiado, R. G. G., Garza-Reyes, J. A., Lona, L. R., & Tortorella, G. (2019). Exploring Industry 4.0 technologies to enable circular economy practices in a manufacturing context: A business model proposal. *Journal of Manufacturing Technology Management, 30*(3), 607–627. https://doi.org/10.1108/JMTM-03-2018-0071

Neergaard, L. (2015). *Medical workers use smartphones as portable labs.* https://www.heraldtribune.com/story/news/2015/05/08/medical-workers-use-smartphones-as-portable-labs/29314367007/

Neumaier, M. (2019). Diagnostics 4.0: The medical laboratory in digital health. *Clinical Chemistry and Laboratory Medicine, 57*(3), 343–348. https://doi.org/10.1515/CCLM-2018-1088/PDF

Portnoy, J., Waller, M., & Elliott, T. (2020). Telemedicine in the era of COVID-19. *The Journal of Allergy and Clinical Immunology: In Practice, 8*(5), 1489–1491.

Rai, N., & Thapa, B. (2015). *A study on purposive sampling method in research.* Kathmandu School of Law.

Rajendran, V. K., Bakthavathsalam, P., Bergquist, P. L., & Sunna, A. (2021). Smartphone technology facilitates point-of-care nucleic acid diagnosis: A beginner's guide. *Critical Reviews in Clinical Laboratory Sciences, 58*(2), 77–100. https://doi.org/10.1080/10408363.2020.1781779

Reddy, M. (2022). *Digital transformation in healthcare in 2022.* 7 Key trends. DAP. https://www.digitalauthority.me/resources/state-of-digital-transformation-healthcare/

Report Linker. (2022). *Medical and diagnostic laboratory services global market.* https://www.globenewswire.com/news-release/2022/02/11/2383472/0/en/Medical-And-Diagnostic-Laboratory-Services-Global-Market-Report-2022.html. Accessed March 25, 2024.

Santos, M. Y., Oliveira e Sá, J., Andrade, C., Vale Lima, F., Costa, E., Costa, C., Martinho, B., & Galvão, J. (2017). A big data system supporting Bosch Braga Industry 4.0 strategy. *International Journal of Information Management, 37*(6), 750–760. https://doi.org/10.1016/J.IJINFOMGT.2017.07.012

Schroeder, A., Ziaee Bigdeli, A., Galera Zarco, C., & Baines, T. (2019). Capturing the benefits of industry 4.0: A business network perspective. *Production Planning and Control, 30*(16), 1305–1321. https://doi.org/10.1080/09537287.2019.1612111

Smith, L. (2021). *Telemedicine improves care across Johns Hopkins medicine.* https://www.hopkinsmedicine.org/news/articles/telemedicine-improves-care-across-johns-hopkins-medicine

Toby Blackburn. (2022, March 16). What are the criteria of a true cloud laboratory? *Drug Discovery and Development.*

Tortorella, G. L., Fogliatto, F. S., Espôsto, K. F., Vergara, A. M. C., Vassolo, R., Mendoza, D. T., & Narayanamurthy, G. (2020). Effects of contingencies on healthcare 4.0 technologies adoption and barriers in emerging economies. *Technological Forecasting and Social Change, 156*. https://doi.org/10.1016/J.TECHFORE.2020.120048

Ugalmugle, S. (2020). *Telemedicine market size & share, growth outlook 2021–2027*. https://www.gminsights.com/industry-analysis/telemedicine-market

von Eiff, M., & von Eiff, W. (2020). The digitalisation of healthcare. *HealthManagement.Org Journal, 20*(6).

Wilson, S., Steele, S., & Adeli, K. (2022). Innovative technological advancements in laboratory medicine: Predicting the lab of the future. *36*(S1), S5–S17. http://mc.manuscriptcentral.com/Tbeq. https://doi.org/10.1080/13102818.2021.2011413

Xu, Y., Li, X., Yao, H., & Lin, K. (2019). Neural networks in drug discovery: current insights from medicinal chemists. *Future Medicinal Chemistry, 11*(14), 1669–1672.

Zaninotto, M., & Plebani, M. (2010). The "hospital central laboratory": Automation, integration and clinical usefulness. *Clinical Chemistry and Laboratory Medicine, 48*(7), 911–917. https://doi.org/10.1515/CCLM.2010.192

Chapter 7
The 3Rs and Industrial Ecology

The "3Rs" in environmental management, Reduce, Reuse, and Recycle, are fundamental waste hierarchy principles and global sustainability initiatives. This strategy seeks to minimize waste production, preserve natural resources, and decrease environmental contamination, promoting a more sustainable relationship between humans and the environment. Industrial ecology is an interdisciplinary framework that seeks to understand the complex interactions between industrial systems and natural ecosystems. It aims to optimize resource utilization by shifting towards more sustainable and eco-efficient practices. The concept of the 3Rs in the industrial ecology framework seeks to minimize waste and promote the efficient use of resources. In the context of industrial ecology, the 3Rs contribute to creating a circular economy, where the life cycles of materials are extended, and waste generation is significantly reduced.

7.1 Introduction to 3Rs

Reduce highlights the significance of reducing the quantity of trash we produce and the resources we utilize. By embracing more efficient and conscientious consumption habits and prioritizing items with a reduced environmental footprint, we can substantially mitigate trash generation and resource extraction from the environment. This phase is essential since it directly tackles the problem at its origin, minimizing unnecessary manufacturing and consumption requirements. A practical method for people and organizations to decrease trash is by reducing the usage of single-use plastic. Opting for reusable water bottles, coffee cups, and shopping bags instead of disposable plastic will substantially minimize plastic waste. On a broader scope, firms can decrease the amount of packaging materials used for their products or use packaging manufactured from recycled materials.

A. Shah et al., *Industrial Ecology*, Industrial Ecology,
https://doi.org/10.1007/978-981-97-3619-5_7

Reuse entails identifying methods to reuse products that would otherwise be discarded, prolonging their lifespan and diminishing the necessity to manufacture new ones (Suchorzewski et al., 2023). This can encompass various activities, including reusing clothes and household items and restoring electronics and machines. Reusing reduces the need for fresh resources and decreases the energy and expenses involved in producing, distributing, and disposing of things. As an illustration, the fashion sector has experienced a significant increase in the appeal of vintage and pre-owned items. This trend offers distinctive design choices and dramatically diminishes the environmental consequences of manufacturing new clothing items. Patagonia and similar companies promote clothing repair by providing repair services and guidelines to clients, enabling them to extend the lifespan of their items.

Recycling involves transforming existing materials into new goods to avoid wasting potentially valuable resources, limit the demand for new raw materials, minimize energy use, and reduce air and water pollution by reducing traditional trash disposal (Shanker et al., 2023). Recycling facilitates the completion of product lifecycles, fostering a circular economy in which materials are utilized for as long as feasible, extracting their utmost value before recovery and regeneration. For instance, some towns have established extensive recycling initiatives that enable recycling paper, glass, metal, and specific types of plastics. These materials can undergo reprocessing to create new products. Recycled paper may be utilized to manufacture fresh paper goods, hence decreasing the need for original wood pulp and conserving trees, energy, and water throughout the manufacturing procedure.

The case study pertains to the three Rs that may be derived from San Francisco's Zero Waste Programme. San Francisco, California, has been leading the way in adopting the principles of reducing, reusing, and recycling with its ambitious zero-waste program. The city implemented compulsory recycling and composting regulations to attain a complete elimination of garbage sent to landfills by the year 2020. While the city has not yet accomplished this objective, it has made substantial advancements. Firstly, homeowners and companies must segregate recyclable and compostable materials from regular waste as part of recycling and composting efforts.

The city's extensive waste management system enables the efficient collection and processing of these materials, thus diverting a substantial proportion of garbage away from landfills. Furthermore, the city has implemented comprehensive education and outreach initiatives to educate individuals about the significance of trash reduction and the correct segregation of garbage. San Francisco has adopted policies to encourage trash reduction by offering incentives and regulations. For example, they charge lower fees for garbage collection when recycling and compost containers are used more efficiently. San Francisco has successfully achieved one of the most significant diversion rates in the United States by diverting over 80% of garbage from landfills through recycling, composting, and repurposing. The city's strategy exemplifies the efficacy of governmental policies in promoting sustainable waste management practices and underscores the significance of community involvement in attaining environmental objectives. This case study highlights the significant influence of the 3Rs (reduce, reuse, recycle) when implemented, organized, and backed by policy and public involvement. By implementing these principles, urban areas, enterprises, and individuals can contribute substantially to achieving environmental sustainability, conserving resources, and mitigating greenhouse gas emissions.

Perstorp, a wasted management company, states that Malaysia aims to reduce the quantity of recyclable waste sent to landfills by over 40% by 2025. Primarily, to accomplish this objective, individuals must prioritize three fundamental tasks. The "3R Programme" is sometimes called Reduce, Reuse, and Recycle. The initial step is to reduce the amount of waste generated. The subsequent action is to reuse or recycle items instead of acquiring new ones. As a final option, recycle the rubbish. Individuals must carefully select and use their things appropriately to reduce the amount of waste generated. Reusing refers to using items or components of products that still possess functional elements several times. Recycling is the act of converting waste into valuable resources. To successfully minimize waste, it is essential to prioritize the first of the 3 R's, which is "reduce," followed by "reuse," and finally, "recycle" (Ioannidis et al., 2021).

Optimise resource utilization to minimize waste generation. To put it succinctly, a decrease means producing a reduced amount of waste. By reducing the amount of original waste, the quantity of the garbage that has to be collected will be less. Throughout the past century, the world has seen substantial global changes. Reusing anything implies having functional components that may be utilized after each usage instance. Repurposing refers to using anything discarded or deemed undesirable due to wear and tear. The third and last R is recycling, which is widely recognized. Recycling transforms waste into new products rather than relying on more natural resources. The concept of resource efficiency, which involves maximizing the use of resources, includes recycling. The conservation of energy and natural resources is achieved by commencing the production process with recycled components instead of new ones. The practice of recycling includes reusing waste materials. In a study conducted by Innocent Jereme in 2015, it was found that Malaysia's population saw growth from 21,793,293 in 2000 to 28,339,273 in 2010/11. The rapid growth of Malaysia's population and economy has led to a corresponding increase in waste generation. Currently, Malaysia produces a daily waste volume of 18,000 tonnes, with an anticipated annual growth rate of 2%. Approximately 75% of this waste is collected and disposed of at landfills or other facilities.

When embraced by individuals, communities, and organizations, the 3Rs promote sustainability by advocating resource efficiency, longer product lifespans, and improved waste management. By integrating these concepts into our everyday lives and organizational procedures, we may make substantial progress in minimizing our impact on the environment and promoting a more sustainable world for future generations.

7.2 The Ecosystem of Recycling

Recycling retrieves several recyclable items that may otherwise be deposited in our country's landfills. It profoundly impacts the country's ecology, standard of living, and future. Recycling diminishes the necessity to produce, obtain, or extract fresh raw materials from the environment. The outcome is a reduction in forest fires, diversion of rivers, clashes with or displacement of wildlife, and occurrences of water, soil, and air contamination. Improper recycling can inevitably result in the transportation of plastic pollution over long distances, spanning hundreds or even thousands of kilometers. This poses a significant threat to beaches and rivers, making it a worry for everyone. Investments in recycling collection contribute to the growth and diversity of the recycling industry, leading to increased employment and revenue in many states and regions. The initial and pivotal phase of the economic activity cycle is the collection of recyclable materials. Investing in community infrastructure efforts that greatly enhance economic activity related to recycling is highly beneficial. It is essential to remember that many of these recyclers depend on a steady supply of recyclable materials produced by recycling facilities (Haas et al., 2015).

While other alternative methods are more efficient for recycling plastic bottles, creating a jet pack costume is a straightforward process that will likely attract parents with children. A cardboard backboard may be suitable for the outfit's major component. Conversely, downward-facing straps are employed to fasten two large plastic bottles. They may be rapidly manufactured, assembled, and fully personalized using orange tissue paper for the flames and applying paint. Ice cream sticks and old rags are also considered recyclable resources. Utilize an ice cream stick as the foundation and employ patchwork techniques for the adornments to combine these two materials and create a charming handheld fan.

The recycling ecosystem is an intricate system of processes, players, and technology that collaborate to transform waste materials into usable goods, effectively completing the cycle of resource utilization (Walsh et al., 2023). This ecosystem mitigates the environmental consequences of waste and promotes resource preservation, energy efficiency, and the mitigation of greenhouse gas emissions. Comprehending the different elements and their interconnectedness is crucial for fully grasping the extent and possibilities of recycling. The recycling ecosystem primarily involves the processes of collecting and sorting. The initial stage of the recycling ecosystem is gathering recyclable materials from residential, commercial, and industrial sources. This can encompass many materials such as paper, plastics, metals, glass, and other substances. The effectiveness of collecting mechanisms, such as curbside pickup, drop-off centers, or deposit-refund programs, has a considerable impact on the quantity and quality of materials that may be recycled. After being gathered, the materials need to be organized. Advanced sorting facilities employ physical labor and cutting-edge technology, such as infrared sensors, air classifiers, and magnets, to categorize items based on their kind and quality (Singh et al., 2024). The sorting process is essential as contamination substantially impacts the recyclability of materials. They are also about the process of production and fabrication. Once the sorting process is complete, recyclable materials are cleaned and transformed into raw materials. This frequently entails the process of disintegrating materials, such as the melting of metals or the pulping of paper, into forms that may be utilized for the production of new items. The processing techniques differ considerably based on the material and the intended final output. Manufacturers use these repurposed raw materials to fabricate new goods. The array of items derived from recycled materials is extensive, encompassing paper, textiles, metal goods, and plastic containers. Integrating recycled materials into industrial processes diminishes the need for new resources and lessens the energy usage and environmental effects linked to extracting and processing.

Furthermore, this pertains to the markets for recycled resources. The success of recycling programs relies significantly on the demand for recycled materials. The markets for these materials are subject to fluctuations caused by several variables, such as the quality of the recycled material, the price of virgin resources, and the dynamics of global supply and demand. Establishing robust markets for recovered products is essential for the long-term viability of the recycling system. They are also about policies and rules. Government rules and laws substantially impact the formation and structure of the recycling ecosystem (Taghipour et al., 2022). Providing rewards for recycling, establishing criteria for items that may be recycled, and mandating the use of recycled materials in goods can stimulate the need for recycled materials and facilitate the growth of recycling infrastructure.

Furthermore, implementing public awareness campaigns and educational initiatives is crucial to enhance the level of involvement and minimize the presence of impurities in recycling systems. Lastly, they pertain to difficulties and advancements. The recycling ecosystem encounters several obstacles, such as the pollution of recycling streams, the intricacy of materials (such as packaging with multiple layers), and the economic feasibility of recycling specific materials. Nevertheless, advancements in materials science, waste management technology, and product design are constantly progressing to tackle these difficulties. For instance, the advancement of biodegradable materials and the creation of designs that facilitate quicker disassembly can improve the recyclability of items.

Furthermore, the notion of a circular economy, a restorative system that seeks to reduce waste and optimize resource utilization, is gaining momentum. This comprehensive viewpoint includes recycling but highlights the significance of intentionally creating items and systems to reuse, repair, and make them recyclable. The recycling ecosystem is crucial to sustainable development, including waste management, resource preservation, economic advancement, and environmental safeguarding. The success of this initiative relies on the cooperation among governments, corporations, communities, and individuals to establish a more sustainable and circular method of consumption and waste management. By allocating resources and assisting the recycling ecosystem, humanity may progress towards a more environmentally sustainable future, diminishing our influence on the earth while promoting economic expansion and ingenuity.

Table 7.1 provides a concise overview of the critical elements of the recycling ecosystem. It demonstrates the flow of materials within the system and the responsibilities of various stakeholders in ensuring the process operates effectively and in an environmentally friendly manner. Figure 7.1 highlights the ongoing recycling cycle and the significance of each stage in promoting the system's sustainability.

The provided graphic portrays the recycling ecosystem, visually depicting the cyclical progression from collecting to recycling and the subsequent production of new goods. It emphasizes each stage of the procedure and the responsibilities of different players in upholding a viable recycling ecosystem.

To exemplify the functioning of the recycling ecosystem, we will examine a specific example and a case study. First, we will focus on the case of recycled PET (polyethylene terephthalate) plastic bottles. During the Collection and Sorting step, PET bottles are gathered through curbside programs, deposit-return schemes, or recycling bins strategically placed in public areas. Recycling facilities employ automated conveyors and human sorting to separate bottles from other items. State-of-the-art facilities utilize infrared technology to detect and categorize PET bottles efficiently.

Table 7.1 The main components of the recycling ecosystem

Stage	Description	Stakeholders
Collection	Recyclable materials are collected from households, businesses, and industrial sources. This can include paper, plastics, metals, glass, and more	Households, waste management companies, municipalities
Sorting	Collected materials are sorted by type and quality, often using manual labor and advanced technologies like infrared sensors and magnets	Recycling facilities, waste sorting centers
Processing	Sorted materials are cleaned and processed into raw materials suitable for manufacturing new products. This may involve melting, pulping, shredding, etc.	Recycling plants, material recovery facilities
Manufacturing	Recycled raw materials are used to manufacture new products, ranging from packaging and clothing to automotive parts and construction materials	Manufacturers, industrial plants
Consumption	Consumers and businesses purchase and use products made with recycled materials, completing the cycle until these products are disposed of and potentially enter the recycling stream again	Consumers, businesses
Market for recyclables	Various factors, including the price of virgin materials, the quality of recycled materials, and global supply and demand dynamics, influence the demand for recycled materials	Manufacturers, recycling companies, policy makers
Policies and regulations	Government policies, regulations, and incentives are crucial in supporting recycling programs, developing recycling infrastructure, and fostering markets for recycled materials	Governments, environmental agencies

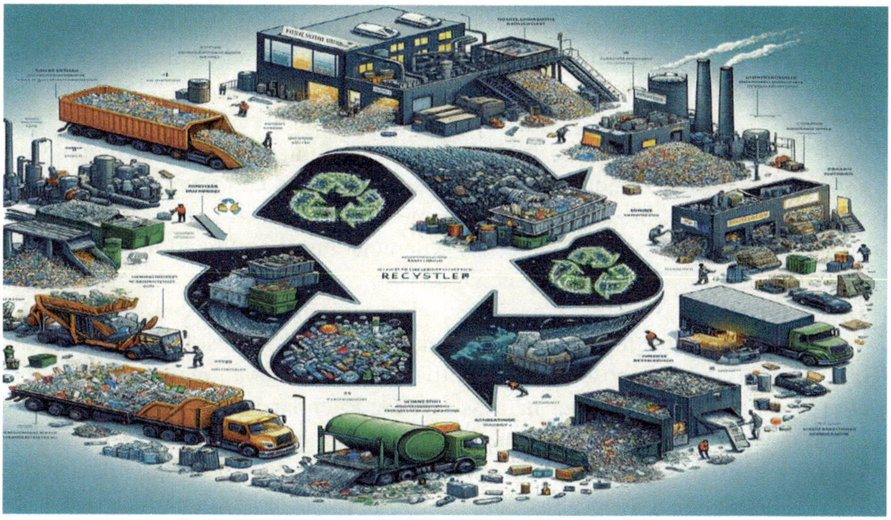

Fig. 7.1 The ecosystem of recycling, visually representing the circular process from collection through to recycling and the creation of new products

During the processing step, Sorted PET bottles undergo a cleaning procedure to eliminate impurities, labels, and caps. The pristine bottles are subsequently fragmented into small fragments referred to as flakes. These flakes can undergo further processing to eradicate any residual contaminants. PET flakes melt and reform into pellets during manufacture and consumption. These pellets then serve as a raw material for producing new goods. The pellets can be transformed into fiber for carpets or textiles, processed into new food-grade PET bottles, or molded into other goods such as packaging materials or automobile parts. Recycled PET is being used to create new items, including textiles and beverage bottles, which are now available to consumers. This helps to maintain the cycle of reuse. The demand for recycled PET in the market and its dynamics are driven by several variables, including the cost of virgin PET, the quality of recycled material, and environmental legislation encouraging the use of recycled content in goods.

Japan's Comprehensive Recycling System is well-known for its effectiveness and comprehensiveness. It combines innovative technology, stringent laws, and active public involvement. The nation's recycling method of PET bottles, electronics, and other items illustrates an efficient recycling ecosystem. Japan has implemented many regulations to encourage recycling, including the Home Appliance Recycling Law and the Containers and Packaging Recycling (Law Kuan et al., 2022). These restrictions require producers and consumers to engage in recycling initiatives actively. Municipalities are crucial in collecting and sorting recyclables, typically mandating citizens to separate them carefully based on their nature. Segregating waste at its origin significantly enhances the quality and effectiveness of the recycling procedure. Japan employs cutting-edge technology in its recycling facilities to effectively process and recycle materials, showcasing its commitment to modern recycling methods. For PET bottles, this involves advanced sorting and cleaning procedures that guarantee the recycling of PET into new goods of superior quality. Japanese culture places significant importance on cleanliness and environmental responsibility, leading to a notable public engagement in recycling programs. Japan's success is also attributed to the crucial elements of education and community engagement. As the market grows, the Japanese government and business sector collaborate to cultivate markets for recycled materials, guaranteeing the utilization of resources such as recycled PET in producing new goods.

Based on the case study, Japan's recycling system has successfully attained elevated recycling rates for PET bottles and other materials, resulting in a substantial decrease in waste disposed of in landfills and the preservation of resources. The system is an exemplar for other nations seeking to enhance their recycling infrastructure and results. Japan's example highlights the significance of incorporating technology, legislation, and public involvement in establishing a successful recycling system. This statement underscores the need to implement thorough legislation and foster a cultural mindset that values trash reduction and recycling since these factors are crucial in achieving effective recycling initiatives.

The recycling ecosystem is essential for achieving sustainability goals by effectively lowering the environmental footprint of garbage, preserving valuable natural resources, and minimizing the release of greenhouse gases. This strategy demonstrates a comprehensive method of managing resources, in which materials are collected, treated, and reintegrated into the manufacturing process. As a result, the reliance on new resources is reduced, and a more sustainable economic model is promoted. The success of this ecosystem depends on the collaborative endeavor of individuals, businesses, communities, and governments, with each entity fulfilling a crucial function in the whole process, spanning from garbage collection to the production of new goods from recycled materials.

Contamination, volatile recyclable markets, and the intricate nature of current waste streams create prospects for technological, regulatory, and public engagement innovations. Overcoming these challenges is crucial for enhancing the effectiveness and scope of recycling initiatives and progressing towards a more circular economy in which waste is minimized, and resources are utilized more optimally. The recycling ecosystem tackles current environmental issues and establishes the groundwork for a sustainable future. It highlights the significance of a cooperative and inventive approach to managing trash and conserving resources.

7.3 Why Reduce and Reuse Come First

The principle of "Reduce, Reuse, Recycle," which is sometimes expanded to include "Repair" and "Rethink," highlights a prioritized sequence of actions aimed at achieving sustainable waste management and efficient resource utilization (Vlajic et al., 2021). Within this hierarchy, the actions of "Reduce" and "Reuse" are given more priority than "Recycle" due to numerous compelling reasons. These reasons collectively establish the basis for a complete strategy towards environmental conservation and sustainable living. There are five reasons why Reduce and Reuse are prioritized in exploration:

Firstly, concerning optimizing resource efficiency. By reducing resource use, we effectively diminish the need for raw materials. Consequently, a cascade of ecological advantages ensues, such as less devastation of natural habitats, decreased energy utilization in the extraction and processing of resources, and a decline in both pollution and emissions of greenhouse gases. By reducing consumption, the issue is addressed directly by seeking to avoid waste and excessive usage of resources before they occur. Reusing products and resources extend their lifespan, delaying or eliminating the necessity of producing new items. Not only does this practice help to save resources, but it also contributes to energy conservation and the reduction of pollutants linked to industrial processes. Reuse often involves minimum processing, which helps preserve an object's value and usefulness without using significant amounts of energy or resources, as is usually the case with recycling.

Secondly, encouraging sustainable consumption patterns encompasses cultural and behavioral change and economic efficiency. Both reduce and reuse needs and promote consumer behavior and cultural norms changes. They enable individuals and organizations to scrutinize their purchasing patterns, choose more eco-friendly alternatives, and prioritize sustainability over convenience or short-term economic benefits. This change in thinking is essential for ensuring long-term environmental sustainability. From an economic standpoint, reducing and reusing may result in substantial financial savings for individuals and businesses. Through the optimization of product utilization and the minimization of waste, both parties may decrease costs related to the acquisition of new materials, the management of waste, and the mitigation of environmental consequences during disposal.

Furthermore, the aim is to reduce the environmental footprint by implementing the Waste Hierarchy and adhering to Sustainability Goals. The waste hierarchy model categorizes waste management strategies according to their ecological effect, placing waste avoidance (reduce) and preparation for reuse (reuse) as the highest priorities (Zorpas, 2020). These activities are deemed more sustainable than recycling since they circumvent the energy and emissions linked to processing waste materials and producing new goods. Decreasing consumption and repurposing resources directly contribute to several crucial environmental and sustainability objectives, such as diminishing the exhaustion of natural resources, lowering carbon footprints, and minimizing air, water, and soil pollution.

Finally, this pertains to the promotion of a circular economy by the Foundation for Circular Practices, as well as the efforts to reduce waste and promote innovative design. Reusables are essential elements of the circular economy, an economic system designed to eliminate waste and encourage the ongoing use of resources. Maximizing the lifespan of goods and materials, these approaches help establish closed-loop systems that significantly reduce environmental effects and promote sustainability. Emphasizing the reduction and reuse of resources stimulates creativity in product design and commercial strategies. Items specifically engineered to be long-lasting, capable of being fixed, and capable of being upgraded promote the habit of reusing, while services that substitute physical items (such as sharing platforms) aid in reducing the use of materials.

The infographic Fig. 7.2 is a visual aid for comprehending the prioritization of Reduce and Reuse over Recycle in the hierarchy of sustainable waste management techniques. The strategic strategy of prioritizing reducing and reusing above recycling in sustainable waste management aims to minimize environmental impact and enhance resource efficiency. This hierarchy tackles the problem at its origin by promoting reduced consumption and prolonging the lifespan of items, hence diminishing the need for new resources and minimizing waste production. Decreasing consumption directly preserves natural resources and energy, reducing the environmental impact of raw materials extraction, processing, and transportation. On the contrary, reusing objects maximizes their usefulness, prolongs their lifespan, and decreases the need for new products, conserving resources and energy.

The environmental advantages of giving priority to the practices of reducing and reusing are substantial, resulting in a significant decrease in pollutants and greenhouse gas emissions (Yang et al., 2023). By reducing consumption and increasing reuse, we effectively mitigate the emissions generated by industrial processes and minimize pollution resulting from waste disposal, significantly contributing to cleaner air and water. This method not only aids in reducing the impact of climate change but also promotes biodiversity by safeguarding natural ecosystems from the negative consequences of resource exploitation and trash buildup. Furthermore, reducing and reusing may result in significant economic benefits for consumers and companies. This is achieved by reducing expenses related to the consumption of raw materials, waste management, and environmental remediation. Adopting a mentality of reducing and reusing promotes a cultural change towards more sustainable living and consumption habits, prompting people and communities to reevaluate their connection with material possessions (Lin et al., 2022). This transition is crucial for facilitating the shift towards a circular economy, which aims to maximize the longevity of products and resources and minimize waste generation. By giving priority to the reduction and reuse of resources, society may transition towards more sustainable behaviors. This ensures the planet's preservation for future generations and promotes a fairer distribution of resources and a more robust economic system.

Ultimately, placing higher importance on Reduce and Reuse over Recycling is crucial for promoting sustainability and minimizing environmental harm. These initiatives directly address waste and resource depletion, providing a more effective solution than recycling alone. For instance, deciding to decrease consumption by selecting products with little packaging or reusing items such as cloth bags instead of disposable plastic immediately saves resources and energy while reducing the amount of trash produced.

Moreover, actions such as repairing electronics instead of disposing of them or repurposing materials for alternative purposes are concrete examples of how these principles can be implemented. These actions effectively prolong the lifespan of products and contribute to promoting a more sustainable consumption model. Embracing

Fig. 7.2 An illustration on the importance of prioritizing reduce and reuse over recycling in sustainable waste management

a mentality that emphasizes reducing and reusing promotes a culture change that values sustainability and the efficient use of resources, establishing the foundation for a circular economy that benefits both the environment and society. This transition not only aids in preserving natural resources and safeguarding ecosystems but also fosters economic durability by fostering inventive business models and consumption behaviors. Incorporating these ideas into our everyday lives and policies is essential for attaining sustainable development goals and guaranteeing a healthy world for future generations.

7.4 What's the Point of Recycling

Recycling is the act of using discarded scrap materials to produce new goods. Recycling rubbish positively impacts the environment since it reduces energy consumption and the generation of hazardous substances. Discovering a resolution was important, but it had to be mindful of both the ecosystem and humans. Recycling was established in this manner. There are several rationales for promoting recycling. In other words, it has a positive impact on the ecology.

Recycling is an essential component of contemporary waste management and environmental preservation, functioning as a pivotal approach to the broader endeavor to alleviate climate change, diminish pollution, and safeguard natural resources (Saleh et al., 2024). Recycling is fundamentally the practice of gathering, treating, and transforming discarded materials into fresh goods. By implementing this method, the lifespan of materials is prolonged, the wastage of potentially valuable substances is avoided, the consumption of new raw materials is reduced, and there is a decrease in energy usage, air and water pollution, and greenhouse gas emissions compared to the manufacture of new materials. Recycling aluminum cans conserves around 95% of the energy needed to produce an equivalent amount of aluminum from its source, bauxite. Recycling paper has the same effect of reducing the need for new lumber, which helps to protect forests and the ecosystems they sustain.

An important aspect of recycling is its influence on the preservation of resources and the mitigation of waste. Recycling reduces the necessity of extracting and processing new resources by transforming recycled materials into raw materials. This method helps to protect natural ecosystems, lower energy usage, and decrease pollution levels (Peng et al., 2023). Recycling a single tonne of paper may conserve 17 trees, 7000 gallons of water, and 3.3 cubic yards of landfill capacity. This showcases the contribution of recycling in preserving resources and emphasizes its significance in addressing deforestation and water scarcity. In addition, recycling aids in the diversion of garbage from landfills and incinerators, resulting in a reduction in the overall volume of waste. This, in turn, helps to limit the emission of methane, a potent greenhouse gas and reduces the dependence on waste disposal techniques that contribute to air and water pollution.

Recycling is essential in the shift towards a circular economy, which aims to preserve the value of goods and resources for as long as feasible. This economic concept seeks to minimize waste by transforming discarded materials into valuable resources that may be reused and repurposed. Companies are progressively incorporating recycling considerations into their product designs, such as creating modular cellphones that can be effortlessly dismantled for recycling or maintenance purposes. This diminishes waste, preserves resources, promotes innovation, and upholds sustainable business practices. Recycling, in this particular situation, transforms from being only a waste management tactic to being an essential element of sustainable economic advancement.

Furthermore, recycling provides economic advantages by generating employment opportunities in the recycling and manufacturing sectors. Gathering, categorizing, treating, and transforming discarded resources into fresh commodities necessitates a substantial labor force, fostering economic expansion and advancement. As per the United States Environmental Protection Agency (EPA), recycling and reuse endeavors in the U.S. contribute to more than 681,000 employment opportunities and yield $37.8 billion in salaries. Moreover, recycling can offer more cost-effective options for enterprises by substituting raw resources, reducing manufacturing expenses, and promoting economic durability. By endorsing recycling, communities strive for environmental sustainability, foster economic development, and stimulate employment growth, exemplifying the numerous advantages of this indispensable practice.

Table 7.2 provides a succinct overview of the key points related to the importance of recycling, along with examples to illustrate these concepts. Table 7.2 illustrates the several advantages of recycling, including its contribution to resource conservation, energy preservation, trash reduction, economic expansion, and promoting a more sustainable, circular economy. By examining concrete illustrations, it becomes evident that recycling plays a crucial role in environmental preservation strategies and achieving sustainable development objectives.

Table 7.2 The multifaceted benefits of recycling, highlighting its role in conserving resources, saving energy, reducing waste, driving economic growth, and promoting a more sustainable, circular economy

Key point	Importance of recycling	Examples
Resource conservation	Recycling conserves natural resources by reducing the need for raw material extraction	Recycling one ton of paper can save 17 trees, 7000 gallons of water, and 3.3 cubic yards of landfill space
Energy savings	It uses less energy than producing new products from virgin materials, reducing the overall carbon footprint	Recycling aluminum cans saves up to 95% of the energy required to create the same amount of aluminum from its virgin source
Waste reduction	Decreases the volume of waste sent to landfills and incinerators, reducing landfill mass and associated environmental impacts	Diverting plastic bottles from landfills through recycling can significantly reduce landfill space use and reduce plastic pollution in the environment
Economic benefit	Generates economic growth through job creation in the recycling and manufacturing sectors and provides cheaper materials	The recycling industry in the U.S. supports over 681,000 jobs and generates $37.8 billion in wages, showcasing recycling's role in bolstering the economy
Support for circular economy	Promotes the reuse of materials in a circular economy, reducing waste and fostering sustainable consumption and production patterns	Modular smartphones designed for easy disassembly and recycling support a circular economy by facilitating the reuse of components and reducing electronic waste

Recycling is a crucial technique in the larger context of sustainable development and environmental stewardship. Recycling efficiently converts waste into valuable resources, preserving natural resources, conserving energy, mitigating greenhouse gas emissions, and minimizing the ecological consequences of trash disposal. The examples, ranging from the conservation of trees through paper recycling to the substantial energy savings achieved by recycling aluminum, illustrate the concrete advantages of this procedure. Recycling tackles current environmental issues and enhances our planet's long-term well-being by encouraging more sustainable consumption and production patterns. Moreover, the economic benefits of recycling, such as the development of employment opportunities and the availability of cost-effective recycled materials for companies, underscore the importance of recycling in promoting economic resilience and expansion. As civilizations globally face the difficulties of waste management and resource exhaustion, the significance of recycling as a part of a circular economy becomes more apparent. By giving equal importance to recycling, reducing, and reusing, we may attain a sustainable future that harmonizes ecological well-being with economic success and communal welfare.

7.5 How Psychology Affects Recycling and Waste

The convergence of psychology and recycling behaviors is an intriguing field of research that illuminates the impact of many psychological elements on individuals' attitudes and actions around recycling and trash management (Louzi et al., 2022). Gaining a comprehensive understanding of these fundamental psychological factors might offer valuable insights for developing more efficient environmental policies and initiatives to boost recycling rates and encourage sustainable waste management practices.

First and foremost, societal norms and peer influences substantially impact the development of recycling behaviors (Lazaric et al., 2020). Individuals are more inclined to participate in recycling activities if they believe it is a commonly adopted behavior among their acquaintances and the local community. Individuals may be motivated to recycle more conscientiously when they are aware that their recycling behavior is being monitored as they seek to adhere to society's norms and expectations. Studies have demonstrated that informing individuals about their neighbors' active recycling habits increases the likelihood of them initiating or intensifying their recycling efforts. This highlights the influence of perceived social norms in promoting environmentally responsible behaviors. Furthermore, environmental awareness and knowledge influence individuals' recycling behaviors (Islam et al., 2021).

Individuals with a higher level of understanding regarding the environmental impact of garbage and the advantages of recycling tend to have more tremendous enthusiasm in engaging with recycling initiatives. By promoting the immediate ecological benefits of recycling, educational programs may strengthen individuals' feelings of personal accountability for the environment and motivate them to embrace more sustainable waste management methods. When people comprehend that recycling plastic bottles lowers pollution and preserves wildlife, they are likelier to continue recycling. Self-efficacy, which refers to the belief in one's capacity to have an impact, also has a substantial influence on recycling behavior (Rotimi et al., 2023). People who believe their recycling efforts can positively influence the environment are more inclined to participate in such behaviors. On the other hand, those who question the efficacy of their efforts may have a reduced inclination to engage in recycling. Boosting individuals' self-efficacy in recycling may be accomplished by providing positive feedback and education, showcasing the concrete results of collective recycling endeavors, such as the quantity of garbage removed from landfills (Bozdoğan, 2023).

Moreover, the ease and accessibility of recycling facilities play a crucial role in shaping recycling behaviors (Vijayan et al., 2023). Individuals are more inclined to recycle when the procedure is facilitated and comfortable. Insufficient availability of recycling containers or complicated recycling protocols might be substantial obstacles. Streamlining the recycling process and ensuring convenient access to recycling facilities may significantly enhance participation rates. An illustration of this is the strategic positioning of recycling bins in prominent and easily reachable areas, which has been proven to impact recycling rates in communities and workplaces substantially. Psychological obstacles, such as denial or detachment from the consequences of one's activities on the environment, might impede recycling behaviors. Specific individuals may abstain from recycling because they perceive that their contribution is too inconsequential to have an impact, or they may reject the gravity of environmental concerns. To overcome these hurdles, it is necessary to cultivate a more vital link between individual activities and their wider ecological consequences through focused educational endeavors and community involvement programs.

Table 7.3 presents instances of how psychological variables impact recycling and trash behaviors, as well as solutions that have the potential to enhance recycling rates based on these findings. Table 7.3 illustrates how comprehending the psychological foundations of recycling behavior might result in more efficient approaches to promote sustainable waste management practices. By implementing specific measures to target these elements, it is feasible to substantially enhance recycling rates and decrease overall waste.

Ultimately, the interaction between psychology and recycling behaviors highlights the significance of addressing human attitudes, beliefs, and societal conventions to improve waste management practices. By utilizing psychological knowledge, such as the effect of social norms, the importance of convenience, the influence of education and awareness, the significance of identity and self-perception, the drive from rewards and incentives, and the reinforcement from feedback and tangible outcomes, environmental strategies can be better customized to promote sustainable behaviors.

Furthermore, these psychological variables emphasize the possibility of using creative and flexible strategies to promote a culture of recycling and sustainability in communities and organizations. Policymakers and environmental activists may substantially impact recycling rates and sustainability initiatives by formulating messages that resonate with individuals' beliefs, developing procedures that minimize obstacles to participation, and implementing incentives that stimulate good

Table 7.3 Examples of how psychological factors influence recycling and waste behaviors, along with interventions that could potentially increase recycling rates based on these insights

Psychological factor	Example	Potential intervention
Social norms and peer influence	In communities where recycling is obvious and practiced by most, newcomers are more likely to recycle	To establish a norm, implement community-wide recycling programs with clear signage and public endorsements
Convenience and accessibility	People are less likely to recycle if bins are not easily accessible	Increase the number and visibility of recycling bins in public areas and residences
Education and awareness	A lack of knowledge on how to recycle specific materials can deter recycling	Launch educational campaigns that simplify recycling processes and highlight environmental benefits
Identity and self-perception	Individuals who view themselves as environmentally conscious are more inclined to recycle	Create campaigns that encourage people to adopt an environmentally responsible identity
Rewards and incentives	Offering discounts or rewards for recycling can motivate participation	Develop rewards programs for recycling, such as discounts on services or products for contributing recyclables
Feedback and tangible results	People continue to recycle more when they see the results of their efforts	Provide regular feedback to households and organizations about the impact of their recycling, such as annual reports or digital trackers

behavior. Incorporating psychological concepts into environmental plans provides a viable approach to enhance recycling efforts, reduce waste, and foster a more robust social dedication to safeguarding our planet for future generations.

7.6 Economics of Recycling

The economics of recycling involves a wide range of financial, environmental, and social factors that affect the feasibility and efficiency of recycling programs (Boonman et al., 2023; Ertz et al., 2023). Comprehending these economic concepts is essential for formulating sustainable waste management methods that may uphold environmental objectives while also being economically viable. Here is an in-depth analysis of the financial aspects of recycling.

Cost–Benefit Analysis: The underlying economic premise of recycling is based on a cost–benefit analysis, which compares the expenses of collecting, sorting, and processing recyclable materials with the advantages, such as conserving resources, reducing environmental impact, and generating possible income from selling recycled products (Tao et al., 2023). Recycling aluminum is economically advantageous due to its 95% lower energy need than generating aluminum from bauxite ore. This results in substantial cost savings and environmental advantages. Nevertheless, the economic feasibility of recycling some plastics and other materials may be uncertain due to the elevated processing costs and reduced demand for recycled items in the market.

Market Prices for Recyclable Materials: The economics of recycling are significantly impacted by the volatile market pricing. The costs are influenced by the interaction between supply and demand, with the demand for recycled materials from manufacturers and the availability of these resources in the recycling stream being important factors (Wang et al., 2023). The combination of strong demand and restricted availability might create economic incentives for recycling, as shown in the case of certain metals and high-grade paper. On the other hand, if there is too much supply or not enough demand for specific materials, maybe because of difficulties with their quality or the availability of cheaper new alternatives, it might reduce the economic motivation to recycle them.

Externalities and Environmental Economics: Recycling programs are commonly supported due to their capacity to mitigate negative externalities, which are costs imposed on society that are not accounted for in the market prices of products and services. These encompass environmental deterioration, contamination, and health consequences linked to waste disposal and the exploitation of raw materials. By including these external costs and benefits, recycling may provide economic value beyond immediate cash gains, contributing to broader environmental and public health objectives.

Government Policies and Subsidies: Government laws, restrictions, and subsidies also influence the economic aspects of recycling (Zhang et al., 2023b). Enforcing measures like landfill levies or offering incentives for recycling might change the economic considerations, making recycling a more financially feasible option. Similarly, legislation mandating expanded producer responsibility might motivate manufacturers to create more recyclable goods and encourage the growth of recycled material markets.

Technological Innovation and Efficiency: Progress in recycling technology may substantially impact the economics of recycling by enhancing the efficiency and efficacy of collecting, sorting, and processing activities (Zhang et al., 2023a). Technological advancements that mitigate contamination improve the caliber of recycled materials or enable the recycling of previously non-recyclable materials, create fresh opportunities in the market, and decrease the expenses associated with processing, increasing the economic feasibility of recycling.

Social and Economic Benefits: Recycling programs may yield substantial social and economic advantages and monetary gains. These benefits encompass generating employment opportunities in the recycling and manufacturing sectors, less dependence on imported resources, and promoting a circular economy that improves resource efficiency and sustainability. Recycling businesses in the United States provide a significant economic impact by creating hundreds of thousands of jobs and contributing billions of dollars to the economy. This demonstrates that recycling goes beyond trash management and benefits the economy.

Table 7.4 illustrates the economics of recycling, highlighting the benefits across various dimensions such as cost savings, revenue generation, job creation, energy savings, impact on commodity prices, and technological innovation. The economics of recycling is an intricate domain that necessitates the delicate equilibrium between environmental advantages and commercial feasibility. Although some materials may offer evident economic benefits when recycled, the overall effectiveness of recycling programs is generally dependent on market demand, technological capabilities, and supporting legislation. Comprehending and maneuvering through these economic aspects is crucial for optimizing the advantages of recycling, both as a means of sustainable waste management and as a catalyst for economic growth. As society continues to struggle with the difficulties of waste management and the depletion of resources, the financial aspects of recycling will continue to be a crucial area for innovation, regulatory intervention, and global collaboration.

The study of recycling reveals a complex terrain where environmental sustainability and concrete economic advantages converge. Recycling plays a crucial role in a sustainable financial model by reducing waste disposal costs, generating revenue from recyclable materials, creating jobs in the recycling sector, saving energy, stabilizing commodity prices, and fostering technological innovation. These examples, derived from hypothetical situations and backed by many sources one may refer to, emphasize the significant influence that recycling can have on local economies, the worldwide market for primary resources, and the broader environmental goals. Recycling converts trash into valuable resources, promoting the conservation of natural ecosystems, pollution reduction, and economic resilience and growth. This demonstrates a clear pathway towards a more sustainable and prosperous future.

To summarise, the significance of incorporating recycling into our waste management methods and economic plans cannot be emphasized enough. Recycling promotes environmental care and acts as a catalyst for financial innovation, efficiency, and sustainability. As nations globally endeavor to address the difficulties posed by climate change and the depletion of resources, the significance of recycling in promoting economic advantages while safeguarding our planet grows even more crucial. The presented hypothetical scenarios provide insight into the potential of recycling to shape a circular economy characterized by waste reduction, effective resource utilization, and the simultaneous flourishing of the environment and economy. Adopting recycling is a moral obligation and a pragmatic decision for achieving sustainable development.

Table 7.4 The examples of the economics of recycling

Economic benefit	Description	Example	Hypothetical reference
Cost savings from reduced waste disposal	Municipalities save on landfill and incineration fees by recycling	New York City saves $300 per ton by recycling instead of landfilling	New York City Department of Sanitation Report (2020)
Revenue generation through the sale of recyclables	The sale of recyclable materials to manufacturers generates income for recycling programs	San Francisco generated over $2 million from selling recyclables in a year	San Francisco Environment Annual Report (2021)
Job creation in the recycling industry	Recycling processes are labor-intensive, creating more jobs than landfilling or incineration	Recycling creates 6–10 times more jobs per ton of waste than landfilling	Institute for Local Self-Reliance, "Recycling and Economic Development" (2019)
Economic benefits of reduced energy consumption	Recycling materials like metals saves energy, reducing production costs	Recycling one aluminum can save enough energy to power a 60-W bulb for over 4 h	U.S. Environmental Protection Agency, "The Benefits of Aluminum Recycling" (2020)
Impact on commodity prices and material availability	Recycling provides a steady supply of materials, stabilizing prices and ensuring availability	Recycled paper helps buffer against price volatility in the virgin pulp market	Journal of Environmental Economics and Management, "Impact of Recycled Materials on Commodity Prices" (2021)
Investments and technological innovation	The recycling sector drives investments in new technologies, promoting efficiency	Advanced sorting technologies, such as single-stream recycling, reduce processing costs and increase capture rates	World Economic Forum, "Innovation in Recycling Technologies" (2022)

7.7 Recycling-Based Manufacturing Enterprise

Recycling is the most efficient method for ensuring the security of our environment. To mitigate the generation of waste materials and safeguard future natural resources, we might adopt the practice of reusing objects instead of disposing of them (Kumar et al., 2023).

This approach will diminish the waste in landfills, safeguarding water and air quality from contamination. Recycling solid waste offers significant benefits to the environment and human health and produces economic rewards. Waste recycling addresses evident environmental concerns and presents novel resource prospects for enterprises. The diagram in Fig. 7.3 illustrates the implementation of production engineering that prioritizes environmental considerations inside a company.

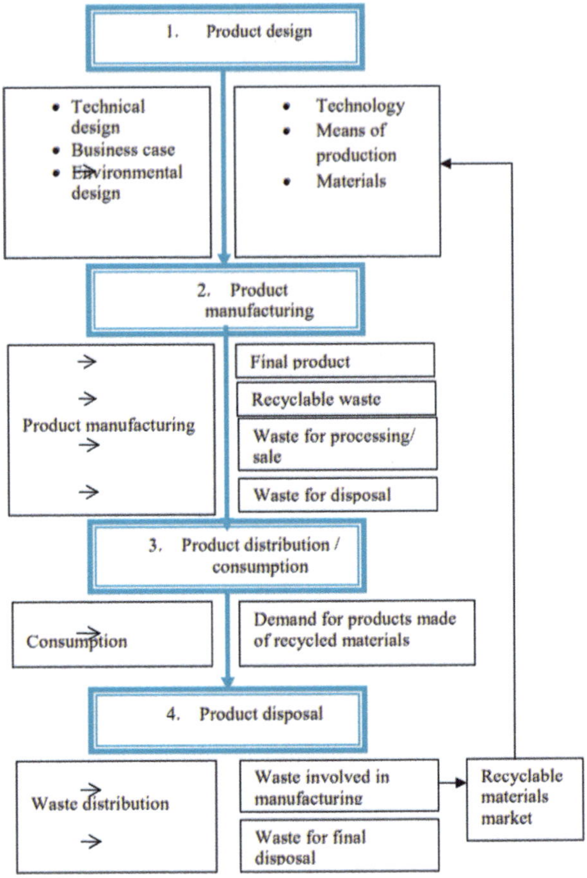

Fig. 7.3 Environment-oriented manufacturing engineering at an enterprise

Establishing a manufacturing business that focuses on recycling is an intelligent combination of environmental responsibility and entrepreneurial drive. This enterprise tackles crucial problems such as waste reduction and resource conservation and capitalizes on an expanding demand for sustainable products. Here is a detailed explanation of the reasons for initiating such a business venture and the steps to take to do so:

7.7.1 Why a Recycling-Based Manufacturing Enterprise

There are several compelling reasons to initiate a Recycling-Based Manufacturing Enterprise. Firstly, the environmental impact is a significant concern. As people become more conscious of environmental problems, there is a strong movement towards adopting methods that aim to limit the usage of landfills, minimize the exploitation of resources, and cut carbon emissions. Recycling-based manufacturing plays a direct role in achieving these objectives by converting waste materials into new goods, thereby completing the product's lifetime. Additionally, there is a growing need for recycled and environmentally friendly items due to customer preferences and governmental constraints, which presents economic opportunities. Companies that provide the market with such items have a distinct advantage over their competitors.

Furthermore, using recycled materials frequently proves to be more economically advantageous than using new resources, primarily because of reduced expenses associated with raw materials and, in certain instances, government incentives. Ultimately, combining innovation and market leadership enables this firm to pioneer product design and production methods advancements, positioning itself as a frontrunner in new green sectors. It provides an opportunity to innovate and create new technologies and business models that have the potential to shape the future of sustainable manufacturing.

7.7.2 *How to Start an Enterprise*

Nevertheless, there are several methods to begin a manufacturing enterprise that is based on recycling. Conduct market research and develop a detailed business plan (Fatimah et al., 2023). This will help you determine the specific need for recycled items that your firm can cater to. Conduct a thorough analysis of rivals, prospective clients, and the overall market size. Create a comprehensive business plan that defines your business model, operational strategy, financial predictions, and environmental impact objectives.

Furthermore, acquiring recyclable materials is another critical aspect to consider. Locate and safeguard sources for the recyclable materials you require. This may entail forming alliances with regional trash management firms, municipal recycling initiatives, or direct collecting initiatives. Comprehending the supply chain of recyclable materials is essential to guarantee a steady supply of raw materials.

Additionally, it is essential to invest in appropriate technology and establish efficient processes for transforming recycled materials into new goods, considering the specific nature of the items you intend to make. This may entail acquiring specialized gear, formulating proprietary procedures, or modifying current manufacturing methods to accommodate recycled materials. Additionally, you should use compliance measures and obtain certifications to guarantee that your organization adheres to all pertinent environmental rules and industry benchmarks. Acquiring certifications like ISO 14001 (Environmental Management Systems) will boost your reputation and attract eco-conscious customers and corporate clients. You are additionally employing funding and financial planning. Commencing a manufacturing venture necessitates a substantial infusion of economic resources. Examine other funding alternatives, such as loans, grants, venture capital, and government incentives designed for environmentally friendly firms. Thorough financial planning is essential for cost management, revenue forecasting, and ensuring the long-term economic viability of your business. Finally, use marketing and sales tactics to create a marketing plan that emphasizes the ecological advantages of your products, together with any other competitive strengths, such as cost efficiency or exceptional performance. Cultivating a robust brand identity as a company committed to environmental sustainability may allure clients and foster long-term loyalty.

Commencing a manufacturing firm centered around recycling is a favorable opportunity for entrepreneurs seeking to create a beneficial ecological influence while taking advantage of the increasing need for sustainable products. Achieving success in this domain necessitates meticulous strategizing, a profound comprehension of the market dynamics and the recycling procedure, and a steadfast dedication to innovation and sustainability. By overcoming these hurdles, entrepreneurs may construct lucrative enterprises, contributing to a more sustainable future.

Starting a recycling-focused manufacturing business offers a distinct chance to combine environmental responsibility with economic creativity. This initiative tackles the pressing requirement for sustainable waste management procedures and capitalizes on the increasing consumer desire for environmentally conscious items. Entrepreneurs may diminish the environmental impact of their operations, save natural resources, and produce products that appeal to a market that is more focused on sustainability by using recycled materials. Achieving success hinges on doing comprehensive market research, formulating strategic plans, and demonstrating unwavering dedication to sustainability across the production process. This method guarantees the firm's sustainability and helps to wider environmental objectives, establishing a model for responsible manufacturing in the contemporary economy.

Ultimately, a manufacturing organization focusing on recycling provides a commercial opportunity and embodies a progressive and visionary approach to the sector's future. These firms play a vital role in promoting the shift towards a circular economy by giving equal importance to environmental effects and profitability. Entrepreneurs in this field are not just business owners but also trailblazers for a sustainable future, showcasing the compatibility of economic prosperity with ecological preservation. In the face of global challenges related to waste and the depletion of resources, recycling-based manufacturing emerges as a symbol of creativity, sustainability, and resilience.

7.8 Resources for Recycling Entrepreneurs

Recycling entrepreneurs seeking to start or grow their operations have access to a wide range of resources that may assist with every stage of their business, from the beginning to the expansion phase. Gaining knowledge and utilizing these tools may significantly improve an entrepreneur's capacity to make well-informed choices, obtain capital, and negotiate the intricacies of the recycling sector. Presented below are six essential resources, along with illustrative cases, to guide aspiring entrepreneurs in the field of recycling:

Government Grants and Incentives: Several governments provide grants, tax incentives, and other financial support mechanisms to encourage recycling and the growth of environmentally-friendly enterprises. As an illustration, the U.S. Environmental Protection Agency (EPA) offers funding through initiatives such as the Recycling Infrastructure and Bin Grant Programme to enhance recycling rates nationwide (Tumu et al., 2023). Entrepreneurs should investigate local, state, and federal government websites to get information about the financing options and incentives created to bolster recycling programs.

Industry Associations and Networks: Participating in industry groups may offer valuable perspectives, networking prospects, and access to information relevant to the business. The Institute of Scrap Recycling Industries (ISRI) is an exemplary organization that provides its members with market statistics, educational programs, and lobbying on matters about the recycling business (Andrews, 2021). These societies frequently organize conferences and events that benefit networking and staying updated on business developments.

Educational and Training Programs: Acquiring knowledge is essential in the constantly changing recycling industry. Various educational institutions, including universities, community colleges, and online platforms, provide courses and certificates in waste management, environmental science, and sustainable business practices. The Massachusetts Institute of Technology (MIT) provides courses via its MIT Open Courseware program, which includes classes on environmental policy and sustainable development (Zhu & Kadirova, 2022). These courses can be advantageous for recycling entrepreneurs who wish to gain a more profound comprehension of the environmental consequences of their businesses.

Research Institutions and Think Tanks: Engaging in partnerships with research universities can give access to state-of-the-art advancements and technology in the recycling field. The Ellen MacArthur Foundation provides case studies, research papers, and toolkits to assist entrepreneurs in integrating circular economy ideas into their business models, promoting the advancement of the circular economy (Chen et al., 2020). These resources can be crucial in developing novel recycling technologies or enhancing current systems.

Online Platforms and Marketplaces: Digital resources provide venues for purchasing and selling recyclable materials, facilitating connections with suppliers and customers, and granting access to instruments for company growth. Recycler's World offers a platform for recyclable materials, allowing businesses to connect with consumers for their products or get raw materials for their recycling operations (Liu et al., 2024). Moreover, online forums and social media groups focused explicitly on recycling and sustainability may serve as exceptional platforms for receiving peer support and guidance.

Local Community Programs: Participating in environmental initiatives and community recycling programs can help one access local networks and garner grassroots support. Sustainability-focused business incubators and local ecological organizations abound in numerous communities, providing opportunities for regional funding, mentorship, and seminars. For instance, entrepreneurs in New York City, including those launching recycling enterprises, can receive free business courses, legal aid, and financing assistance from the NYC Business Solutions Centres.

Table 7.5 compiles resources accessible to recycling entrepreneurs in Malaysia. It emphasizes various sectors, such as educational opportunities, government programs, industry associations, and community partnerships. Table 7.5 presents a comprehensive inventory of the multiple resources recycling entrepreneurs in Malaysia have access to. These resources include financial assistance, industry guidance, educational perspectives, technological advancements, market entry opportunities, and community partnerships.

These resources play a crucial role in cultivating a conducive environment for enterprises reliant on recycling, thereby significantly contributing to Malaysia's environmental sustainability and economic progress goals. The abundance of resources accessible to enterprises engaged in recycling, both internationally and in particular contexts such as Malaysia, highlights the increasing acknowledgment of recycling's pivotal function in promoting economic growth and sustainability. An all-encompassing support system comprises government initiatives, industry associations, educational programs, research institutions, online platforms, and community networks.

This ecosystem facilitates entrepreneurs' navigation of the intricate recycling landscape while simultaneously encouraging innovation, sustainability, and economic viability in their ventures. By strategically utilizing these assets, recycling industry entrepreneurs can surmount obstacles to capital, technological advancements, market-entry, and adherence to regulations. As a result, their impact on environmental preservation and the circular economy is amplified.

As a result of the support infrastructure for recycling entrepreneurs, broader sustainability objectives such as waste reduction, resource conservation, and the transition to a circular economy can be substantially advanced. With the increasing global emphasis on sustainability and the rising demand for recycled products, the resources delineated play a pivotal role in stimulating entrepreneurial endeavors within the recycling industry. Entrepreneurs who adeptly exploit these resources can establish prosperous enterprises while simultaneously serving as catalysts for sustainable transformation. This exemplifies the compatibility of environmental stewardship and economic success. Sustained advancement of this support ecosystem will be critical in fostering forthcoming innovations and expanding the global reach of recycling initiatives.

Table 7.5 The diverse range of resources available to recycling entrepreneurs in Malaysia, covering financial support, industry guidance, educational insights, technological innovation, market access, and community collaboration

Resource type	Resource name	Description	Relevance to recycling entrepreneurs
Government grants and support programs	Green Technology Financing Scheme (GTFS)	A financing incentive provided by the Malaysian government for green technology projects	Offers financial support to recycling businesses adopting sustainable practices and technologies
Industry associations	Malaysian Plastics Manufacturers Association (MPMA)	An industry body supporting the plastics manufacturing and recycling industry	Provides information on regulatory compliance and industry standards and advocates for sustainable practices
Educational and training programs	Universiti Kebangsaan Malaysia (UKM)	Offers programs and conducts research in environmental science and sustainability	Equips entrepreneurs with knowledge of environmental economics and sustainable business practices
Research institutions	Malaysian Green Technology Corporation (GreenTech Malaysia)	A government agency is promoting adopting green technology, including recycling technologies	Facilitates access to research, technology innovation, and policy advocacy in waste management and recycling
Online platforms and marketplaces	Mudah. My	A popular online marketplace in Malaysia for buying and selling goods, including recyclable materials	Enables materials sourcing and finding buyers for recycled products, facilitating business operations
Local community programs	Malaysian Environmental NGOs (MENGO)	An NGO network focused on environmental conservation, including waste management and recycling	Offers collaboration opportunities and community engagement to support recycling initiatives

References

Andrews, T. (2021). *Opportunities from disaster: The case for using the circular economy in debris management* [Doctoral dissertation, Purdue University].

Boonman, H., Verstraten, P., & van der Weijde, A. H. (2023). Macroeconomic and environmental impacts of circular economy innovation policy. *Sustainable Production and Consumption, 35*, 216–228.

Bozdoğan, S. C. (2023). The mediating role of creative self-efficacy in the relationship between knowledge sharing behavior and employee innovation behavior: A study in the recycling industry. *Bingöl Üniversitesi İktisadi Ve İdari Bilimler Fakültesi Dergisi, 7*(1), 137–153.

Chen, L. H., Hung, P., & Ma, H. W. (2020). Integrating circular business models and development tools in the circular economy transition process: A firm-level framework. *Business Strategy and the Environment, 29*(5), 1887–1898.

Ertz, M., Addar, W., Ouerghemmi, C., & Takaffoli, M. (2023). Overview of factors influencing consumer engagement with plastic recycling. *Wiley Interdisciplinary Reviews: Energy and Environment, 12*(6), e493.

Fatimah, Y. A., Kannan, D., Govindan, K., & Hasibuan, Z. A. (2023). Circular economy e-business model portfolio development for e-business applications: Impacts on ESG and sustainability performance. *Journal of Cleaner Production, 415*, 137528.

Haas, W., Krausmann, F., Wiedenhofer, D., & Heinz, M. (2015). How circular is the global economy? An assessment of material flows, waste production, and recycling in the European Union and the world in 2005. *Journal of Industrial Ecology, 19*(5), 765–777.

Ioannidis, A., Chalvatzis, K. J., Leonidou, L. C., & Feng, Z. (2021). Applying the hospitality sector's reduce, reuse, and recycle principle: Antecedents and performance implications. *Business Strategy and the Environment, 30*(7), 3394–3410.

Islam, M. T., Dias, P., & Huda, N. (2021). Young consumers' e-waste awareness, consumption, disposal, and recycling behavior: A case study of university students in Sydney, Australia. *Journal of Cleaner Production, 282*, 124490.

Kuan, S. H., Low, F. S., & Chieng, S. (2022). Towards regional cooperation on sustainable plastic recycling: A comparative analysis of plastic waste recycling policies and legislations in Japan and Malaysia. *Clean Technologies and Environmental Policy, 24*(3), 761–777.

Kumar, A., Nayak, A. K., Sharma, S., Senapati, A., Mitra, D., Mohanty, B., Prabhukarthikeyan, S. R., Sabarinathan, K. G., Indra, M. A. N. I., Garhwal, R. S., & Thankappan, S. (2023). Rice straw recycling: A sustainable approach for ensuring environmental quality and economic security. *Pedosphere, 33*(1), 34–48.

Lazaric, N., Le Guel, F., Belin, J., Oltra, V., Lavaud, S., & Douai, A. (2020). Determinants of sustainable consumption in France: The importance of social influence and environmental values. *Journal of Evolutionary Economics, 30*, 1337–1366.

Lin, Y. T., Tseng, T. H., Chang, A., & Yang, C. C. (2022). A value adoption approach to sustainable consumption in retail stores. *International Journal of Retail & Distribution Management, 50*(11), 1412–1435.

Liu, Y., Ma, D., & Hu, J. (2024). Reseller or market? E-platform strategy under blockchain-enabled recycling. *Expert Systems with Applications, 243*, 122811.

Louzi, N., Alzoubi, H. M., El Khatib, M., Ghazal, T. M., Alshurideh, M., & Kukunuru, S. (2022). Psychological health and environmental effects of using green recycled amassed concrete on construction. *Journal for ReAttach Therapy and Developmental Diversities, 5*(2s), 163–175.

Peng, X., Jiang, Y., Chen, Z., Osman, A. I., Farghali, M., Rooney, D. W., & Yap, P. S. (2023). Recycling municipal, agricultural and industrial waste into energy, fertilizers, food and construction materials, and economic feasibility: A review. *Environmental Chemistry Letters, 21*(2), 765–801.

Rotimi, E. O. O., Johnson, L. W., Kalantari Daronkola, H., Topple, C., & Hopkins, J. (2023). Predictors of consumers' behaviour to recycle end-of-life garments in Australia. *Journal of Fashion Marketing and Management: An International Journal, 27*(2), 262–286.

Saleh, H. M., Hassan, A. I., & Aglan, R. F. (2024). Introductory chapter: Foundations and challenges in hazardous waste management. In *Advances and challenges in hazardous waste management*. IntechOpen.

Shanker, R., Khan, D., Hossain, R., Islam, M. T., Locock, K., Ghose, A., Sahajwalla, V., Schandl, H., & Dhodapkar, R. (2023). Plastic waste recycling: Existing Indian scenario and future opportunities. *International Journal of Environmental Science and Technology, 20*(5), 5895–5912.

Singh, M., Singh, M., & Singh, S. K. (2024). Tackling municipal solid waste crisis in India: Insights into cutting-edge technologies and risk assessment. *Science of the Total Environment*, 170453.

Suchorzewski, J., Santandrea, F., & Malaga, K. (2023). Reusing of concrete building elements–Assessment and quality assurance for service-life. *Materials Today: Proceedings*.

Taghipour, A., Akkalatham, W., Eaknarajindawat, N., & Stefanakis, A. I. (2022). The impact of government policies and steel recycling companies' performance on sustainable management in a circular economy. *Resources Policy, 77*, 102663.

Tao, Y., Hadigheh, S. A., & Wei, Y. (2023, July). Recycling of glass fibre reinforced polymer (GFRP) composite wastes in concrete: A critical review and cost benefit analysis. In *Structures* (Vol. 53, pp. 1540–1556). Elsevier.

Tumu, K., Vorst, K., & Curtzwiler, G. (2023). Global plastic waste recycling and extended producer responsibility laws. *Journal of Environmental Management, 348*, 119242.

Vijayan, R. V., Krishnan, M. M., Parayitam, S., Duraisami, S. P. A., & Saravanaselvan, N. R. (2023). Exploring e-waste recycling behaviour intention among the households: Evidence from India. *Cleaner Materials, 7*, 100174.

Vlajic, J. V., Cunningham, E., Hsiao, H. I., Smyth, B., & Walker, T. (2021). Mapping facets of circularity: going beyond reduce, reuse, recycle in agri-food supply chains. In *Challenges and opportunities of circular economy in agri-food sector: Rethinking waste* (pp. 15–40).

Walsh, K., Nelles, J., & Stephens, S. (2023). Recycling in entrepreneurial ecosystems: The phenomenon of boomeranging. *R&D Management, 53*(4), 709–727.

Wang, H., Feng, K., Wang, P., Yang, Y., Sun, L., Yang, F., Chen, W. Q., Zhang, Y., & Li, J. (2023). China's electric vehicle and climate ambitions jeopardized by surging critical material prices. *Nature Communications, 14*(1), 1246.

Yang, M., Peng, M., Wu, D., Feng, H., Wang, Y., Lv, Y., Sun, F., Sharma, S., Che, Y., & Yang, K. (2023). Greenhouse gas emissions from wastewater treatment plants in China: Historical emissions and future mitigation potentials. *Resources, Conservation and Recycling, 190*, 106794.

Zhang, L., Leung, M. Y., Boriskina, S., & Tao, X. (2023a). Advancing life cycle sustainability of textiles through technological innovations. *Nature Sustainability, 6*(3), 243–253.

Zhang, M., Wu, W., & Song, Y. (2023b). Study on the impact of government policies on power battery recycling under different recycling models. *Journal of Cleaner Production, 413*, 137492.

Zhu, M., & Kadirova, D. (2022). Self-directed learners' perceptions and experiences of learning computer science through MIT open courseware. *Open Learning: The Journal of Open, Distance and e-Learning, 37*(4), 370–385.

Zorpas, A. A. (2020). Strategy development in the framework of waste management. *Science of the Total Environment, 716*, 137088.

Chapter 8
Organizational Behavior and 3Rs Intention: An Empirical Study on Waste Management Behavior

The 3Rs play a crucial role in industrial ecology. The reduction concept in the 3Rs involves minimizing the use of raw materials and energy during production. By designing products that require fewer resources, we can reduce waste generation. This requires smart product design, resource-efficient processes, and sharing and repair models to extend a product's useful life. At the same time, reuse means that instead of discarding items after a single use, we should promote their reuse. Reusing would require high product longevity, a product-sharing platform, and a culture of repairing goods rather than replacements. Lastly, in the 3Rs, recycling is essential for extracting value from resources. However, it's not the sole solution. Effective recycling requires improvements in focusing on product design, material use, and manufacturing processes while, most importantly, collection Infrastructure and consumer and organizational behavior. The chapter focuses on understanding organizational behavior and its practices in the context of the 3Rs.

The 3Rs concept is now a norm in developed countries; however, if we look at the global supply chain from a holistic point of view, The evidence in developing countries suggests questioning the sustainability performance of the whole chain. A manufacturing firm in Pakistan was selected to understand the upstream behavior and the intention of the 3Rs. The study will inform how organizations act regarding green practices, their plans to use the 3Rs, and how waste is handled. For empirical testing, the study looked at 106 responses from Pakistani manufacturing companies, mainly in Karachi, the country's financial hub. The findings showed that organizations in Pakistan were more likely to practice the 3Rs when they had a positive attitude. COVID-19 has influenced industrial ecology; thus, moderation of the pandemic was also checked, but results suggested that no moderation was found. The study's results helped us learn more about organizations' plans to use 3R and trash management practices and the factors that motivate them to do so.

A. Shah et al., *Industrial Ecology*, Industrial Ecology,
https://doi.org/10.1007/978-981-97-3619-5_8

8.1 Introduction

Plastic pollution is the result of excessive plastic production and use. Plastic pollution is a problem all over the world since it has a significant impact on people and the environment. There are no safe methods for getting rid of plastic waste because plastic is made of a material that is not biodegradable. The toxic material caused harm to people, animals, and the environment (Koushal et al., 2014). Denmark pioneered the implementation of taxes on plastic packaging manufacturing, preceding nations such as Israel, Bangladesh, Albania, and China, which prohibited single-use plastic shopping bags. A fee of 0.05 pounds sterling was levied by the United Kingdom per plastic bag, while the two largest retailers in Australia declined to furnish complimentary plastic packaging (Klein, 2018). With an excess of 8 million tons of plastic entering the sea yearly, humankind should reexamine how we make and use plastics so they don't become squandered.

Modest, light, and flexible plastics are the predominant materials of our cutting-edge economy. Their creation is supposed to be twofold throughout the following twenty years. However, just 14% of all plastic bundling is gathered for reusing after use, and massive amounts escape into the climate. This is not just an outcome in that frame of mind of $80 billion to $120 billion yearly to the worldwide economy; however, assuming the latest thing proceeds, there could be more plastic than fish by weight in the seas by 2050.

8.2 Problem Statement

One of Pakistan's major socioeconomic and environmental problems is plastic pollution. Geyer et al. (2017) estimate that 8300 metric tons of plastic were manufactured globally between 1950 and 2017, of which 6400 kg tons were consumed and discarded, 79% of which ended up in landfills, 9% was recycled, and the remaining 12% was burned. The rising problem of plastic garbage is severely harming our ecology. If not addressed, this issue will endanger both people and animals. Being a growing nation, Pakistan is no stranger to the problems brought on by plastic garbage. High rates of urbanization, a population that is expanding at an exponential rate, and an increase in consumption have all considerably exacerbated the problem. Most plastic trash is found in plastic bags, often used for everyday tasks like grocery shopping and in packaging industrial raw materials and finished goods. According to the Environmental Protection Agency, Pakistan uses over 55 billion throwaway plastic bags yearly (Mansoor, 2019). This garbage has built up in the water canals and reservoirs, clogging the equipment used to clean wastewater. Due to this, Pakistan's big centers like Karachi and Lahore frequently and severely flood. There is currently insufficient control of plastic garbage disposal in less developed countries like Pakistan. According to the US Census Bureau, Pakistan is the sixth most populous country in the world (Census, 2023). Pakistan does not place any restrictions on the use of plastic products; hence, the issue of trash management gets worse every year.

Given the COVID-19 pandemic, this study aims to investigate how Pakistani citizens behave regarding waste management and the factors that affect consumer recycling behavior patterns in relation to plastic waste. It focuses, in particular, on identifying and evaluating factors that influence people's intention to participate in waste management. Understanding these consumer characteristics might affect how people behave in managing garbage disposal.

8.3 Background of the Study

Plastics are expected to be a massive element in supportable activities' ecological, cultural, and cheap components. Plastics are light, sturdy, spotless, and adaptable; used to make bundling, auto, constructing, and electronic is greater electrical gadgets. For instance, Americans use a hundred billion plastic sacks in 12 months, producing around 12 million barrels of oil; all things considered, 10 billion paper sacks each year implies slicing down 14 million timber. The utilization of unrefined petroleum for turning in plastics consumes a scant asset (power); however, the utilization of paper means the lower capability of the planet Earth to hold CO_2 (Wong & University of Hull Business School and Logistics Institute, 2010).

Recognizing the importance of plastics and the reality that plastics are crafted from scarce assets, many efforts were made in studies and improvements to make plastics reusable and recyclable. According to the United Kingdom Authorities' Waste Strategy 2007, the Department for Environment, Food and Rural Affairs has set a goal of forty-five percent recycling with the aid of 2015 (Wong, 2010).

In 2008–09, local authorities gathered 27.3 million tons of municipal waste; however, three percent were dispatched to landfills, nine percent were recycled or composted, and 12.2% were incinerated for energy recovery. Despite understanding that plastics are difficult to degrade clearly, the United Kingdom discards 4 out of five plastic bottles. There seems to be a lack of awareness and research (Wong, 2010).

8.3.1 Waste Volume in Pakistan

Pakistan generates more than 49.6 million tons of solid waste annually, with this figure experiencing an annual growth rate exceeding 2.4%. Like developing countries, Pakistan undergoes substantial environmental challenges due to its inadequate waste management infrastructure see Table 8.1 for city-wise waste generation and Fig. 8.1 for a type of waste. Plastic waste produces a total of 8% of all waste in Karachi. Therefore, 1320 tons per day of plastic waste is produced.

Table 8.1 Solid waste generation concerning cities of Pakistan (US Department of Commerce)

City	Population in millions	Solid waste generation/day in tons
Lahore	10,000,000	7690
Hyderabad	5,500,000	3973
Karachi	20,500,000	16,500
Faisalabad	7,500,000	5017
Rawalpindi	5,900,000	4500
Gujranwala	4,800,000	3480
Multan	5,200,000	3680
Sargodha	4,500,000	3072
Quetta	600,000	716
Peshawar	2,900,000	2048

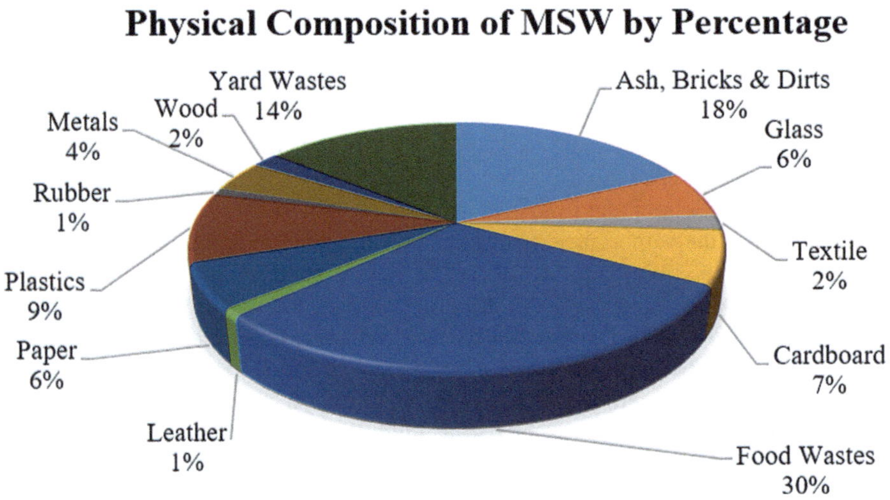

Fig. 8.1 Physical composition of MSW by percentages

8.4 Objective of the Study

Green climate supportability is a primary issue of security around the world. Implementing the 3Rs of trash management—Reduce, Reuse, and Recycle—is one way to prevent the maximum amount of waste from entering landfills. Environmental sustainability is a severe problem for safety on a global scale. Countries all across the world are going toward recycling. Unfortunately, not much has been done in Pakistan to promote recycling since there are inadequate sources for the supply of recyclable garbage and because most people are still ignorant of its potentially dangerous consequences. This project will examine the 3Rs environmental strategy and its implications on the plastic industry because it takes about 450–1000 years for plastic to decompose. This research will use a quantitative or qualitative approach to examine 3R and its impact on plastic businesses in Pakistan and compare them to those in developing and emerging nations. It will also explore the role of COVID-19 in waste management.

The study's primary goal is to determine what makes people want to reduce, reuse, and recycle (R2R) and how that affects how they handle the waste. In addition to looking at how financial slack affects 3R intentions, the paper will also discuss how attitudes, subjective norms, and behaviors are seen. The study also looks at how these intentions affect how waste is managed in real life and how people's feelings about COVID-19 might change this relationship. While the pandemic is ongoing, the main goal is to fully comprehend how 3R intentions affect environmental behaviors.

8.5 Literature Review

8.5.1 Relationship Between Attitude and 3R Intention

An attitude is a collection of thoughts, feelings, and actions towards a situation or an event (Chaiklin, 2011). Although attitudes and behavioral intentions to use less plastic have a strong positive association, the link is weak (Hasan et al., 2015). According to Lin et al. (2017), there is a favorable correlation between attitude and behavioral intention on pro-environmental programs that are both mandated and optional. However, the required program's attitude score toward its behavioral objective was lower than the optional program's. Additionally, Chang and Chou (2018) concluded that attitudes had a favorable and direct impact on customers' behavioral intention to "Bring Your Shopping Bags" (BYOB) during grocery shopping in Taiwan. According to Wan et al. (2012), attitude impacts behavior intention about recycling. Chun T'ing et al. (2020) concur that attitudes are one of the critical indicators of intentions to recycle. It suggests that attitude significantly influences 3R behavior intention favorably. The tendency to have a favorable attitude toward engaging in 3R conduct will grow along with the positive intention to do so (Susanto et al., 2019).

As per the study's outcomes, attitude is the most crucial characteristic and correlates highly with the 3Rs' behavior intention. However, subjective norms, behavioral control perception, and environmental elements are irrelevant. The 3Rs conduct and intention are strongly and significantly correlated. The government ought to launch a 3R campaign to raise awareness and foster a more positive attitude among those with unfavorable views about the concept, helping to define the subjective standards for 3R conduct. The government must also offer training for recycling garbage (Susanto et al., 2019).

H1: *Attitude has a significant effect on 3R intention*

8.5.2 Relationship Between Subjective Norm and 3R Intention

Emotional standards are the conviction that a vast individual or gathering would recognize and back a particular behavior. The apparent prevalent burden from others to act a specific way and the desire to concur with others' perspectives comprise emotional standards. In earlier examinations, it was shown that the impact of mentality was usually more grounded than that of abstract principles on making goals. As was noted at that point, regulating sees led to what is known as "abstract standards," or saw prevalent difficulty. In summary, an individual's perception of pervasive resistance from crucial individuals (such as family, friends, and coworkers) regarding their decision to act (or refrain from acting) in pursuit of a specific objective and their inclination to adhere to the perspectives of others are both associated with affective standards (for a scientometric approach, consider abstract standards and other hypotheses incorporating this term) (Eckhardt, 2009). Tong et al. (2018) found that subjective nos have a significant and positive relationship with the intention of households to recycle and concluded that support from family, friends, and society influences the level of intention to recycle.

Moreover, Lee and Tanusia (2016) concentrate on uncovering a vast and good relationship between the saw standard and the 3R expectations. The outcomes indicate that families, companions, and the local area might encourage them to 3R expectations (Chun T'ing et al., 2020). As per the study by Park et al. (2014), subjective standards also encourage recycling intention when social expectations are being conveyed. Another research by Zainol et al. (2012) showed that subjective norms are favorably associated with families' desire to recycle. In contrast, 3R intentions in Sweden were not affected by other individuals, including family and friends, according to research by Hage et al. (2009) and Ali & Yusof (2018).

H2: *Subjective Norm has a significant effect on 3R intention*

8.5.3 Relationship Between Perceived Behaviour and 3R Intention

The perception of controlled behavior. Preferred behavioral control refers to the perception of the difficulty of a behavior. According to the Theory of Reasoned Action, perceived behavior control is the primary distinction between the two. The "waste management hierarchy," which establishes priority for the most desirable choices for the management of waste within the context of sustainability, is the foundation of the current paradigm for waste management. Wastes are viewed as misapplied resources; hence, every effort should be made to improve resource exploitation efficiency (Ali & Yusof, 2018).

Therefore, preventing waste should be the priority for our production and consumption systems, followed by preparing materials for reuse, recycling, and composting, high-efficiency energy, and finally, only a limited, non-recoverable portion of the waste stream should be disposed of. A person's perceived behavior control is their belief in their ability to manage their internal state and behavior, stimulate and respond to their environment, and produce the intended results (Wallston et al., 1987).

Al Mamun et al. (2018) explored the variables that impact reusing expectations and ways of behaving in low-pay families. As per the exploration, I saw social control and want to reuse decidedly affect 3R aim. Latip et al. (2018) explored the impacts of demeanor and emotional standards. They saw social control toward the expectation to conform to natural administration rehearses and resolved that profound saw conduct control will have a higher goal to consent (Chun T'ing et al., 2020).

H3: *Perceived Behavior has a significant effect on 3R intention*

8.5.4 Relationship Between Financial Risk and 3R Intention

The potential for monetary misfortune in a venture or business undertaking. Liquidity risk, credit risk, and functional gambling are some of the most ordinary and different monetary dangers. This might bring about state-run administrations letting go over money-related arrangements completely and defaulting on securities or other monetary commitments. Organizations might bomb in an undertaking that overwhelms the organization, notwithstanding the risk of defaulting on obligations they take on. Modern waste is an "unloaded mine" with the development of consideration from researchers and modern firms' past natural vested parties (Zhang et al., 2018).

It is crucial to comprehend the financial and credit risk your firm confronts in light of our current challenging global economic climate. Businesses cannot afford the ambiguity and financial risk associated with prolonged payment delays, delinquencies that erode cash reserves, or, even worse, write-offs of bad debt. Many waste recycling companies have traditionally used the "Relationship Approach" to control their credit risk. But those days are long gone when a handshake sufficed. Businesses can't afford to assume they understand a firm just because a customer or executive told them so. Long-standing partnerships with a strong personal connection that lack a good, unbiased analysis frequently have an incorrect impact on many loan choices. This method of managing credit risk is incredibly ineffective. Poor financial outcomes are specific when credit choices are made exclusively based on intuition or a gut feeling. The decision-making process for credit must be free of emotion. Decisions must be based on factual information that provides a comprehensive picture of a customer's credit profile. Hence, theories have a positive relationship between financial risk and 3R intention (Dworin, 2012).

H4: *Financial Slack has a significant effect on 3R intention*

8.5.5 *Relationship Between Waste Management Behaviour and 3R Intention*

Most foundations and networks don't deal with their waste utilizing the 3Rs; there is a gamble to individuals and other living things, which uplifts the risks of an unnatural weather change. Because of the way that advanced education foundations should consider the territorial, public, and worldwide targets laid out to lessen CO_3 and CH_4 gas emanations, these organizations are expected to encounter more prominent administrative and political strain. As per EPA (2012), reusing blended paper might save what could be compared to 185 gallons of petroleum for each reused ton. Everybody should take care of the strategic execution of waste decrease and reuse so that the 3Rs can be appropriately carried out. At request for the essential execution of waste decrease, reuse, and reusing to be appropriately embraced in HEIs to save expenses and increment execution, members' way of life of discarding results should be changed by utilizing schooling and practice. This may be achieved by updating how strong waste administration is currently finished at HEIs, changing from customary to incorporated ones for a superior, cleaner climate, and expanding framework proficiency and cost viability. The preparation program should be carried out in HEIs so that staff and students can make their own school-based, coordinated, strong waste administration program (Jibril et al., 2012).

The 3Rs thought, an essential waste administration technique, is utilized in this review for examinations. Decrease means to abstain from using. A few states in Malaysia ran a mission to illuminate, increment mindfulness about, and persuade individuals to utilize items that create less rubbish. Steel straws should be utilized instead of single-utilized plastic straws with bio-sacks or reusing packs (Ahmadi, 2017).

Reuse can be achieved by purchasing, selling, giving, or retouching things; subsequently, it can limit squandering. For example, plastic packs can be reused for garbage assortment, reusable cups rather than expendable ones, and drink bottles should be topped off with water instead of purchasing new ones (Ahmadi, 2017).

Reusing involves assortment, arranging, and handling of trash. The recyclable materials incorporate paper, plastic, metal, and glass (Ahmadi, 2017). Studies indicate that following the 3R's conduct assists with addressing ecological issues (Nindyati, 2014). Subsequently, there is a positive connection between Waste Management Behavior and 3R Intention (Chun T'ing et al., 2020).

H5: *3R intention has a significant effect on Waste management behavior*

8.5.6 *Relationship Between COVID-19 and 3R Intention*

As a result of the WHO's inability to control the spread of the coronavirus disease 2019 (COVID-19), which was officially designated a pandemic on March 11, 2020

(1), over 126 million confirmed cases, including over 2.7 million fatalities, have been documented globally as of March 28, 2021(Tran et al., 2021).

The characterizing issue of our day is the COVID-19 disease (COVID-19) pandemic. The global local area's obligation to propel the 2030 Agenda for Sustainable Development and the Decade of Action to accomplish the Sustainable Development Goals will be scrutinized as we rise out of the fiasco and spotlight on remaking better (SDGs). If we are to involve the troubles as a springboard for a street toward comprehensive, versatile, and harmless to the ecosystem, reasonable development and multilateral activity are required. Territorial cooperation and coordination will be critical to understanding this potential and can lead worldwide endeavors toward an essential shift in economies and society. Moves initiated at the provincial level during and after the emergency can act as a springboard for longer-term cooperation on issues expected to turn out to be more squeezing as the world looks for elective strategies to understand the 2030 Agenda.

8.5.7 Relationship Between COVID-19 and Waste Management

The COVID-19 urges us to think about our communication with the climate. Because of environmental change, deforestation, and sea contamination, biological systems are debasing, biodiversity is vanishing, and sicknesses are spreading rapidly. Worldwide carbon dioxide discharges are anticipated to drop by 8% in 2020, in contrast to 2019, because of the pestilence, which is returning to levels from a decade earlier. The transitory declines in energy interest and fossil fuel byproducts during the pandemic have given the climate space to breathe. If monetary impetuses and speculation to restore the economy are not engaged toward a cleaner and stronger energy foundation, the recuperation in emanations may be more prominent than the fall. Low oil and gas costs ought to be purposefully utilized as a fantastic chance to execute environmental change changes and approaches to decarbonize our economies, progressively eliminate petroleum derivative appropriations, coordinate more sustainable power, and upscale energy effectiveness across financial areas for moderation and variation to the environment crisis, even though they increment the gamble of relapse temporarily A Green Economic Reset Will Deliver an Inclusive, Lasting Recovery for Asia and the Pacific (Asian Development Bank, 2022; Zachariadis et al., 2021).

Without more robust social protection policies, the epidemic would further increase the disparities in racial and social equality. The most vulnerable populations must receive investments made possible by the creative use of modern technology for social protection measures. Developing mechanisms and instruments, such as extending digital ID, is necessary to enable improved data and statistical availability and to bring to light the hundreds of millions of individuals who lack legal identification in some locations. To secure the safety of the whole population and to

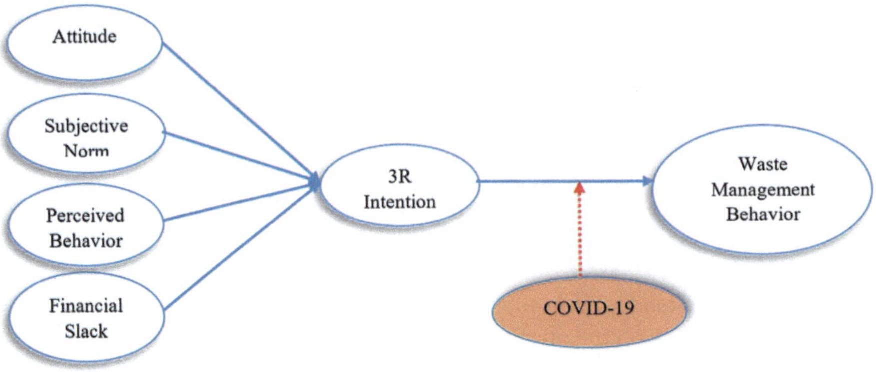

Fig. 8.2 Theoretical framework

effectively combat the epidemic, it is now imperative to alleviate the terrible situation of migrants, IDPs, and refugees. Even if increasing public spending on health and social protection systems may be difficult given the drop in tax revenue, doing so is unquestionably essential to a full-fledged and long-lasting recovery. There must be strong and comprehensive social security to maintain consumption and the demand for economic production. These initiatives help improve community resilience and crisis readiness, including natural catastrophes, armed conflict, and climate change. Hence, there is a positive relationship between COVID-19 and Waste Management (Asian Development Bank, 2022; Zachariadis et al., 2021).

H6: *COVID-19 has a significant moderating effect on 3R intention and Waste management behavior*

8.6 Theoretical Framework

See Fig. 8.2.

8.7 Methodology

The methodology for this study incorporates research design, data collection scales, scale, validity and reliability, sampling strategy, research procedures, and sample size used to evaluate the hypotheses. This section of the chapter will be a road map as a guide to understanding better how the research was conducted, what methods and methodologies were employed, and what significance they have for this particular research study, which is performed to analyze the relation 3R Intention, Waste Management Behaviour, and COVID-19 (as moderation) for Pakistan.

Research Design

The quantitative approach is employed in the current study because of its dependable, exact, and accurate measurement based on statistical analysis, which can confirm the truth. Statistical analysis is carried out to check the hypothesis described in the prior chapter. The primary data collecting approach employed in this study because the data is being collected for a specific purpose on the employees of companies here from the population of Karachi to evaluate variables depending on customer satisfaction; the data was collected in the form of Google form with the help of already developed construct in the past literature. A deductive approach is carried out in this research, as already-developed theories are used to carry out the analysis in companies to measure waste management behavior.

Sampling Technique and Sample Size

The non-probabilistic sampling technique was used in this study, and the research objective was exploratory research to determine the approximation of reality. For this research, respondents of 100 samples size are selected as we are using Smart PLS based on the ten times rule; as per this rule, the minimum required sample size is ten times the number of inner or outer links that point to the variable in the model (Hair et al., 2011).

Scale and Measure

Kataria and Saini (2020) and Tran et al. (2020) adopted the scale utilized in the present research. The total number of scales used in this research is 28, based on a five-point Likert scale (see Table 8.2). These scales have questions according to the variables for which they are opted. These instruments range from 1 to 5, where one represents strongly disagree, and five represents strongly agree. The reliability of these items is already established in previous papers, which are now retested in the current research for measuring the waste management behavior of Pakistani organizations based on the variables 3R intention, COVID-19 19, attitude, subjective norm, perceived behavior, and financial risk. All Variables Items are measured using a 5-point Likert scale from 1 to 5 where one expresses strongly disagree and five expresses strongly agree.

Table 8.2 Construct use for measuring the variables

Construct and item	Citation
AT 1–5	Attitude (Chun T'ing et al., 2020)
SN 1–4	Subjective norm (Chun T'ing et al., 2020)
PBC 1–6	Perceived behavioral control (Chun T'ing et al., 2020)
BI 1–5	3R intention (Chun T'ing et al., 2020)
3RB 1–6	3R behavior (Chun T'ing et al., 2020)
FS 1–6	Financial slack (Mwathi, 2014)
COV 1–5	COVID-19 (Khan et al., 2021)

Data Analysis Method

After collecting primary data, a quantitative research method is used to evaluate the hypothesis developed in the previous chapter. The test is performed on Smart PLS4. A minimum sample size of 60 was determined using g power software analysis. PLS-SEM (Partial Least Square Structural Equation Modeling) has several benefits in research. First, using PLS-SEM with a limited sample size is wis (Reinartz et al., 2009). Thirdly, the most frequently cited justifications for employing PLS-SEM are utilizing various scale types in data, abnormal data, dimensions beyond the mean, and the requirement for a minimum sample size (Cassel et al., 1999). Second, PLS-SEM provides highly efficient results, even when dealing with complex models and sample sizes, and makes no practical assumptions about the underlying data (Hair et al., 2012). PLS-SEM is well-suited for managing formative and reflective measurement models because it can handle multivariate absence and avoid generating a typical data set for distributional assumptions. Finally, a composite-based method was devised, and replacements were used to represent indicator variables and constructs weighted for a particular construct (Sarstedt et al., 2014). Reliability, validity, correlation, and regression tests are run after ensuring that the data is normal. These tests are covered in more depth below.

Reliability Analysis

Cronbach alpha is utilized as the tool to determine whether or not items are internally consistent (Sekaran & Bougie, 2016). The internal consistency or reliability of the responses submitted by respondents presents the homogeneity of the collected data (Fink & Litwin, 1995). Cronbach alphas greater than 0.8 are ideal; however, Cronbach alphas of 0.6–0.7 are also acceptable (Sekaran & Bougie, 2016). According to Fink and Litwin (1995), it is measured as a correlation coefficient and represented as r. Moreover, reliability is considered good when it has a value equal to or greater than 0.7.

Validity Analysis

Validity refers to the interpretations, actions, conclusions, or judgments drawn from a score on a test rather than the test itself. This relates to how much the data collected backs up the desired assessment of test scores for the suggested the creation. Furthermore, validity is concerned with whether the conclusion made is acceptable, relevant, and valuable in light of the sample with whom one works and the setting where the test user and independents operate (Hubley & Zumbo, 2011).

8.8 Results and Discussion

The research approach for this research study was mentioned in the previous chapter. The results of the data collected are analyzed in detail and explained thoroughly in this chapter. Below is Table 8.3, mentioning the response rate; for responses, 150

Table 8.3 Response rate analysis

Employees approached	150
Responses received	106
Response rate	71%
Response error	6
Sample size	100
Minimum sample size criterion	(6 inner links) × 10 = 60

employees from different organizations are approached, out of which 106 responses are received. Six of the responses received have response errors. So, the calculation is being carried out on Smart PLS 4 using 100 responses.

8.8.1 Construct Reliability and Validity

Interpretation for Reliability

The Measurement Model on Waste Management Behavior for Organizations in Pakistan by Using Smart PLS Software is shown in the Reliability Table. Cronbach's Alpha findings in the table must be greater than 0.6; however, these results are obtained and are acceptable see Table 8.4.

Interpretation for Construct Validity

The Measurement Model of Waste Management Behavior for Organizations in Pakistan by Using Smart PLS Software is shown in the validity Table. AVE value

Table 8.4 Construct reliability and validity

	Cranach's alpha	Composite reliability (rho_a)	Composite reliability (rho_c)	Average variance extracted (AVE)
3R intention	0.830	0.834	0.881	0.599
Attitude	0.821	0.843	0.874	0.582
COVID-19	0.867	0.875	0.904	0.653
Financial slack	0.862	0.894	0.896	0.591
Perceived behavior	0.702	0.714	0.799	0.490
Subjective norm	0.758	0.797	0.842	0.573
Waste management behavior	0.818	0.826	0.868	0.526

findings in the table must be greater than 0.4 (Hair et al., 2009). However, these results are obtained and are acceptable as all values are greater than 0.4.

8.8.2 *Discriminant Validity*

Statistical test of discriminant validity analysis quantifies a variable's uniqueness. It can be determined by comparing the square of each correlation coefficient's value to the square root of the explained variance; the latter should be greater (Fornell & Larcker, 1981).

Interpretation of Discriminant Validity

From Table 8.5, it is seen that the top of each column value (i.e., square root of AVE) is higher than the lower values (i.e., the shared variance between construct); hence, it shows the variable uniqueness and the Fornell and Larcker criterion is achieved.

Multicollinearity

Outer Model Multicollinearity

Variance Multicollinearity in the indicator is accessed via variance factor statistics (Fornell & Bookstein, 1982). Multicollinearity is not a severe concern in VIF if it is less than 5.

Inner Model Multicollinearity

Variance Multicollinearity in the variables is accessed via variance factor statistics (Fornell & Bookstein, 1982). Multicollinearity is not a severe concern in VIF if it is less than 5, which indicates the values are below the threshold level.

Table 8.5 Discriminant validity

	3RI	A	CO	FS	PB	SN	WMB
3R intention	0.774						
Attitude	0.542	0.763					
COVID-19	0.369	0.164	0.808				
Financial slack	0.347	– 0.004	0.609	0.769			
Perceived behavior	0.612	0.362	0.334	0.377	0.633		
Subjective norm	0.478	0.326	0.384	0.264	0.501	0.757	
Waste management behavior	0.637	0.401	0.410	0.360	0.601	0.481	0.725
CO × 3RI	0.072	0.094	– 0.025	– 0.101	– 0.124	0.048	0.129

8.8.3 Theoretical Model Testing

Interpretation of Hypothesis Testing

Table 8.6 reflects the direct relation between two variables on which the model is described in the above chapters as a hypothesis. The relation between Attitude and 3R intention is significant as the t-value is greater than 1.96, the *p*-value is less than 0.05, and the beta value is 0.376 (Tran et al., 2021). The relation between Perceived Behavior and 3R intention is significant as the t-value is greater than 1.96, the *p*-value is less than 0.05, and the beta value is 0.357 (Tran et al., 2021). The relation between Financial Slack and 3R intention is insignificant as the *p*-value is greater than 0.05, and the t-value is less than 1.96 (Tran et al., 2018). The relation between Subjective Norm and 3R intention is insignificant as the *p*-value is greater than 0.05, and the t-value is less than 1.96 (Schwab et al., 2012). The relation between waste management behavior and 3R intention is significant as the *p*-value is less than 0.05 (Hair et al., 2010), the t-value is greater than 1.96, and the beta value is 0.571. The moderating relation of COVID-19 on Waste management behavior and 3R intention is insignificant as the *p*-value is greater than 0.05, and the t-value is less than 1.96. Due to the insignificant relation of the moderating variable, slope analysis will not be performed here.

Mediation Analysis

Table 8.7 reflects that the relation between 3R intention and WMB is significant as the *p*-value is less than 0.05. Similarly, the relation between the attitude and 3R Intention, Financial Slack and 3R Intention, Perceived behavior, and 3R Intention is also significant, which is mentioning the significant mediating relation of attitude and Waste Management Behavior through 3R intention, Perceived behavior, and Waste Management Behavior through 3R intention as mediating, as the range of percentile bootstrap does not contain the 0 which also confirms the relationship from the literature describe in an earlier chapter.

Table 8.6 Testing hypothesis

	Beta	Standard deviation	T-value	*P*-value
3RI → WMB	0.571	0.086	6.609	0.000
A → 3RI	0.366	0.077	4.766	0.000
CO → WMB	0.203	0.096	2.121	0.034
FS → 3RI	0.120	0.077	1.553	0.121
PB → 3RI	0.357	0.083	4.326	0.000
SN → 3RI	0.140	0.084	1.661	0.097
CO × 3RI → WMB	-0.056	0.078	0.722	0.470

Table 8.7 Mediation test

	Beta	Standard deviation	T statistics	*P* values
A → 3RI → WMB	0.209	0.054	3.872	0.000
PB → 3RI → WMB	0.204	0.057	3.605	0..000
SN → 3RI → WMB	0.080	0.054	1.490	0.137
FS → 3RI → WMB	0.068	0.046	1.493	0.136

Table 8.8 Moderation analysis

	Beta	Standard deviation	T statistics	*P* values
CO → WMB	0.203	0.096	2.121	0.034
CO × 3RI → WMB	– 0.056	0.078	0.722	0.470

The relation between the subjective norm and the 3R intention is insignificant as the *p*-value is less than 0.05, which also mentions the indirect relationship between subjective norm and waste management behavior. Whereas the relation between Financial Slack and the 3R intention is significant as the *p*-value is greater than 0.05 from the above table, it also mentions the insignificant indirect relation between subjective norm and Waste management behavior as mediating, as the range of percentile bootstrap does not contain the 0.

Moderation Analysis

Table 8.8 reflects the moderation results. COVID-19 has a significant relation with waste management behavior as the value of *p* is less than 0.05, the t-value is greater than 1.96, and the beta value of beta is 0.203. However, the moderating relation of COVID-19 on 3R intention and waste management behavior is also insignificant as the *p*-value of COVID-19 and 3R intention with waste management behavior is 0.47, which is greater than 0.05. As the moderating relation has been proven insignificant, there will be no need for slope analysis due to the absence of significant moderation.

8.9 Conclusion

All of the hypotheses align with prior research on the same variables. All of the tests were executed in the previous chapter and met the requirements for each analysis. The importance of the hypothesis will be examined and discussed in detail in the present chapter.

Hence, from the above Discussion on results, it can be concluded that to develop waste management behavior in organizations, organizations must have the employees have the right attitude and perceived behavior towards the 3R intention, which can be achieved either by training or hiring people with the required attitude towards waste management behavior. Furthermore, the significance of financial slack concerning waste management clearly shows that the financial situation influences the waste management behavior of people within an organization. Moreover, the COVID-19 breakdown could not influence the relationship between 3R intention and waste management behaviors.

8.9.1 Recommendations

- 3R intention can be developed by managing attitudes, which significantly impact 3R intention.
- Attitude to buy products that can be recycled, reduced, and reused to develop in the people of Pakistan more so that waste management behavior could be accomplished.
- Attitudes towards recycling products are to be developed to influence waste management behaviors in Pakistani culture drastically.
- Renewable sources are used more to achieve Industrial Ecology and manage the reduction of hazardous greenhouse gases, such as replacing Diesel and petrol engines and turbines with renewable resources like solar electric power, water turbines, and wind turbines. This step can reduce emissions and significantly contribute to the next generation's 3R intention, perceived behavior, and attitudes towards waste management activities.
- More variables affecting 3R intentions could be searched; this study finds that waste management behavior can be significantly affected by COVID-19; therefore, more variables like these can be searched in the future to estimate their impacts on waste management behavior.
- Using IR 4.0 reduces human error to rely more on efficient machines as the implementation of Industrial Revolution 4.0 can bring machine learning and artificial intelligence, which brings efficiency to the system by reducing human error.
- Developing societal pressure regarding waste management behavior in the next generation can be achieved by managing the upcoming generation's attitude so that they can bring societal pressure on each other in the future.
- Attitude to adjust regarding Industrial ecology for lining output and input with global carrying capacities.
- Literate people regarding the financial benefits of Waste Management will directly attract industries to invest more in sustainable products to avail themselves of these financial benefits.
- Knowledge of Pandemics (COVID-19) can bring waste management behavior as it directly impacts society, so literature on pandemics and diseases can impact waste management behavior in the corporate culture.

References

Ahmadi, M. (2017). Evaluating the performance of 3Rs waste practices: Case study-region one municipality of Tehran. *Advances in Recycling and Waste Management, 2*(130), 2.

Al Mamun, A., Mohiuddin, M., Bin Ahmad, G., Thurasamy, R., & Ali Fazal, S. (2018). *Recycling intention and behavior among low-income households*. Mdpi.Com. https://doi.org/10.3390/su10072407

Ali, M. S. M., & Yusof, R. N. R. (2018). Intention to practice reduce, reuse & recycle (3R) among expatriates working in Malaysia. *International Journal of Academic Research in Business and Social Sciences, 8*(3). https://doi.org/10.6007/ijarbss/v8-i3/3928

Asian Development Bank. (2022, June 3). *A green economic reset will deliver an inclusive, lasting recovery for Asia and the Pacific*. Asian Development Bank. https://www.adb.org/news/features/green-economic-reset-will-deliver-inclusive-lasting-recovery-asia-and-pacific

Cassel, C., Hackl, P., & Westlund, A. H. (1999). Robustness of partial least-squares method for estimating latent variable quality structures. *Journal of Applied Statistics, 26*(4), 435–446.

Census. (2023). Population. www.census.gov

Chaiklin, H. (2011). Attitudes, behavior, and social practice. *The Journal of Sociology & Social Welfare*. HeinOnline.

Chang, S. H., & Chou, C. H. (2018). Consumer intention toward bringing your own shopping bags in Taiwan: An application of ethics perspective and theory of planned behavior. *Sustainability, 10*(6), 1815. https://doi.org/10.3390/SU10061815

Chun T'ing, L., Moorthy, K., Yoon Mei, C., Pik Yin, F., Zhi Ying, W., Wei Khong, C., Zhao Chern, G., & Zin Lin, T. (2020). Determinants of 3Rs behaviour in plastic usage: A study among Malaysians. *Heliyon, 6*(12), e05805. https://doi.org/10.1016/j.heliyon.2020.e05805

Dworin, J. (2012). Managing credit risk in scrap recycling—Google Search. Business of Recycling. https://www.google.com/search?q=Managing+credit+risk+in+scrap+recycling&oq=Managing+credit+risk+in+scrap+recycling&aqs=chrome..69i57j69i60l2.455j0j7&sourceid=chrome&ie=UTF-8

Eckhardt, A. (2009). The significant others of subjective norm-A scientometric study of subjective norm in IS top-journals over two decades.

EPA. (2012). Land, waste, and cleanup topics. US EPA. Municipal solid waste. Overviews & factsheets. https://www.epa.gov/environmental-topics/land-waste-and-cleanup-topics

Fink, A., & Litwin, M. (1995). How to measure survey reliability and validity.

Fornell, C., & Bookstein, F. L. (1982). Two structural equation models: LISREL and PLS applied to consumer exit-voice theory. *Journal of Marketing Research, 19*(4), 440–452.

Fornell, C., & Larcker, D. F. (1981). Evaluating structural equation models with unobservable variables and measurement error. *Journal of Marketing Research, 18*(1), 39–50.

Geyer, R., Jambeck, J. R., & Law, K. L. (2017). Production, use, and fate of all plastics ever made. *Science Advances, 3*(7), e1700782.

Hage, O., Söderholm, P., & Berglund, C. (2009). Norms and economic motivation in household recycling: Empirical evidence from Sweden. *Resources, Conservation and Recycling, 53*(3), 155–165. https://doi.org/10.1016/J.RESCONREC.2008.11.003

Hair, J. F., Anderson, R. E., Babin, B. J., & Black, W. C. (2010). *Multivariate data analysis: A global perspective* (Vol. 7). Pearson.

Hair, J. F., Black, W. C., Babin, B. J., Anderson, R. E., & Tatham, R. L. (2009). *Análise multivariada de dados*. Bookman editora.

Hair, J. F., Ringle, C. M., & Sarstedt, M. (2011). PLS-SEM: Indeed a silver bullet. *Journal of Marketing Theory and Practice, 19*(2), 139–152.

Hair, J. F., Sarstedt, M., Pieper, T. M., & Ringle, C. M. (2012). The use of partial least squares structural equation modeling in strategic management research: A review of past practices and recommendations for future applications. *Long Range Planning, 45*(5–6), 320–340.

Hasan, S. N. M. S., Harun, R., & Hock, L. K. (2015). Application of theory of planned behavior in measuring the behavior to reduce plastic consumption among students at Universiti Putra

Malaysia, Malaysia. *Procedia Environmental Sciences, 30*, 195–200. https://doi.org/10.1016/J.PROENV.2015.10.035

Hubley, A. M., & Zumbo, B. D. (2011). Validity and the consequences of test interpretation and use. *Social Indicators Research, 103*(2), 219–230. https://doi.org/10.1007/S11205-011-9843-4/FIGURES/1

Jibril, J. D. azimi, Sipan, I. Bin, Sapri, M., Shika, S. A., Isa, M., & Abdullah, S. (2012). 3Rs critical success factor in solid waste management system for higher educational institutions. *Procedia—Social and Behavioral Sciences, 65*, 626–631. https://doi.org/10.1016/J.SBSPRO.2012.11.175

Kataria, S., & Saini, V. (2020). The mediating impact of customer satisfaction in relation of brand equity and brand loyalty: An empirical synthesis and re-examination. *South Asian Journal of Business Studies, 9*(1), 62–87. https://doi.org/10.1108/SAJBS-03-2019-0046

Khan, S. A. R., Ponce, P., Thomas, G., Yu, Z., Al-Ahmadi, M. S., & Tanveer, M. (2021). Digital technologies, circular economy practices and environmental policies in the era of COVID-19. *Sustainability, 13*(22), 12790. https://doi.org/10.3390/SU132212790

Klein, A. (2018). New Zealand becomes the latest country to ban plastic bags.

Koushal, V., Sharma, R., Sharma, M., Sharma, R., & Sharma, V. (2014). Plastics: Issues challenges and remediation. *International Journal of Waste Resources, 4*(1), 1–6.

Latip, M., Sharkawi, I., Sharifuddin, J., & Mohamed, Z. (2018, undefined). Investigating owner-manager's intention to adopt environmental management practices among food processing SMEs in Malaysia. *Ijemp.Com, 1*(3), 1–11.

Lee, J. W. C., & Tanusia, A. (2016). Energy conservation behavioural intention: Attitudes, subjective norm and self-efficacy. *IOP Conference Series: Earth and Environmental Science, 40*(1), 012087. https://doi.org/10.1088/1755-1315/40/1/012087

Lin, S. C., Nadlifatin, R., Amna, A. R., Persada, S. F., & Razif, M. (2017). Investigating citizen behavior intention on mandatory and voluntary pro-environmental programs through a pro-environmental planned behavior model. *Sustainability, 9*(7), 1289.

Mansoor, A. (2019). *Environment: How plastic is killing us.*

Mwathi, E. M. (2014, undefined). Factors influencing effective management of electronic waste: A case of cyber cafes in Nairobi central business district, Kenya.

Nindyati, A. (2014). Green behavior (reuse, reduce, recycling—3R) at home and at workplace. https://doi.org/10.13140/RG.2.1.3229.7365

Park, J., & Ha, S. (2014, undefined). Understanding consumer recycling behavior: Combining the theory of planned behavior and the norm activation model. *Wiley Online Library, 42*(3), 278–291. https://doi.org/10.1111/fcsr.12061

Reinartz, W., Haenlein, M., & Henseler, J. (2009). An empirical comparison of the efficacy of covariance-based and variance-based SEM. *International Journal of Research in Marketing, 26*(4), 332–344.

Sarstedt, M., Ringle, C. M., Henseler, J., & Hair, J. F. (2014). On the emancipation of PLS-SEM: A commentary on Rigdon (2012). *Long Range Planning, 47*(3), 154–160.

Schwab, N., Harton, H. C., & Cullum, J. G. (2012). The effects of emergent norms and attitudes on recycling behavior. *46*(4), 403–422. https://doi.org/10.1177/0013916512466093

Sekaran, U., & Bougie, R. (2016). *Research methods for business: A skill building approach.* John Wiley & Sons.

Susanto, N., Davidesyta, L., Nurkertamanda, D., & Putranto, T. T. (2019). The influence of behavioral prediction factors and intention in improving 3R (reduce, reuse, recycle) household behavior in Tanjung Mas, Semarang, Indonesia. *AIP Conference Proceedings, 2114*(1), 030002. https://doi.org/10.1063/1.5112406

Tong, X., Nikolic, I., Dijkhuizen, B., van den Hoven, M., Minderhoud, M., Wäckerlin, N., … & Tao, D. (2018). Behaviour change in post-consumer recycling: applying agent-based modelling in social experiment. *Journal of Cleaner Production, 187*, 1006–1013.

Tran, K. T., Nguyen, P. V., & Nguyen, L. M. (2018). The role of financial slack, employee creative self-efficacy and learning orientation in innovation and organizational performance. *Administrative Sciences, 8*(4), 82. https://doi.org/10.3390/ADMSCI8040082

Tran, Q. A., Nguyen, H. T. T., Bui, T. V., Tran, N. T., Nguyen, N. T., Nguyen, T. T., Nguyen, H. T., & Nguyen, S. H. (2021). Factors associated with the intention to participate in coronavirus disease 2019 frontline prevention activities among nursing students in Vietnam: An application of the theory of planned behavior. *Frontiers in Public Health, 9*, 699079. https://doi.org/10.3389/FPUBH.2021.699079

Tran, V. D., Vo, T. N. L., & Dinh, T. Q. (2020). The relationship between brand authenticity, brand equity and customer satisfaction. *Journal of Asian Finance, Economics and Business, 7*(4), 213–221. https://doi.org/10.13106/JAFEB.2020.VOL7.NO4.213

Wallston, K. A., Wallston, B. S., Smith, S., & Dobbins, C. J. (1987). Perceived control and health. *Current Psychology, 6*(1), 5–25. https://doi.org/10.1007/BF02686633

Wan, C., Cheung, R., & Shen, G. Q. (2012). Recycling attitude and behaviour in university campus: A case study in Hong Kong. *Facilities, 30*(13), 630–646. https://doi.org/10.1108/02632771211270595/FULL/XML

Wong, C. (2010). *A study of plastic recycling supply chain*. The Chartered Institute of Logistics and Transport, University of Hull Business School and Logistics Institute. https://www.ciltuk.org.Uk/Portals/0/Documents/Pd/Seedcornwong.Pdf

Zachariadis, T., Giannakis, E., Taliotis, C., Karmellos, M., Fylaktos, N., Howells, M., Blyth, W., & Hallegatte, S. (2021). A science-policy framework for a green economic recovery after the COVID-19 pandemic.

Zainol, Z., Pembaziran, P., Galakan, D., Hijau, A., & Tih, S. (2012). Minimizing waste and encouraging green practices. *Researchgate.net, 46*(1), 157–164.

Zhang, B., Lai, K.-H., Wang, B., & Wang, Z. (2018). Financial benefits from corporate announced practice of industrial waste recycling: Empirical evidence from chemical industry in China. *Resources, Conservation and Recycling, 139*, 40–47. https://doi.org/10.1016/J.RESCONREC.2018.07.019

Chapter 9
Social Economical Benefit of Industrial Ecology

The concept of industrial ecology is a holistic framework that seeks to redefine and restructure industrial processes to promote sustainability and efficiency. This framework incorporates a distinctive environmental, social, and economic amalgamation. The present chapter comprehensively examines the socio-economic advantages of industrial ecology, emphasizing the substantial favorable effect that sustainable industrial practices may exert on society and the economy. By integrating ecological principles into industrial processes, firms have the potential to decrease their environmental impact and enhance their socio-economic standing, thus making a valuable contribution towards the progress of a more sustainable and equitable society. This chapter will explore how industrial ecology accelerates socio-economic growth, providing numerous advantages beyond environmental sustainability. The chapter seeks to offer a comprehensive understanding of how industrial ecology is transforming modern industries, guiding them toward an economically sustainable, socially accountable, and environmentally friendly future. This is achieved by combining theoretical perspectives with real-life illustrations.

9.1 Introduction

Industrial ecology has a broader scope than environmental conservation, as it tackles various socio-economic concerns. The introduction of green supply chain methods has positively impacted multiple aspects, including waste reduction, emissions reduction, worker safety, community health, and local development. Companies such as Patagonia and Unilever have established standards in incorporating sustainability into their operational frameworks, showcasing concrete advantages concerning their market standing, consumer confidence, and financial outcomes.

In addition, industrial ecology promotes the establishment of environmentally and socially responsible industries, stimulating economic growth and the creation of

A. Shah et al., *Industrial Ecology*, Industrial Ecology,
https://doi.org/10.1007/978-981-97-3619-5_9

green jobs. The adoption of renewable energy sectors, exemplified by firms such as Tesla and Vestas, demonstrates how the principles of industrial ecology may foster innovation, generate employment opportunities, and enhance economic resilience.

Industrial ecology is crucial in identifying and mitigating hazards related to resource scarcity, volatile commodity prices, and societal unrest through a systemic approach. Companies that actively adopt industrial ecology concepts are more adept at navigating the global market's intricacies, guaranteeing their long-term viability and financial success.

9.2 Social and Economic Concerns of Supply Chain Sustainability

Supply Chain Sustainability (SCS) is a crucial component of industrial ecology that focuses on implementing ecologically and socially responsible practices across the supply chain. The subject matter encompasses many issues, including but not limited to labor rights, fair trade, community engagement, and economic inclusiveness. Implementing supply chain sustainability (SCS) is not solely a matter of ethics but a strategic decision that allows organizations to improve their risk management methods, strengthen their reputations, and develop more resilient and robust supply chains.

Labor Rights and Fair Trade: SCS prioritizes the preservation of labor rights and the advancement of equitable trade. This entails implementing measures to guarantee secure working environments, equitable remuneration, and eradicating child or coerced labor throughout the supply chain. Companies such as Fairphone, which try to obtain cell phone materials free from conflicts, are prime examples of how firms may include labor rights and fair trade in their fundamental operations. This enables them to establish industry benchmarks and influence wider market patterns.

Community Engagement and Economic Inclusion: Engaging with local communities and fostering economic inclusion are fundamental components of SCS. This includes the allocation of resources towards local development initiatives, the provision of assistance to small firms, and the promotion of equitable distribution of economic advantages. A notable example is the partnership between Starbucks and indigenous coffee cultivators, wherein the corporation not only procures coffee grown sustainably but also allocates resources toward farmer education and initiatives to foster community advancement. This exemplifies the company's dedication to both ecological preservation and socio-economic advancement.

Enhancing Risk Management and Reputation: SCS plays a crucial role in contemporary risk management and reputation improvement. Organizations that actively embrace supply chain sustainability (SCS) concepts are more adept at managing the intricacies of worldwide supply chains, predicting and addressing risks associated with environmental and social governance, and fostering confidence among consumers and stakeholders. This strategy is seen in companies such as Patagonia, which has garnered acclaim for its dedication to ecological sustainability and ethical conduct, enhancing its market standing and customer allegiance.

Building Resilient Supply Chains: A sustainable supply chain possesses inherent resilience. Companies may enhance the adaptability and resilience of their supply chains by implementing strategies such as diversifying sources, minimizing reliance on non-renewable resources, and strengthening supplier relationships. Tesla's commitment to localizing battery production and sustainably sourcing materials exemplifies the integration of resilience into the supply chain, following the principles of industrial ecology while guaranteeing uninterrupted business operations.

SCS stands at the crossroads of social responsibility, environmental stewardship, and economic sustainability, playing a crucial role in shaping the future of industrial practices. When companies integrate SCS principles into their operations, they adhere to ethical standards and lead toward a more sustainable and just global economy. Such commitment propels them to contribute positively to the planet and its inhabitants and positions them advantageously in a market that increasingly values conscientious business practices. SCS is instrumental in guiding industries toward a future where economic growth, environmental health, and social well-being are inextricably linked and mutually reinforcing, underscoring the transformative power of SCS in realizing a sustainable industrial future.

9.3 Socioeconomics in Supply Chain Management

Including socio-economic factors in supply chain management is crucial, which entails adopting a holistic approach that acknowledges the complex interplay between economic activities and social frameworks inside supply chains. Industrial ecology promotes the integration of various aspects, including efficiency, profitability, social well-being, equitable resource allocation, and economic inclusivity, within the supply chain paradigm. Implementing this comprehensive strategy ensures that supply chain management improves corporate performance and actively contributes to sustainable and equitable economic progress, aligning with the core principles of industrial ecology.

Social welfare within supply chains extends beyond the mere assurance of equitable labor practices, encompassing value generation for all parties involved, such as employees, communities, and customers. Companies such as Ben and Jerry's exemplify how supply chains can bolster local economies and promote sustainable agriculture by demonstrating their dedication to fair trade ingredients. Similarly, corporations such as Tesla demonstrate resource equality by responsibly procuring materials for batteries and addressing concerns over the impact of mineral extraction on local communities and surroundings.

Economic inclusion in supply chains aims to expand the financial advantages to a more diverse range of participants, focusing on individuals in disadvantaged circumstances. An example is the partnership between significant retailers such as

Walmart and small local suppliers, which broadens the range of products the retail behemoth offers and fosters local economic expansion.

Through the implementation of these integrated methodologies, supply chain management surpasses conventional commercial aims, thereby aligning with the overarching objectives of industrial ecology to cultivate a global economy that is more resilient, sustainable, and egalitarian. This heightened viewpoint on supply chain management highlights the capacity of industrial operations to contribute to social advancement and environmental conservation.

9.3.1 Impact on Social Welfare

Supply chain decisions substantially impact social welfare, encompassing the working conditions within factories and the overall health, well-being, and economic stability of entire communities. By incorporating socio-economic factors into their supply chain management plans, firms have the potential to convert their operations into positive entities, thereby improving the well-being of persons and communities linked to their commercial operations. Progressive firms such as Patagonia and H&M are at the forefront of adopting ethical labor standards in the garment business, often subject to scrutiny due to labor breaches. In addition to guaranteeing equitable remuneration and secure labor conditions, they also allocate resources towards community development initiatives to enhance the localities where their goods are manufactured.

Other sectors besides the textile business are also making significant advancements in this domain. Apple, a prominent technology company, has implemented substantial measures to conduct audits of its supply chain to identify labor infractions and assess its environmental impact. The objective is to establish a supply chain that is more transparent and accountable. Starbucks' Coffee and Farmer Equity (CAFE.) Practices exemplify how companies can promote social welfare by implementing procurement policies that guarantee fair compensation and satisfactory working conditions for coffee growers. This, in turn, fosters economic development in coffee-producing regions.

These instances exemplify a more extensive pattern in which firms acknowledge their responsibility to advance societal well-being through responsible supply chain management. By engaging in such actions, organizations not only bolster their standing and foster customer confidence but also make valuable contributions to developing sustainable and resilient local economies. This demonstrates the compatibility of ethical business practices with financial profitability and corporate prosperity.

9.3.2 *Promoting Equitable Resource Distribution*

Facilitating fair allocation of resources within supply chains is crucial for attaining socio-economic equilibrium, guaranteeing that the advantages resulting from economic endeavors are distributed equitably among all parties involved, especially in underdeveloped regions. This notion is crucial for tackling issues such as the exhaustion of resources, deterioration of the environment, and disparities in economic distribution. The chocolate sector, which is frequently subject to scrutiny due to its supply chain disparities, has witnessed the emergence of companies such as Tony's Chocolonely that have taken the lead in implementing programs to promote equitable trade practices. The organization demonstrates a strong commitment to eradicating child labor and ensuring that cocoa farmers receive a sustainable income, thereby establishing a standard for fair allocation of resources.

Moreover, the coffee business presents an additional noteworthy illustration through enterprises such as Fair Trade Coffee, which significantly emphasizes direct and ethically-driven procurement arrangements. These agreements guarantee equitable remuneration for coffee producers, fostering community advancement and promoting sustainable agricultural methods. Furthermore, within the technology industry, endeavors such as the Fairphone prioritize the significance of ethically obtained resources to enhance the well-being of those in the mining sector and mitigate environmental consequences.

Companies may promote long-term sustainability and resilience in global marketplaces by including fair resource allocation in their supply networks. Not only does this have advantages for producers and communities engaged in the initial phases of the supply chain, but it also ensures that corporations are in line with the growing consumer demand for ethical and sustainable corporate conduct. In essence, promoting equitable resource distribution extends beyond mere fairness, establishing a sustainable framework for global trade that fosters the welfare of all parties concerned.

9.3.3 *Enhancing Economic Inclusivity*

Promoting economic inclusion in supply chain management is paramount in fostering a more equitable economic environment wherein diverse stakeholders, including small-scale producers and marginalized groups, are afforded equal possibilities. Adopting this method is crucial in mitigating poverty and promoting comprehensive economic resilience. For example, prominent technology corporations like Microsoft have played a critical role in incorporating inclusion into their supply chains through active collaboration with local suppliers in developing economies. These collaborations enhance the supply chain by incorporating varied inputs and stimulating local economies by introducing new opportunities and technical developments.

In addition to the technology sector, the agricultural business provides significant perspectives on economic inclusion. Companies such as Starbucks, through their ethical sourcing initiatives, guarantee equitable remuneration and assistance to smallholder coffee farmers, making a valuable contribution to the long-term viability of rural economies. Similarly, the fashion industry attempts to foster more inclusivity throughout supply chains through various efforts such as the Ethical Trading Initiative. These initiatives aim to promote the rights and equitable treatment of people employed in garment factories in economically disadvantaged areas across the globe.

The abovementioned instances highlight the diverse advantages of economic inclusion within supply chains, which go beyond individuals and stimulate wider social and economic advancements. By cultivating an environment that embraces diversity and inclusivity, businesses may unleash various unexplored possibilities, encouraging creativity and expansion while positively impacting a fair and secure global economy.

9.3.4 Community Engagement and Relationship Building

Community engagement and relationship building are fundamental practices in supply chain management, emphasizing the significance of establishing connections with local communities to understand their requirements and foster strong, mutually beneficial relationships. The importance of this involvement extends beyond corporate responsibility, as it represents a strategic necessity that has the potential to enhance trust, loyalty, and a company's social acceptance of its operations. The extractive industry, which is frequently criticized for its environmental and social consequences, has observed the implementation of extensive community engagement techniques by businesses such as Rio Tinto. These projects are formulated to mitigate adverse effects and contribute to local communities, showcasing a dedication to sustainable and ethical business principles.

The agricultural business is an additional illustration of successful community engagement, extending beyond the extractive sector. Nestlé and similar companies have established initiatives to collaborate closely with farmers, providing them with training and resources to enhance crop productivity and promote sustainable practices. This approach guarantees the long-term sustainability of their supply chains and contributes to local agricultural communities' well-being.

Within the retail domain, IKEA's strategy for procuring resources for its merchandise illustrates robust community involvement. IKEA promotes local economies and crafts by collaborating directly with local artists and small-scale producers, especially in developing nations. This approach ensures that IKEA's products are sourced sustainably.

The examples above highlight the diverse advantages of involving the community in managing supply chains. Companies can enhance their operational resilience market reputation and contribute to broader societal and economic development by cultivating transparent, respectful, mutually beneficial relationships with local communities. This approach aligns their business objectives with the principles of industrial ecology.

9.3.5 *Conclusion: A Call for Integrated Supply Chain Practices*

Incorporating socio-economic issues in supply chain management goes beyond being a simple strategic advantage; it becomes necessary within industrial ecology. The integration of various components within supply chains not only enhances operational efficiency and profitability but also promotes social responsibility and inclusion. By actively involving themselves in socio-economic factors, businesses can contribute to sustainable economic growth, improve the well-being of communities, and contribute to the development of a fair and resilient global economy.

The implementation of inclusive supply chain policies by enterprises can foster economic stability and growth at the local level, particularly in developing regions. This, in turn, can contribute to the overall strength of the global economy. For example, the Africa Progress Platform, established by the World Economic Forum, seeks to enhance the influence of small and medium firms throughout Africa. This effort showcases the potential of integrated supply chains to foster economic progress.

Community Well-being: Unilever and similar companies have exemplified the efficacy of incorporating community health and education into their supply chain strategy, resulting in enhanced living standards. This demonstrates a deep dedication to promoting community well-being. Corporate corporations such as Patagonia prominently show the convergence of socioeconomics and environmental sustainability. This company prioritizes equitable labor practices and is committed to mitigating environmental harm throughout its supply chain, exemplifying a holistic approach to sustainable business.

Incorporating socio-economic factors into supply chains promotes innovation and facilitates collaboration among enterprises, non-governmental organizations (NGOs), and governmental entities. Using blockchain technology by IBM in conjunction with smallholder farmers to augment crop traceability is a prime illustration of how collaborative efforts can yield inventive resolutions that yield advantages for all parties involved.

Companies' integration of socio-economic factors into supply networks contributes to the enhancement of their resilience and adaptability. An example is Toyota's cooperative strategy with its suppliers, which enabled it to rebound from the 2011 tsunami swiftly, showcasing the strength of resilient and interconnected supply networks.

For firms dedicated to industrial ecology, it is crucial to incorporate socioeconomics into supply chain management. This method guarantees supply chains' long-term viability and moral uprightness while establishing firms as frontrunners pursuing a more sustainable, fair, and robust global economy. Employing a sequence of illustrative instances and comprehensive examination, this section has expounded upon the transformational capacity of socio-economic integration within supply chains, emphasizing its pivotal function in stimulating favorable worldwide transformation.

9.4 An Economic Perspective

An Economic Perspective on Industrial Ecology explores the intricate relationship between economic systems and ecological sustainability, focusing on the potential for restructuring industrial activities to achieve improved financial and environmental results. This viewpoint examines the ideas of industrial ecology, which proposes a shift in economic models from linear to circular, focusing on optimizing resources, reducing waste, and converting by-products into valuable inputs.

Central to this economic perspective is recognizing that industrial processes, conventionally perceived as sequential trajectories leading to inefficiency, can be reconceptualized as closed loops in which outputs are reintegrated as inputs. This approach not only alleviates environmental consequences but also presents economic benefits by reducing material expenses and creating other sources of income. The method is consistent with the overarching principle of the circular economy, which aims to preserve the value of products, materials, and resources within the economy for an extended period while minimizing waste output.

From a microeconomic standpoint, companies that embrace industrial ecology methods have the potential to improve their operational efficiency, save expenses related to waste disposal and raw material usage, and stimulate creativity by creating novel products and processes. At the macroscopic level, the systematic implementation of industrial ecology principles has the potential to foster economic expansion through the establishment of industries focused on resource recovery and sustainable production. This, in turn, can generate employment opportunities and promote economic diversification.

9.4.1 Economic Rationale of Industrial Ecology

Industrial ecology offers a paradigm shift in economics by suggesting that integrating environmental sustainability with industrial efficiency might lead to a symbiotic association. This perspective emphasizes that the strategic utilization of resources and the reduction of waste go beyond just ordinary environmental stewardship and become crucial economic imperatives.

- Resource Optimization and Cost Reduction: Industrial ecology's fundamental basis is optimizing resource efficiency, resulting in substantial financial benefits. By optimizing material and energy utilization, industries have the potential to

decrease costs related to the acquisition of raw materials and the management of waste. An illustrative example of this idea is the implementation of industrial symbiosis, wherein waste generated by one industry serves as the raw material for another. One notable example is the collaborative efforts observed within the Kalundborg Eco-Industrial Park in Denmark, wherein enterprises share resources, leading to cost reduction and improved environmental effects.
- Waste Valorization and Circular Economy: The field of industrial ecology presents a counterargument to the conventional perception of trash as a valueless thing, calling for its reintegration into the production cycle to revalue what was previously regarded as waste. This methodology follows the tenets of the circular economy, which aims to preserve the value of goods, materials, and resources for an extended duration. A modern example of this phenomenon is the emergence of biorefineries, which are establishments that transform waste biomass into energy and other valuable commodities. This exemplifies how waste valorization can generate fresh income sources while mitigating environmental consequences.
- Innovation and Competitive Advantage: Implementing industrial ecology principles can stimulate innovation, creating novel products, services, and processes that confer competitive advantages within the market. Organizations at the forefront of adopting these concepts establish benchmarks within the sector and attract an expanding segment of environmentally aware consumers. For instance, the automotive industry has witnessed businesses such as Tesla get a significant edge in the market through their innovative efforts in electric automobiles and sustainable energy solutions. This exemplifies how industrial ecology can propel technological progress and establish a unique position in the market.
- Enhancing Economic Stability and Growth: Industrial ecology plays a significant role in promoting economic stability and growth by facilitating the development of industries prioritizing sustainability on a larger scale. In addition to stimulating economic diversification, this phenomenon generates employment opportunities within burgeoning industries, prioritizing recycling, renewable energy, and sustainable manufacturing. The expansion of the renewable energy industry, specifically in the domains of solar and wind power, is a prime illustration of how using industrial ecology concepts can stimulate the emergence of novel sectors, thus fostering economic advancement and fortitude.

In summary, industrial ecology's economic rationale is significant and transformational, offering a strategic framework for enterprises to achieve financial success while maintaining environmental sustainability. This strategy highlights a crucial change, urging industries to adopt the mutually beneficial connection between industrial productivity and environmental sustainability, thus promoting a model where economic expansion does not compromise ecological well-being.

Industrial ecology encourages industries to go beyond conventional operational frameworks, promoting a comprehensive incorporation of sustainability into the fundamental aspects of corporate operations. In addition to addressing environmental concerns, this integration presents opportunities for economic growth, innovation, and a competitive edge. By adopting a perspective that regards trash as a valuable resource and incorporating closed-loop systems, enterprises have the potential to identify novel avenues for generating value, thereby propelling the advancement of a circular economy that exhibits resilience, adaptability, and sustainability.

Furthermore, the tenets of industrial ecology follow the increasing worldwide focus on sustainable development, which resonates with consumer demands and regulatory patterns. Companies leading the way in this transformation get a competitive advantage and significantly contribute to the broader societal transition towards sustainability, thereby playing a pivotal role in the worldwide pursuit of a sustainable future.

Incorporating industrial ecology concepts allows firms and economies to enhance resource efficiency, commit to environmental responsibility, and foster economic vitality. This path requires cooperation, creativity, and a dedication to achieving long-term sustainability objectives. By adopting this comprehensive approach, industries and economies not only enhance their economic success but also promote the health and well-being of the planet. This sets the stage for a future where economic and environmental goals are mutually supportive, guaranteeing a prosperous future for future generations.

9.4.2 Enhancing Efficiency and Reducing Costs

Implementing industrial ecology principles at the microeconomic level provides a robust approach for companies seeking to enhance efficiency and save expenses. This method entails a careful and precise optimization of resource utilization and a deliberate decrease in waste production, which are essential activities that support environmental responsibility and demonstrate economic prudence.

- Process Redesign for Resource Optimization: Process redesign is a fundamental aspect of industrial ecology that allows organizations to reorganize their processes to achieve the highest level of resource efficiency. This entails conducting a comprehensive examination and reorganizing production procedures to reduce input materials and energy usage while optimizing output quality and quantity. These redesigns have the potential to result in significant decreases in raw material expenses, providing a distinct economic motivation in addition to environmental advantages.
- Technological Innovation for Waste Minimization: Technological innovation is crucial for achieving industrial ecology's aims, especially in waste reduction. Companies can convert waste materials into valuable resources by implementing advanced technologies, such as recycling, repurposing, or energy recovery. These innovations decrease the expenses linked to garbage disposal and create new sources of income from materials considered "waste," so fundamentally questioning the conventional perception of waste as an economic burden.

The Interface Case: A Paradigmatic Example

1. Interface Inc. serves as an exemplary model in this context. With its 'Mission Zero' initiative, the company set forth an ambitious goal to eliminate its negative environmental impact by 2020. Through comprehensive process redesign and integration of innovative recycling technologies, Interface significantly reduced waste generation and resource consumption. This propelled the company toward its environmental objectives and resulted in marked cost savings and enhanced profitability, showcasing the economic viability of industrial ecology principles.
2. Latest Example of Circular Economy in Action: A more recent example can be seen in the electronics industry, where companies like Fairphone are redefining product design and lifecycle management. By focusing on modular design and promoting repairability, upgradability, and recycling, Fairphone is minimizing waste and extending the lifecycle of its products, which in turn reduces the long-term costs for both the company and its customers, illustrating the economic and environmental advantages of industrial ecology.

In a nutshell, integrating industrial ecology principles into the operational structure of companies presents a significant benefit, allowing them to enhance their efficiency and reduce operational expenses while positioning themselves as leaders in sustainable business practices. The combination of economic and environmental advantages highlights the crucial importance of industrial ecology in today's business climate, offering a strategic opportunity for organizations to achieve financial success while promoting environmental responsibility.

Companies undertake a transformative journey by adopting the principles of industrial ecology, which involves reinventing their production methods, resource utilization, and waste management strategies. This shift facilitates financial savings by enhancing the efficiency of materials and energy usage and stimulates innovation by promoting novel products, services, and market prospects that adhere to sustainable standards. Consequently, companies reduce their ecological footprint and bolster their market competitiveness and adaptability in an ever more environmentally aware global economy.

In addition, implementing industrial ecological practices establishes organizations as frontrunners in corporate responsibility, enhancing their standing among consumers, investors, and regulatory entities. The improved corporate image can yield concrete advantages such as heightened consumer loyalty, augmented investment, and advantageous regulatory treatment, bolstering the firm's economic prosperity.

Industrial ecology is a comprehensive approach combining economic goals with environmental needs, providing a plan for enterprises to succeed in a future where sustainability is paramount. Adopting these methods by an increasing number of firms contributes to a wider transition towards a circular economy, thereby promoting a more sustainable and prosperous global environment for future generations. This comprehensive investigation reveals that industrial ecology is crucial in achieving individual business success and establishing global economic and environmental resilience.

9.4.3 *Fostering Innovation and Competitive Advantage Through Industrial Ecology*

Industrial ecology emerges as a framework for sustainability and a powerful catalyst for innovation, compelling companies to create and improve goods, services, and processes that align with sustainability principles. The pursuit of innovation, deeply anchored in industrial ecology, not only corresponds with environmental mandates but also establishes a notable competitive advantage within the market.

- Innovation as a Core Principle: The core principle of industrial ecology revolves around innovation, which involves reconsidering and restructuring resource utilization and product creation processes to reduce environmental harm and optimize efficiency. The scope of this notion surpasses gradual enhancements and encompasses transformative innovations capable of redefining entire product categories or business practices.
- Competitive advantage in a sustainable economy: In the context of the dynamic green economy, enterprises that actively adopt sustainable innovations can secure a significant portion of the market, thereby attracting an expanding consumer base that places a high value on environmental stewardship. Moreover, these advancements can give organizations a competitive advantage over their rivals, particularly in light of the escalating stringency of global ecological rules and the growing inclination of consumer preferences toward sustainability.

Case Study: Tesla's Market Disruption

Tesla Inc. is a paradigm of how industrial ecology principles can catalyze industry-wide innovation. By reimagining the automobile around electric power, Tesla challenged the conventional automotive industry, driving a sector-wide shift towards electric vehicles (EVs). This shift is not confined to product design alone. However, it extends to the entire value chain, including battery production, energy sourcing, and vehicle end-of-life management, embodying the holistic approach advocated by industrial ecology.

Latest Example: Circular Economy in Fashion

A more recent example of industrial ecology-driven innovation can be found in the fashion industry, which is notorious for its environmental footprint. Companies like Patagonia are pioneering the circular fashion movement, designing clothing with recyclability and durability in mind and promoting repair and reuse. This approach reduces waste and resource consumption and establishes a new competitive parameter in the fashion industry, attracting consumers who value sustainability.

Companies are not only responding to the requirements of environmental sustainability by incorporating the ideas of industrial ecology into their strategic frameworks, but they are also leveraging innovation to establish a unique position in a highly competitive global market. The dedication to innovation, guided by the principles of industrial ecology, enables companies to introduce new products, services, and processes that effectively decrease environmental impacts while stimulating economic development, improving market distinctiveness, and strengthening long-term business sustainability.

The field of industrial ecology catalyzes fostering innovation, compelling organizations to critically reassess and adapt their goods and processes from a sustainable perspective. This reconceptualization not only aids in preserving the environment but also reveals fresh prospects for generating wealth and gaining a competitive edge. Tesla's innovative progress in electric vehicles and battery technologies is a prime example of how a company can utilize industrial ecology principles to disrupt conventional industries, establish new opportunities for industry dominance, and meet the increasing consumer desire for environmentally friendly choices.

Moreover, Patagonia's dedication to sustainability is apparent through its utilization of recycled materials and its efforts to mend and recycle clothing. This exemplifies how industrial ecology can catalyze innovation in product creation and business strategies. This strategy not only improves the brand reputation of Patagonia but also appeals to a consumer demographic that places importance on corporate responsibility, bolstering the company's competitive advantage.

Furthermore, adopting industrial ecology promotes a culture of ongoing enhancement and adjustment, essential when technical progress and consumer demands change swiftly. Businesses that actively incorporate sustainable innovations into their activities are more likely to predict and adjust to forthcoming regulatory modifications, environmental obstacles, and market fluctuations, guaranteeing long-term sustainability and prosperity.

Incorporating industrial ecology concepts provides businesses with diverse benefits, allowing them to attain environmental objectives while fostering economic prosperity and establishing themselves as leaders in the industry. By demonstrating a dedication to sustainable innovation, corporations have the potential to not only make a positive impact on the worldwide need for environmental conservation but also position themselves as progressive frontrunners in the sustainable economy. This positioning positions them for long-lasting prosperity in a global landscape where the equilibrium between economic progress and ecological guardianship is becoming increasingly crucial.

9.4.4 Stimulating Economic Growth and Transitioning to a Circular Economy

Industrial ecology transcends traditional environmental management by advocating for a systemic shift towards a circular economy at the macroeconomic level. The current paradigm shift significantly emphasizes extending the materials' lifespan and optimizing the resources' value. This change contributes to resource conservation and the reduction of ecological footprints and plays a crucial role in driving economic growth and fostering innovation. Integrating industrial ecology concepts into economic systems leads to new industries, transforming current sectors and numerous job possibilities, promoting a strong and resilient economy.

- Enhancing Resource Efficiency and Economic Value: The circular economy concept, advocated by the field of industrial ecology, aims to optimize resource efficiency by promoting product and material reuse, recycling, and remanufacturing. This strategy serves the dual purpose of prolonging the lifespan of materials, resulting in resource and energy conservation while also enhancing the value of previously considered garbage. Industries that embrace this approach frequently experience notable cost reductions, less reliance on raw material extraction, and mitigated environmental consequences while generating economic value.
- Fostering Innovation and Industry Transformation: Adopting a circular economy requires and promotes innovation in multiple industries. Organizations are motivated to engage in the advancement of novel technologies, the creation of products with extended lifespans and recyclability, and the reconsideration of business models to align with the concepts of circularity. This invention facilitates economic expansion and stimulates industry metamorphosis, establishing a comprehensive sustainable development foundation.
- Creating New Industries and Job Opportunities: Applying industrial ecology concepts has developed novel sectors, specifically recycling, renewable energy, and sustainable materials. The recycling sector has grown substantially as enterprises recognize the benefits of recuperating resources from used items and waste streams. The rise of this business makes a significant contribution to the economy. It creates diverse employment prospects, encompassing various tasks such as material recovery and processing and advanced positions in research and development driven by sustainability.
- Example: The Rise of Green Technologies: A contemporary exemplar of industrial ecology is the burgeoning green technology sector, which includes advancements in renewable energy, sustainable materials, and eco-efficient technologies. Companies like Beyond Meat, which produces plant-based meat substitutes, epitomize how industrial ecology principles can give rise to new market segments. By offering sustainable alternatives to traditional meat products, Beyond Meat addresses environmental issues such as resource-intensive agriculture and greenhouse gas emissions while tapping into new consumer markets and driving economic growth.

Ultimately, implementing industrial ecology concepts and shifting towards a circular economy has significant economic consequences beyond simply meeting environmental regulations. This paradigmatic shift aligns with the broader goals of global sustainability. It has notable economic prospects, paving the way for a future where economic well-being and environmental preservation are not in conflict but work together harmoniously.

Businesses and economies can achieve a range of positive results by incorporating the principles of industrial ecology into their operating frameworks. They have the potential to encourage innovation, facilitate the emergence of novel market segments, and propel technological progress that forms the foundation of sustainable practices. This transformation enables a reconceptualization of resource utilization, transforming previously deemed trash into valuable assets, fostering resource efficiency, and mitigating environmental consequences.

Furthermore, the transition towards a circular economy signifies a novel phase in economic progress, distinguished by heightened adaptability and environmental sustainability. It provides opportunities for employment in developing sectors, such as renewable energy and sustainable materials, and presents a solid solution to the urgent issues of resource exhaustion and environmental deterioration. Economic restructuring, which draws inspiration from the concepts of industrial ecology, enhances the stability and growth of individual firms and the overall economy, enhancing its ability to navigate future difficulties effectively.

Hence, incorporating industrial ecology into economic planning and business strategy is more than acknowledging environmental responsibility; it embodies a progressive mindset toward economic advancement. It entails a dedication to constructing an economy that utilizes innovation and sustainability as foundations for expansion, guaranteeing that the current prosperity does not compromise future welfare. Industrial ecology is vital in promoting a sustainable, fair, and economically prosperous future, highlighting the inherent connection between ecological awareness and economic prosperity.

Case Studies Demonstrating Economic and Environmental Benefits

- Zero-Waste Initiatives: Companies like Subaru have implemented zero-waste initiatives in their manufacturing plants, significantly reducing landfill waste and cutting costs associated with waste disposal and raw material acquisition.
- Energy Co-Generation: Facilities that utilize waste heat for energy co-generation, such as those in the Kalundborg Eco-Industrial Park in Denmark, showcase how industrial symbiosis can reduce energy costs and decrease reliance on fossil fuels.
- Sustainable Supply Chains: The fashion industry, with pioneers like Patagonia, demonstrates how sustainable supply chain practices can reduce environmental impacts, enhance brand loyalty, and open up new market segments focused on sustainability.

Chapter 10
Realized Social Economic Benefits of Industrial Ecology: A Study on Textile Industry

This chapter goes into great detail about socio-economic benefits and industrial ecology in the context of the textile industry. This study would shed light on the cultural, regulatory, market, and technological hurdles that prevent the textile industry from fully embracing industrial ecology and how these adaptations have succeeded. Industrial ecology makes ecological and human systems work together in a good way for the economy, the environment, and society. After that, an in-depth look at the previous research lays the theoretical groundwork for the study. Additionally, it explores the difficulties that make it hard to recycle textiles and handle trash. Qualitative research is being conducted; data have been collected from various representatives via surveys, interviews, and open-ended questions.

Moreover, a thematic analysis is conducted to identify the codes and subsequently examine the themes to present the key indicators contributing to the various advantages of ecological practices. The analysis provides evidence that the export of textiles and garments has adverse ecological and socio-economic consequences, which is also an eye-opener for big retail brands in developed countries, mainly the USA and Europe. It depletes natural capital through the sustained degradation of vital ecosystem functions, and it does not contribute to the overall economic welfare of society. The chapter concludes with reflections on the findings and recommendations for future research.

10.1 Socio-economic Benefits and Industrial Ecology

Socio-economic benefits in sustaining the industrial ecology of the export sector are influential in the local/regional industrial ecosystem, which is envisioned as having environmental, economic, and social "wins." It is advised to base the ecosystems metaphorically while implementing them domestically or globally or as "a circular put system" comprising material and energy fluxes. The system produces

A. Shah et al., *Industrial Ecology*, Industrial Ecology,
https://doi.org/10.1007/978-981-97-3619-5_10

less waste and emissions thanks to the major stakeholders involved, which include the government, the public, and the private sectors. They all should work together to use wastages efficiently, including (cascading of energy, recycling of matter, and long-term use of locally renewable energy and matter resources) (wastes are used instead of virgin resources, and nonrenewable resources are replaced with renewables). Economic gains are possible by lowering costs associated with wastages, carbon emission, organic matter and energy use, transportation, and costs associated with implementing environmental legislation requirements, as well as by enhancing the system's "green market situation" and public perception of the environment. Investors and other capitalists might be persuaded to infuse in this local structure because it is "risk-free." In this idealized representation of the domestic and globalized ecosystem, the collective benefit results from increased use of territorial resources and increased local economy, making it self-sufficient and creating employment possibilities for residents.

Local administration of the flow of materials and energy can also result in the emergence of new commercial and economic sectors, such as trash or recycling businesses. Corporate social responsibility (CSR) can be achieved through collaboration and involvement of all the parties involved. The flow of wastage should be controlled to apply the concept of IE so that there can be much better integration and the desired solutions can be achieved while keeping in view the aspects of sustainability and biodiversity.

A branch of ecology called "industrial ecology" tries to redesign the industrial environment by managing and planning closed-loop industrial fabrication and dissipating systems. Industrial ecology aims to create harmonious relationships between the ecological and human systems to benefit communal, ecological, and economic sustainability in this period of significant environmental change and environmental ambiguity, which are highlighted by the outspread consequences on energy use, resource-draining, water wastage, global warming, and viable usage of material and production. The wider adoption of the Sustainable Development Goals (SDGs) has made industrial ecology (IE) far more appealing.

IE is a branch of ecology that focuses on managing and constructing closed-loop industrial systems to rebuild the industrial environment. Industrial ecology aims to create harmonious relationships between the ecological and human systems to benefit social, environmental, and economic sustainability. Applying such processes can help operate effectively and accurately, or they can be amalgamated in these ways to create a wealth of research opportunities from the management perspective. The term "industrial ecology" has come to refer to all aspects of how environmental requirements, particularly for materials, affect both natural and human capital. As a result, there is an urgent need to create methods, structures, and tools for the industrial ecosystem to support natural and human capital. A conceptual framework must be created to assess the prospects for manufacturers within the industrial ecosystem to contribute to the ongoing development of IE.

Numerous academic studies and commercial initiatives investigate fresh approaches to producing greener and more ecologically responsible chemicals and materials for various applications. Creating, manufacturing, recycling, and disposing of green plastics, polymer composites, adhesives, mixes, and several other industrial items made from renewable resources constitute a sizable portion of this endeavor. Particularly in the automobile business, natural fibers have the potential to provide better-added value, renewability, sustainability, and lower prices. Tons of fiber crops will be the subject of further study. A growing body of scholarly literature extols the advantages of natural fibers from a technological, economic, and ecological standpoint. This excitement often extends to the socio-economic advancement of the farmers of fiber crops and their local communities. However, there isn't much discussion or proof for claims about the alleged benefits for the impacted population in rural areas.

Despite the anticipated increase in natural fiber demand, it is unlikely that this will improve agricultural fiber growers' and their communities' standard of living. For tens of thousands of natural fiber growers, social equity and improved environmental quality can only be attained through consistent political will and cooperation between consumer groups, governments, scientists, industry, and local communities, as well as a suitable economic strategy like local subsidies should internalize a project's externalities by incorporating planning, execution, and implementation processes into a larger sustainability framework. Current project life cycle management (LCM) approaches poorly address sustainable development objectives. Then, for South Africa, a procedure to enhance the consideration of environmental factors in project LCM is introduced. A case study from the process industry is used to illustrate the procedure. Furthermore, a framework of social sustainability standards that apply to initiatives in the process industry is put forth.

10.2 Socio-economic Benefits in Textile Industry

The industrial revolution around the world brought severe threats that we, as the citizens of our planet Earth, are facing today. Our natural resources are depleted rapidly, and similarly, we are seeing rapid climate changes. Some parts of the world have started seeing heavy rainfalls and big fires erupting in some of the most enormous forests worldwide, all due to increased global warming. Various industries and their plants emit carbon, and industrial effluents hamper our seas, rivers, and lakes.

This study concentrates on the socio-economic benefits of sustaining industrial ecology in the textile sector. Industrial ecology represents production and consumption as closely linked with the natural environment. The textile industry worldwide shares a market size of US $993.6 Billion and is expected to grow by 1032.1 Billion in 2022. For many countries, this industry can be seen as the backbone of their economy, specifically in Asia. In Pakistan, the industry contributes to about 8.5% of the country's GDP. In the past two years, especially during the pandemic, we have seen a rise in exports, mainly because of the textile industry. However, as said earlier, industrialization has brought severe environmental damage, particularly in the textile industry; the consumption of various chemicals, the considerable use of water, and greenhouse emissions have created adverse effects. The annual amounts of water used (93 billion cubic meters), greenhouse gas emissions (1200 million tons), fertilizers for cotton (8 million tons), pesticides for cotton (200,000 tons), chemicals (42 million tons), and dyestuffs (1 million tons) around the world are all given.

The research reports that the chemical waste released in the environment accounts for 45% because of chemical and dyeing wastage. In addition to some recent industrial reports, $400 Billion worth of clothing is wasted annually.

As the number speaks for itself, it can be seen that this industry, responsible for the increase in global carbon footprint and various other wastages, needs to adopt behaviors based on sustainability and recycling. The increasing global population has doubled the consumption in the last two decades. Socio-economic factors can help with sustainability because every step of the textile waste recycling process—collecting, sorting, transporting, and recycling—creates jobs and opens up business opportunities for small or family-run companies.

10.3 The Ecosystem of the Textile Industry

Only the function of socio-economic advantages in preserving the industrial ecology of the textile industry's export sector will be discussed in this paper. According to critical analysis, globalization is merely the beginning of a period of capitalist expansion. As per Aina (1996), exploitation, inequality, accumulation, and polarization in international politics continue to be hallmarks of globalization. Dembele (1998) succinctly noted that globalization will strengthen the current division of labor. This means Africa can only supply raw materials and commodities and buy manufactured goods from wealthier countries. Globalization, on the other hand, would make it much harder for African governments to set goals for national development or even make it impossible for them to do so.

Any country's textile sector is crucial to the growth of that nation's economy since it fosters commerce, generates jobs, and is a significant source of foreign direct investment. This industry has been significant ever since the Industrial Revolution. However, because of resource depletion and climate change, it needs to adapt its production methods and eliminate the throw-away culture, which can help achieve ecological and environmental benefits. Recycling textiles is so urgently needed. The availability of varying degrees of statistical data and effect measures has led to an increase in the consideration of the environmental effects of textile production, use, and disposal in commercial, societal, and academic disputes. Concerns have been raised about the environmental impacts of cotton production at different stages of the textile value chain. These impacts include biodiversity loss due to the use of pesticides and other harmful chemicals and habitat loss (Chapagain et al., 2006).

The textile business has a detrimental effect on ecosystems, particularly those that produce freshwater. According to estimates, the dyeing and processing of textiles contribute to 20% of all industrial water pollution worldwide (Kant, 2011). The production of textile fibers is projected to reach 130 million tons by 2025, up from the planned 70.6 million tons in 2007 to 85.5 million tons in 2013 and 90.8 million tons in 2014 (Dahlbo et al., 2017). Although there have been various attempts to quantify the global amounts of textile waste, this endeavor has proven challenging due to the inconsistent approaches used for identifying, measuring, and reporting trash. An estimated 15 to 25 percent of all textile waste is collected separately and entered into possible recycling channels (Zamani, 2014).

It is standard to recognize the difference between mechanical and chemical reusing (Ragaert et al., 2017). Be that as it may, reusing courses habitually combines mechanical, chemical, and warm forms. To resolve this vulnerability, they separate the taking after-reusing strategies: texture reusing, fiber reusing, polymer/oligomer reusing, and monomer reusing. Reusing utilized clothing and materials into unused filaments, feedstock, components, and items ensures total circularity for things that can't be reused. Michaud et al. (2010) frequently categorize three sorts of material reusing: wiper generation, fiber materials, and reaction filaments utilized to make modern texture items. After being stripped of metallic components, such as zips, post-consumer materials are cut into fine shreds to sell to the industry.

Materials from both the pre-and post-consumer markets can be utilized as feedstock. These sorts of physical and mechanical material reusing drop beneath the category of open-loop frameworks, where items are dismantled for utilization in a distinctive, auxiliary item framework. By substituting virgin materials, fabric costs can be decreased at this point. In any case, regularly, no arrangement is made for re-collecting these materials at their moment of end-of-life arrangement, and in case they were to be collected, rehashed preparation would render the high-quality materials futile. Both pre-and post-consumer materials can be mechanically reused into filaments, which can then be utilized to make unused yarns and textures, clothing, covers, and sewn merchandise (Payne, 2015).

Shorter fiber lengths deliver materials of lower quality; in this way, modern and recycled fibers must be combined to achieve the required qualities. A squandered parcel of moo quality, unusable-for-reuse materials comes from sorting and evaluating the assembled materials. Even though there are no demonstrated markets for these items at the minute, there's a chance to occupy these materials for use as feedstock in cutting-edge chemical reusing forms presently being created. Material item reusing has been recommended as a potential reply to these issues as the portion of the more significant move from a direct to a circular economy. Environmental weakening and climate change are happening more rapidly around the globe. By 2050, junk generation is expected to have expanded by 70% yearly, and fabric utilization will have multiplied over the past 40 years. Moreover, businesses cannot incorporate natural concerns, activities, and targets into their vital arrangements. They steadily receive circular commerce models due to open weight on natural issues, thorough confinements, and tall levels (Bartlett et al., 2013).

10.3.1 Industrial Symbiosis

A division of IE is industrial symbiosis. It explains how a network of many organizations can promote environmental innovation and long-term cultural change, produce and share mutually beneficial transactions, and enhance operational and technical procedures. IS works with a network of organizations to promote eco-innovation by generating and sharing knowledge for new input sources, destinations for non-product outputs with added value, and enhanced business and technical processes. Synergies are transformed into appealing business prospects through the mutual benefits of higher revenues, cost reduction, and competitive advantages. Technologies that offer increased levels of efficiency and transparency can encourage the use of CSCs. The design and production of products are better suited to three-dimensional printing and augmented reality. Further, businesses can lessen the uncertainty associated with environmental collaboration and international sourcing management thanks to IBM's artificial intelligence technology, Supply Chain Insights.

10.3.2 Barriers to Textile Recycling

Regarding obstacles, recycling textiles is typically complex, and the recycling rate is poor (Sandin & Peters, 2018). Few countries, if any, can recycle 50% or more of their potential trash today. For example, less than 12% of recyclable materials are collected in Brazil, compared to rates of 10%, 11%, and 20%, respectively, in Russia, South Africa, and China. Developed countries have higher rates, at 32, 40, and 43%, respectively—including Canada, France, and Italy (Conke, 2018).

The following is a discussion of some of the main obstacles to textile recycling optimization:

- Economic Viability
- Poor coordination, weak policies and standards
- Composition of textile products
- Technological limitations
- Non-availability of recyclable textile materials
- Lack of information and limited public participation.

Businesses find it challenging to value textile waste. The sale of undesirable goods that are still in good condition has the most value. The projected size of the global used and worn clothing export market is $4 billion (Sharma, 2018). However, there is a finite and declining market for these goods.

Even though the methods of recycling differ from region to region, it is also clear that a large amount of textile waste cannot be processed using any of these methods because the materials have been worn down, may be dirty or damp, may have branding that prevents them from being sold again, may be made of a mixture of different fiber types, or may contain metal or plastic components.

These items may never cover the underlying collection, sorting, and transportation costs. Businesses that want to promote sustainability have worked to get harmful chemicals out of their supply chains. Even though much work still needs to be done, this has had an enormous influence. Pesticides and water use issues specific to cotton are mainly unaffected. Brands are becoming more eager to introduce products that include recycled fibers, which is undoubtedly a step in the direction of the circular economy. Secondhand retailing and product take-back programmers were found to be the two main methods that retailers are now assuming post-retail responsibility by Hvass (2014). The most critical channel for educating consumers about sustainability and the circular economy seems to be take-back.

The interdisciplinary industrial ecology (IE) field studies how natural and economic ecosystems are connected. The idea uses the relationship between natural and industrial ecosystems as a metaphor to fix industrial systems that can't last. Multiple techniques and strategies have helped the development of the IE in particular. The main goal of industrial ecology is to figure out how economic activity affects the environment and find ways to lessen those effects. Tourism is an economic activity that affects the environment, so it should be treated like any other business. This research suggests using a theoretical analysis of IE to find the best way to combine IE with tourism activities. Seeking an in-the-IE idea sounded interesting.

10.3.3 Socio-economic Factors and Textile Recycling

The textile industry worldwide is essential as it involves huge investments and revenues and provides employment opportunities. Like every industry, this industry is also responsible for wastages, which impact the environment and humans (Leal Filho et al., 2019). To protect and preserve the environment, recycling waste is the need of the hour, and with this, there are many benefits associated with socio-economic factors. Textile recycling can be recognized as one of the main aspects of achieving sustainability. It is vital that the industry itself starts recognizing the increasing global population needs and shrinks down the use of raw materials (Watson et al., 2017).

Talking about the processes involved in textile recycling (i.e., collection, sorting, transport, recycling) creates jobs and opportunities for small or family businesses (Zamani, 2014). Textile recycling is often considered the last process, but it can be seen as an instrument of circular economy. This could be one of the many solutions needed for building sustainability. Recycling helps minimize the manufacturing cost and can be cost-effective for achieving sustainability. For instance, Ecosign (2017) found that recycling used clothes could cut greenhouse gas emissions by 53%, pollution from chemical processing by 45%, and water eutrophication by 95%. The materials are of great value as they can be reused in different sectors of the economy, such as the automobile sector, fishing sector, etc. (Ecosign, 2017). Various wastages can be used in further processes like spinning and weaving (Nunes et al., 2018). The imperative for market expansion of recycled grades, such as closed-loop fiber-to-fiber recycling, has been underscored by the volatile fluctuations in the textile industry and the unpredictability surrounding specific crucial export markets (WRAP, 2016).

The wastage of the textile industry can be taken as an opportunity for new ventures as they can make a profit by designing them into new products (Rosa, 2016). Ehrenfeld (1995) asserts that the evolution of environmental management paradigms is bringing about the emergence of a brand-new idea known as industrial ecology. The desire to apply sustainability concepts to environmental and economic systems drives it (Ehrenfeld, 1995). The limitations of the industrialized world's reliance on command-and-control management are reflected in recent environmental theory, which has focused on an awareness of the profound and essential linkages between human behavior and the natural world. The main problem is that the economy functions, for the most part, as an open system that draws resources from the environment and returns massive amounts of garbage and pollutants that were not used.

10.4 Socio-economic Factors and COVID-19

The pandemic immediately halted all the economic activities that international organizations carried out. Events with a low probability of happening but a significant impact, including natural disasters, can cause supply chain disruptions. Nearly every nation has taken action to halt the spread of this epidemic by enforcing lockdowns, limiting social interactions, and enforcing "smart lockdowns" while keeping in mind the need to maintain economic activity to prevent any adverse effects on the country's economy. According to the International Labor Organization (ILO), the COVID-19 problem might lead to the loss of up to 25 million jobs due to the high labor cost in the textile industry, particularly in Asian countries.

The word "sustainability" comes from scientific literature, classifying resource management as sustainable if not used over a critical level of natural capital. Industrial equilibria (IE) and industrial symbiosis (IS) have implications for sustainable company management techniques. The IE examines the logical connections between societal and environmental issues and organized settlement's environmental and economic requirements. IS refers to the network of waste, byproduct, and product exchanges that reduces the environmental impact of industrial locations. Several different models for the symbiotic expansion of industrial networks are available globally.

10.5 Research Map

This research has opted for the deductive approach as the authors have been testing the theory; however, there is no complete familiarity regarding the socio-economic benefits, especially in third-world countries. However, there is research regarding the environment and a comprehensive study of ecological benefits in regards to society and environment of third world country so, so the authors have gathered the information from various articles, found a common point or theme from the articles, and did a research on the topic. In this research, qualitative research is being conducted as the data have been gathered through interviews and surveys, and open-ended questions from various representatives have been asked. A thematic analysis has been done to identify the codes and then review the themes to report the prominent indicators of the causes of different benefits obtained from ecological practices; see Fig. 10.1 for a comprehensive map illustration of the study.

To separate knowledge from information and information from data, qualitative methodology offers a variety of methods and analysis techniques (Gupta et al., 2015), either by assisting quantitative analysis or particularly in situations where other methods, such as quantitative survey, pose limitations (Ghanshala et al., 2013). One of these methods is content analysis, which is conceptualized and used in this chapter.

A general definition of content analysis is the "method for concluding by systematically and objectively identifying specified properties of messages" (Holsti, 1969).

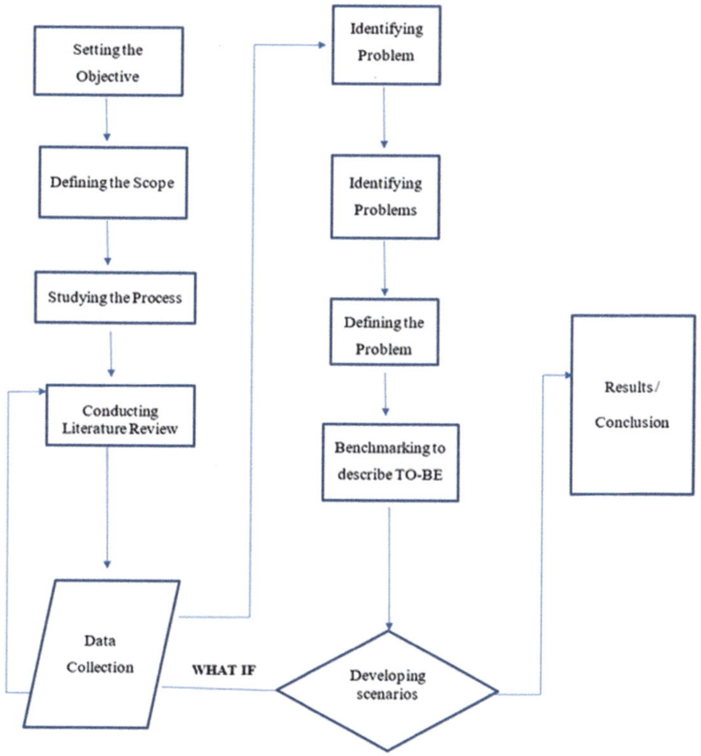

Fig. 10.1 Road map for the research

Content analysis is used for systematic and impartial written, visual, and spoken communications investigation.

However, others who have focused on its methodology have described it as a "tool for objective, methodical, and quantitative description of manifest content communication" (Berelson, 1952) and "a research strategy that uses a series of processes to draw meaningful conclusions from the text" (Weber, 1990). Content analysis gives a short but complete explanation of a phenomenon in the form of categories and concepts. This helps make a model or conceptual map (Elo & Kyngäs, 2008). One benefit of content analysis is its ability to generate dependable and replicable results from contextual data.

This, in turn, facilitates the generation of novel data and enhances the body of existing knowledge. Primarily, it facilitates the development of practical action guides, which precisely align with the objective of this research. Additionally, because it is sensitive to content and context, it allows for flexible research design that deals with meaning and intention and finds key processes (Downe-Wamboldt, 1992; Harwood & Garry, 2003). Thus, content analysis is the foundation for many studies that use text as their primary data source. These studies include everything from newspapers (Mason & Poyatos Matas, 2016) to magazines (Döring et al., 2016) to websites (Bazzini et al., 2015) to religious texts (Graneheim & Lundman, 2004). The positivist, qualitative-deductive method of content analysis is covered in this chapter.

10.5.1 Data Collection

To kick start the data collection phase, this study has cascaded down the respondents to different tiers to have the say of most of the employees. The stakeholders are divided into three tiers; first, it revolves around the three tiers with whom the authors have been able to collect the data.

Tier 1: In this category, employees and organizational workers fall.
Tier 2: In this tier, the clients were asked to have their valuable feedback.
Tier 3: In this zone, stakeholders and regulatory bodies can ask for their feedback.

Since the research is qualitative so, the data can be gathered through interviews and interactions with the associated parties of the Towel Manufacturing Company and even their partners or competitors to have an industry overview; as well, the employees participated working in a factory to have the specific detailing and to forecast the future patterns. Even clients were keen to save the date for the interview. Extra information was extracted from research articles (Aluko et al., 2004).

In this research, the authors came across different research articles to extract various case studies to identify the linkages and patterns and figure out the objectives of this study. In this regard, the following people were subjected to in-depth interviews:

- Customers
- Personnel Managers at the Locations
- Owners/Shareholders.

To diversify the scope of the research, the authors agreed to obtain data collected from all the stakeholders to provide an overview of both sides. In this regard, around seventy respondents were met and have added their feedback to this research. The questions they asked were also open-ended to identify the linkages, if any, in the processes.

This research study had been able to gather information from case studies as well, and interviewing the internal stakeholders has provided the leverage to make the production processes even more handy and technology usage to be also encouraged, which the authors have observed during the visit as well case studies shows how company deals the situation.

Thematic Analysis

Thematic analysis is a strategy for analyzing subjective information. It can be in different forms, such as writings or transcripts. The analyst then does the analysis to find out the typical topics that come up frequently. There are usually six steps involved in the thematic analysis process; however, the primary purpose is to focus on any of the themes found in qualitative data to figure out the promptest dataset.

Thematic analysis is generally used in situations where there are tons of different data and does not have any figurative identification to which it pertains, so in this regard, this study tries to specify and cascade down the data into smaller bits called codes, which are a subjective form of information (Aluko et al., 2004). See the extracted code through thematic analysis in Table 10.1.

Table 10.1 Improvement strategies for textile industry

S. No.	Codes	Extracted code
1	Increase efficiency	Lack of knowledge about the new technology. Unable to enhance the machinery. Also, there is a lack of qualified human capital
2	Process reengineering	Best practices of textiles are not adopted. This results in a lack of efficiency in operations
3	Wastage control	Recycling knowledge, usage, optimization of material usage, and safe waste disposal are to be partially adopted. There is much room for improvement and adoption of lean manufacturing
4	Cost controlling	Old technology, partial wastage utilization, high energy costs, compliances, and other overheads
5	Less environmental damage	Water wastage un-controlling factors

10.6 Thematic Analysis: Familiarization

Thematic analysis, as discussed in the Research Map section, gathers data in qualitative research to identify the main codes, that is, items that lay the foundation of any effect caused by them. So, it is pivotal to have a Hawkeye over the items that are being gathered. It can be through interactions or audio sources and can be written in books or quoted in international journals. This leads to familiarity with the data for further usage in concluding.

By keeping a specific focus on the textile operations of the topic, which is "Socio-economic benefits of industrial ecology in the textile sector of Pakistan", this study will mention the data in the familiarization tab of thematic analysis, which will further help in determining the core issues faced by the industry.

As is already mentioned in the introduction of this report about the company which is the focus of this study, the process of how the operations go on concerning the topic will be discussed. The authors interviewed a person from the Towel Manufacturing Company to assess the theme of the discussion, identify the familiarity, and summarize the discussion.

Since its inception, the core competency of the company has been ecology as it relates to the manufacture of textiles. The normal production of clothing and other items has numerous, diverse, and often detrimental environmental effects. Pollution of soil and aquifers, escalating climate change, disregard for species diversity, resource waste, chemicals getting into our clothing, and other issues are all present. Environmental preservation and consumer safety remain two of the primary focal points of our association and our members, even though our concentration is less on clothing and more on other textile products. Only we can protect this one planet that we have, the Earth. Similarly, we consumers are showing more and more respect for our health.

10.6.1 Production Ecology: Familiarization

Impact of Production Processes

The process starts with yarn production, the core material used in all textile products. The process involved collecting, reprocessing, and controlling waste in the raw material. The core social issue was the fluff generated due to the shuffling and cotton processing. Breathing issues may affect the workers around. Regulations regarding this issue are not being followed. This could be the problem, along with the problem statement of this study. The same fluff and production environmental issues were raised at the weaving level of operations, where warping the loom beam is required for the production process to continue.

After these processes, the core of the textile operations involves using chemicals in washing, dyeing, and bleaching the grieve-produced fabric/towel. This also involves water and energy consumption. Waste water and waste treatments, as well as the generation of noise, are also considered. The washing, bleaching, and dyeing process involves considerable water and chemicals, which may pose a risk to the environment and the humans involved. In an interview, it was observed that high-grade SOPs were created and implemented under monitoring. The monitoring is only involved in the production process, but the wastage seems uncontrolled due to a lack of awareness. Chemical-contained process water is being released into the sewerage, which may harm someone in case of overflow. This could cause damage to the sewerage line and result in administrative issues in the industrial area. The textile industry's main issue in ecology is adopting production techniques that can help reduce costs and become more environmentally friendly. For this reason, recycling has also become an integral part of production, saving materials and controlling production. Ways to be adopted to control and to lessen the amount of wastage. The commonality is that production processes and wastages must be deep-dived to find the socio-economic benefits of process/reprocessing and waste.

10.6.2 Processes Familiarization

De-sizing

It is alarming that water resources are depleted rapidly; the foremost reason behind this is the process of de-sizing. This reaches the point where the water is not usable anymore.

Bleaching

Some bleaching agents are considered hazardous, like the oldest industrial bleaching agent, hypochlorite, which is problematic for humans and the environment.

Neutralization

Acetic acid is used as needed in the neutralization process, which has the same consequences as hypochlorite.

Dye

This process contains the maximum amount of pollution as it involves dyeing chemicals. It is amongst the main textile processes. The wastewater contains hydrolyzed reactive dyes, which are hazardous mixtures of dyeing chemicals of about two g/L.

Stitching

Stitching involves the energy issues of the most energy-consuming motors, whereas servo motors are also available, which consume less electricity than conventional motors. However, due to the owners' low intentions, such things cannot be implemented. This stage of textile production also involves noise pollution.

In addition to the above requirement, the customers are making these things mandatory for their vendor to get approved on their approved vendor list. All these points must be confirmed to gain the maximum market share and customers for the business's ongoing concern. This is to control and ensure customers reserve the right to a third-party audit of vendors.

10.7 Results and Discussion

Amongst them are the challenges of noise pollution and water issues from bleaching and directly going into the central holes without being treated, creating issues for human health through the fluff, mainly generated during the weaving and raising process. This area of this study will first discuss the issues that are still untapped in the said industry, and the second part of the chapter will consist of the solution that has been put forward to the industry owner at first sight. The different types of looms generate the base and noise. One power looms. These old technology looms are the 1940s technology and are still being used.

These are low in production and may produce 100 lbs. of the towel and fabric. The main disadvantages are energy consumption, noise, and high wear and tear. The second type of loom is shuttleless, which generates a large ground base and sound, along with a considerable amount of yarn fluff that spreads around the whole production area. Workers working in have been found with severe respiratory disease, which becomes chronic over time. This issue is continued with all looms, be it the power loom, shuttleless, or high-tech Air jets. The second phase of the identification is waste handling management, recycling with proper processes involved, and waste destruction.

The lack of a comprehensive and well-coordinated structure significantly impacts the sector. A missing legal affirmation would also increase the effectiveness of refurbishing the virgin material in the ecosystem.

The supply chain adds value to various parts that operate independently, occasionally with competing interests. It will be possible to create unified policies and regulations that address specific industry, technological, categorization, and test standards by centralizing the operations of these individual components.

The dyeing process is another area that needs concentration in terms of waste handling. The dyeing process is one of the most energy-consuming processes. The average dyeing unit consumes about 1200–1500 tons per day. In this regard, there needs to be proper research and development to minimize the toxicity of the chemicals being used, energy consumption needs to be reduced, and equipment needs to be upgraded. They need to set up an Effluent Treatment Plant (ETP). The water recycling process also needs to be carried out, especially the cooling water, as it is comparatively less uncontaminated and can be used for rinsing.

Chemical consumption is also one of the main wastages in this industry. Minimizing chemical consumption will lower the treatment costs. Companies should avoid using chemicals while updating themselves on the new dyestuffs entering the market. They also need to control the dosages and spillages in the processes by minimizing the chemical use centrally while pumping them into machines. They need to measure the mixing processes manually, and automatic dispensers should be used to calibrate the system correctly.

Similarly, wastage is involved in every process, from dyeing to bleaching to finishing. At each process, multiple areas require focus, training, and educating the workers that are part of these processes and making them aware of the wastages that can be recycled. Automation is the key to ensuring that means better controls to minimize wastage.

The findings highlight the trade's adverse effects, for instance, on soybeans. Numerous choices and legislative initiatives are accessible or already in place to lessen the adverse effects. This study offers a series of, but not limited to, measures that might be taken to drive the international venues towards a more comprehensive and confined stage without attempting to solve this complicated problem. The measures range from being preventive to being curative. They can be taken domestically and internationally, specifically for a place using resources from different avenues as a joint obligation.

There are three main methods by which the recommended activities communicate with the market. First, using alternatives and encouraging sustainable consumption are some activities that can lower demand. Second, certain activities can lessen the environmental harm caused by the production. Numerous subsidies encourage harmful behavior. Thirdly, conservation initiatives, financial rewards, and payment plans for environmental services could be used to offset unavoidable environmental harm. Economic value makes "polluting" items less desirable by charging buyers for the costs of externalities (attaching more cost to it). The criteria for calculating the value can lead to determining fines regarding environmental harm.

The analysis demonstrates that exports have detrimental ecological and socio-economic repercussions, depleting natural capital and resulting in the long-term loss of crucial ecosystem functions while offering no economic advantages to society. The ecosystem services concept provides a method to show actual effects on present and future populations worldwide and on the inhabitants within the boundary to purify the contagious water due to wastage of fertilizers (e.g., international climate regulation decrease by deforestation). The measurement of these consequences in monetary terms for this particular global trade market enabled the theory to intricately understand the opposite relation associated with the growth of a worldwide marketplace.

10.8 Conclusion

A branch of ecology called "industrial ecology" tries to redesign the industrial environment by managing and planning closed-loop industrial systems. The main motto of this ecological mechanism is to create harmonious relationships between the ecological and human systems to benefit social, environmental, and economic sustainability. The enormous consequences of resource usage, resource decrease, water scarcity, climate change, and excessive virgin raw material usage have brought about a decade of enormous material utilization change and climatic devastation. The wider adoption of the Sustainable Development Goals (SDGs) has made industrial ecology (IE) far more appealing. IE tools and practices can be applied to address the difficulties of global sustainability.

According to research, interest in conversations about long-term apprehension has grown through the past 30 years. Beyond the rising interest in sustainability, an international focal point objective to address comprehensive sustainability issues to support nations tends to emerge. Nonetheless, a resurgence to captivate these objectives, little has been accomplished. The industrial sectors worldwide have seen modern commercial activity's economic and financial rewards. However, there has been a significant change in how business is conducted due to recent ecological and climate challenges. It asks for innovative corporate tactics and strategies to turn current global problems into the driving force behind sustaining the long-term agenda by utilizing ecological frameworks and instruments to advance those aims.

Determining a comprehensive and long-lasting ecological and human development potential depends significantly on industrial ecology. However, IE could specifically help with waste reduction, promoting closed-loop manufacturing and material consumption in networks that rely little or bulk on the manufacturing systems of particular organizations or businesses that particularly benefit from local and international cooperation. Industry can obtain benefits for sustainable development by converting efficient material and energy systems in a lean manufacturing framework into closed-loop flows by utilizing industrial ecology today. Besides strategy, the economic system in a cyclical system has been crucial in improving industrial ecology performance. Sustainable development is increasingly a central issue in the literature on ecology as industrial ecology becomes more pervasive than ever.

The inter-firm environmental management system can be understood from industrial ecology and sustainable production perspectives. It also helps define the system and illustrate that it arises instead of interacting with various parties in energy-efficient loop situations. To bring about technical development, public and private entities need to collaborate. A healthy and long-lasting production system links other stakeholders of associated industries (Hart, 1995). Although inside and outside participants cannot provide lean manufacturing, they could take part in developing and providing energy-efficient and waste-free solutions. Capacity plays a pivotal role in the creation of sustainable solutions for the manufacturing of goods and processes. Industrial ecology is crucial for identifying well-designed.

The research analysis demonstrates that import and export harm socio-economic repercussions, depleting natural capital by permanently losing crucial ecosystem functions while offering no financial advantages to society. A complete resource value and comprehensive cost pricing of resources are made using the economic technique. Environmental expenses must allegedly be considered in management and policy discussions. A technique to account for consequences of human activities that are otherwise not considered is the economic assessment of ecosystem services (e.g., wastage and polluted environment). That promotes drastic inclusion of dialogue, and the generations and places aren't currently represented in these talks. It is crucial to remember that the financial numbers should not be taken exclusively as an absolute value but rather as a conservative and rough approximation. The ecosystem services concept provides a method to show actual effects on present and future populations worldwide and domestic localities, for example, purified water contaminated through fertilizer wastage) (e.g., international climate regulation decreases due to deforestation). The measurement of these consequences in monetary terms for this particular global trade market enabled delusional talks to intricate the mismatch associated with growth around worldwide marketplaces. This variety of initiatives to improve Europe's protein self-sufficiency and lessen the obstructive effects of worldwide trade are then presented.

References

Aina, T. A. (1996). Globalization and social policy in Africa. *Research Network Codesira Bulletin, 4.*

Aluko, M. A. O., Akinola, G. O., & Sola, F. (2004). Globalization and the manufacturing sector: A study of selected textile firms in Nigeria. *Journal of Social Sciences, 9*(2), 119–130.

Bartlett, C., McGill, I., & Willis, P. (2013). *Textiles flow and market development opportunities in the UK*. Waste & Resources Action Programme.

Bazzini, D. G., Pepper, A., Swofford, R., & Cochran, K. (2015). How healthy are health magazines? A comparative content analysis of cover captions and images of women's and men's health magazine. *Sex Roles, 72*, 198–210.

Berelson, B. (1952). *Content analysis in communication research.*

Chapagain, A. K., Hoekstra, A. Y., Savenije, H. H. G., & Gautam, R. (2006). The water footprint of cotton consumption: An assessment of the impact of worldwide consumption of cotton products on the water resources in the cotton producing countries. *Ecological Economics, 60*(1), 186–203.

Conke, L. S. (2018). Barriers to waste recycling development: Evidence from Brazil. *Resources, Conservation and Recycling, 134*, 129–135.

Dahlbo, H., Aalto, K., Eskelinen, H., & Salmenperä, H. (2017). Increasing textile circulation—Consequences and requirements. *Sustainable Production and Consumption, 9*, 44–57.

Dembele, D. M. (1998). Africa in the twenty-first century. *Codesria Bulletin, 1.*

Döring, N., Reif, A., & Poeschl, S. (2016). How gender-stereotypical are selfies? A content analysis and comparison with magazine adverts. *Computers in Human Behavior, 55*, 955–962.

Downe-Wamboldt, B. (1992). Content analysis: Method, applications, and issues. *Health Care for Women International, 13*(3), 313–321.

Ecosign. (2017). *Textile recycling as a contribution to circular economy and production waste enhancement.*

Ehrenfeld, J. R. (1995). Design for environment: A new framework for strategic decisions. *Environmental Quality Management, 4*(4), 37–51.

Elo, S., & Kyngäs, H. (2008). The qualitative content analysis process. *Journal of Advanced Nursing, 62*(1), 107–115.

Ghanshala, K. K., Pant, D., & Pandey, J. (2013). A gaps approach to access the efficiency and effectiveness of IT-initiatives in rural areas: Case study of Samalta, a village in the central Himalayan region of India. ArXiv Preprint 1303.0490.

Graneheim, U. H., & Lundman, B. (2004). Qualitative content analysis in nursing research: Concepts, procedures and measures to achieve trustworthiness. *Nurse Education Today, 24*(2), 105–112.

Gupta, M., Ganguli, S., & Ponnam, A. (2015). Factors affecting employee engagement in India: A study on offshoring of financial services. *The Qualitative Report, 20*(4), 498–515.

Hart, O. (1995). *Firms, contracts, and financial structure.* Clarendon Press.

Harwood, T. G., & Garry, T. (2003). An overview of content analysis. *The Marketing Review, 3*(4), 479–498.

Holsti, O. R. (1969). *Content analysis for the social sciences and humanities.* Addison-Wesley (Content Analysis)

Hvass, K. K. (2014). Post-retail responsibility of garments–A fashion industry perspective. *Journal of Fashion Marketing and Management, 18*(4), 413–430.

Kant, R. (2011). *Textile dyeing industry an environmental hazard.*

Leal Filho, W., Ellams, D., Han, S., Tyler, D., Boiten, V. J., Paço, A., Moora, H., & Balogun, A.-L. (2019). A review of the socio-economic advantages of textile recycling. *Journal of Cleaner Production, 218*, 10–20.

Mason, S., & Poyatos Matas, C. (2016). Language teacher supply: A content analysis of newspaper coverage across the 'Anglobubble.' *Issues in Educational Research, 26*(3), 446–463.

Michaud, J.-C., Farrant, L., Jan, O., Kjær, B., & Bakas, I. (2010). *Environmental benefits of recycling–2010 update.* Waste and Resources Action Programme.

Nunes, L. J. R., Godina, R., Matias, J. C. O., & Catalão, J. P. S. (2018). Economic and environmental benefits of using textile waste for the production of thermal energy. *Journal of Cleaner Production, 171*, 1353–1360.

Payne, A. (2015). Open-and closed-loop recycling of textile and apparel products. In *Handbook of life cycle assessment (LCA) of textiles and clothing* (pp. 103–123). Elsevier.

Ragaert, K., Delva, L., & Van Geem, K. (2017). Mechanical and chemical recycling of solid plastic waste. *Waste Management, 69*, 24–58.

Rosa, R. (2016). *Circular economy in the clothing industry: Challenges and strategies.* KTH Industrial Engineering and Management.

Sandin, G., & Peters, G. M. (2018). Environmental impact of textile reuse and recycling–A review. *Journal of Cleaner Production, 184*, 353–365.

Sharma, S. (2018). *Global market scenario for used clothing.* Fibre2Fashion.

Watson, D., Elander, M., Gylling, A., Andersson, T., & Heikkilä, P. (2017). *Stimulating textile-to-textile recycling* (Vol. 2017569). Nordic Council of Ministers.

Weber, R. P. (1990). *Basic content analysis* (Vol. 49). Sage.

WRAP. (2016). *Textiles market situation report.*

Zamani, B. (2014). *Towards understanding sustainable textile waste management: Environmental impacts and social indicators.* Chalmers Tekniska Hogskola (Sweden).

Chapter 11
Climate Change and Its Impact on Supply Chain: A Sourcing Case Study to Effect Industrial Ecology

The chapter is based on a research study to assist a sourcing company located in Hamburg in devising its long-term strategy regarding cotton-based products. Considering the importance of indigenous cotton yield within a country, any sourcing company continues to hunt for suitable supply locations to buy products and ensure supply chain responsiveness and efficiency. The research is based on the impact of climate change on textile manufacturing companies in the context of industrial ecology. Future climatic conditions are projected to foresee the severity of climate change, gathering climate-related data from governmental and global climate sources. The study analyzes the impact of climate change on cotton yield and compares the climatic conditions in all five sourcing countries of the sourcing company. The analysis also incorporates price and proximity to market factors as necessary for the company's long-term strategy. The factor rating method is deployed to reach a conclusive decision for comparing the feasibility of the five sourcing countries: Pakistan, India, China, Vietnam, and Bangladesh. Finally, concludes the chapter with some recommendations.

11.1 Background

Climate change is one of the most critical challenges facing the world at present. It refers to long-term temperature and weather shifts (Malhi et al., 2021). From economics and socio-ecological perspectives, climate change is one of the most debated topics of the twenty-first century. Soil degradation, environmental population, and natural resource degradation are mainly caused by anthropogenic activities, which are significant ecological problems related to climate change. It has been found that the primary drivers of the current climate change problems are human-induced changes (Solomon, 2007). Global warming, sea-level rise, glacier melting, ocean

A. Shah et al., *Industrial Ecology*, Industrial Ecology,
https://doi.org/10.1007/978-981-97-3619-5_11

acidification, and extreme weather events indicate that climate change has become a reality (Solomon, 2007).

According to the Sixth Amendment report of the United Nations Intergovernmental Panel on Climate Change (IPCC), regional temperature and drought directly impact human health, ecosystem functioning, food security, and the livelihoods of resource-dependent populations. It also mentions that the average land surface temperature in 2003–2012 increased by 0.78 °C compared to 1850–1900, and the atmospheric temperature is projected to increase by 4.3 °C by 2100. South Asia, with almost a quarter of the world's population, is highly vulnerable to these changes because most of the region's population depends on agriculture as a source of income (Nadiruzzaman et al., 2021). Moreover, NASA has claimed the significant effects of climate change on crop production. Record-breaking heat waves, shifts in rainfall patterns, intense droughts, and rising average temperatures are all due to human-caused climate change. Crop production has been stressed due to high-temperature rise due to climate changes (Gray, 2021).

In developing countries, the impacts of climate change are found to be more severe because of their agriculture-based economy, and extreme weather events have made it challenging to persist. Developing countries also need more money to adopt the new method, which is a challenge they face (Morton, 2007). The two foremost needs to minimize the impact of global warming are adaptation and climate change mitigation (Song & Ye, 2017). Major causes for such socio-ecological costs are the excessive burning of fossil fuels, inappropriate use of natural resources, and agro-chemicals. Adopting high-yielding varieties (HYVs) and mono-cropping patterns is also considered a significant cost for socio-ecological impact.

Due to its vast climatology and intricate geographical structures, Pakistan is more vulnerable to extreme hydrogeological events such as storm surges, heat waves, and famines. The agriculture sector is one of the most essential parts of Pakistan's economy and is facing significant challenges due to climate change (Habib-ur-Rahman et al., 2022). According to the Climate Risk Country Profile (CRCP), as of 2019, agriculture was still the largest employer with 42.6% of the workforce, and so variations in temperature and water availability could have a drastic impact on both the crop's productivity and the economy (Chapman et al., 2021). Pakistan is the world's fifth largest cotton producer, contributing almost 0.6% to GDP and 2.4% of the value added in agriculture. During 2021–2022, the cropped area has reduced by 6.8%, from 2,079 thousand hectares to 1937 thousand hectares (Iqbal et al., 2023). This reduction is due to excessive utilization of irrigation water, pesticide and fertilizer application, climate change, and insufficient cotton production management practices (Imran et al., 2018).

11.2 Climate Change and Industrial Ecology

Industrial ecology was an effort made a decade ago to make sense of the correlation between the natural atmosphere and the human system. By analyzing this correlation, major understandings were created on how humans took from the natural environment, transformed it according to their needs for the product, and put things back into the environment. The industrial ecology community has made remarkable advances in data sciences to gather the analysis that was impossible a decade ago (Chester, 2020). Throughout the study of industrial ecology, examining the cyclization of the resources revealed how multiple organizations exploit particular resources in the quest for their survival and growth (Volik et al., 2004).

According to the research, gases such as CO_2, N_2O, CH_4, and halocarbons cause significant damage to the climate. An increase in pollution and industrial activities has significantly affected the climate. The growing world population and their growing demand have forced industries to produce more, resulting in emissions of more harmful substances that are primarily responsible for the increased temperature and negative climate changes. Industries worldwide need to do more to control these industrial activities in terms of environment-friendly policies. Therefore, innovative policies in this sector are necessary (Ojeaga & Posu, 2017). Introducing a continuous monitoring system to monitor the emissions of industries will help minimize the impact of global warming. Industries' and environmental evaluation methods and rules must be strictly followed (Wadanambi et al., 2020).

11.3 The Effects of Climate Change on Agriculture

In an economic city, agriculture is dependent upon climate and weather. Spatial climate variability and significant temporal are subjected to agricultural regions worldwide. The production of food crops and the variability of the impact on yield are essential in each case. To achieve optimum conditions in a specific soil climate sector for sustainable productivity, understanding climate parameters and agriculture and how they link with each other is a prerequisite for tailoring management practices and cropping systems.

11.3.1 Cotton Industry

Cotton Production

Cotton is the world's most used natural fiber, meeting 1/3rd of its fiber needs (EPA, 2021). It contributes 40% to the global market and is cultivated in over 100 countries. Over 350 million people are indirectly or directly linked to the cotton production industry. Cotton holds a significant share of the global agricultural economy, but its yield and production are compromised by various climate factors (Achparaki et al., 2012).

With the advent of COVID-19, many agricultural-related businesses were temporarily closed, and fertilizers were unavailable to farmers, which resulted in delays in cotton sowing. Unfortunately, farmers were receiving lower than the average price for their cotton, so they could not employ as many laborers during the period. The uncertainty in the international market continued to impact the local markets (Better Cotton, 2020). Cotton production worldwide remained stable starting from 2018/2019 with 25.9 million Tonnes; it remained stable in 2019/2020 with 26.1 and reported a 7% decrease in 20/21 and 25.7 million Tonnes in 21/22. The cotton dropped to 22.7 Million Tonnes, the lowest in eight years.

Nevertheless, global cotton demand increased by 25.6 Million Tonnes and 26 Million Tonnes in 21/22. On the other hand, cotton imports decreased to 8.7 million Tons in 19/20 from 9.2 million Tons in 18/19, and an increase of 10.1 million tons was reported in 20/21 and 10.5 million Tons in 21/22. Cotton ending stocks for the year reported 22.8 million Tonnes compared to 19.1 million Tonnes in 18/19, and for the years 20/21 and 21/22, cotton ending stocks were 20.2 and 19.9 million, respectively (Hughes, 2021).

The Effect of Climate Change on Cotton

A study shows climate change is leading to a reduction in cotton production of 2330 kg per farmer (Soviadan, 2019). Achieving high cotton yields and maintaining the sustainability of cotton production requires more precise water, plant growth management, and nutrition. In addition to this cultivator, management practices, pests, and climate conditions influence and determine the fruiting of cotton plants. These are some of the critical factors that have a significant effect on the production of cotton. To improve the cotton yield, applying suitable cultural practices will minimize the harmful effects of these factors (Sawan et al., 2002).

One of the most important environmental factors is Temperature, which affects plant reproductive processes such as pollen growth and germination. The most significant climatic factor influencing cotton output is air temperature. Reduced fruit retention is the main effect of heat stress on cotton, which can lower total lint yields, delay crop maturity, and impair lint quality (Cetin & Basbag, 2010). For germination, cotton needs an air temperature of 15 °C; for vegetative growth, it needs 21–27 °C; and for fruiting, it needs 27–32 °C. Currently, available commercial cultivars typically require temperatures over 15 °C to yield a crop, becoming inactive below 15 °C, and are killed by cold temperatures (Cetin & Basbag, 2010; Waddle, 1984).

Rainfall is another factor necessary for cotton production. Cotton is said to be rainfed. Therefore, it is proclaimed that the higher cotton yield requires a minimum of 500 mm rainfall, and from below that, an economic yield cannot be achieved. Moreover, a minimum of 850–1100 mm of well-distributed rainfall is required for a profitable yield, but its quality is damaged if it rains in a boll bursting period. Water continues to be a production barrier for cotton. In addition to the natural rainfall, irrigation is employed when it is available to sustain viable crop production. Increased water collection would decrease the need for irrigation, increase water usage efficiency, and lower input costs. Therefore, identifying the crucial water scarcity periods for cotton productivity could assist farmers in choosing when to plant when to irrigate, when to operate in the field, etc.

Relative humidity tells the air's moisture ratio to the moisture that saturated air can hold at a specific temperature. With temperature, moisture-containing capability rises, and relative humidity decreases until moisture increases. This is another factor necessary for cotton production. Relative humidity (RH) should be between 60 and 80% for cotton production (iKisan, 2020). Disaster risk is a highly unfavorable and terminal adverse factor for the cotton supply chain (Chari & Ngcamu, 2019). Disasters, including floods, droughts, storms, etc., are lethal for cotton growth, and thus, the risk of such disasters is incorporated in this study.

11.4 Climate Change Projections

The degree to which exposure to CC is influenced by shifting global, regional, and even local socioeconomic situations can significantly impact the size of these effects (Lehtonen et al., 2021). So, to investigate these effects, researchers have tested and applied many scenarios to study these effects, one of which was SRES (Special Report on Emission Scenarios) by Carter in 2005. Soon, it was replaced by a new framework introduced by (Moss et al., 2010) that addressed a more comprehensive set of research questions focusing more on regional aspects of CC.

The IPCC did not develop these frameworks like their predecessors; instead, it was created by the international research community. They were recently reviewed by O'Neill et al. (2020), and they included two main factors: one RCP-representative concentration Pathway and one SSP-shared socioeconomic pathway.

11.4.1 RCP Pathway

RCP is a method for evaluating the assumptions within scenarios to find possible future evolutions. These assumptions are necessary to predict change and help discover the economic, social, and physical environmental variations that will affect CC. In comparison to pre-industrial levels, RCPs describe greenhouse gas concentrations that will cause the total radiative forcing to increase by a target amount by 2100. The difference between the radiation entering from above and leaving it is known as total radiative forcing. 2.6, 4.5, 6.0, and 8.5 Watts per square meter (Wm^{-2}) are set radiative forcing targets for 2100 to predict from a wide range of possible future scenarios, and each pathway leads to a different range of global mean temperature increase over twenty-first century (see Table 11.1) (UK Met Office, 2018).

Table 11.1 RCP, UK Met Office (2018)

RCP	Change in temperature (°C) by 2081–2100
RCP2.6	1.6 (0.9–2.3)
RCP4.5	2.4 (1.7–3.2)
RCP6.0	2.8 (2.0–3.7)
RCP8.5	4.3 (3.2–5.4)

The RCP pathways are neither forecasts nor suggestions for policies but rather a wide range of potential climatic outcomes. Numerous presumptions about population expansion, economic progress, technological advancement, and environmental and social sustainability perspectives are included. A mix of various socioeconomic presumptions can satisfy each pathway.

11.4.2 SSP Pathway

The other main factor introduced by O'Neill et al. (2014) was Shared Socioeconomic Pathways—SSPs, which were developed to provide a system for researchers to carry out integrated analyses. They are reference pathways that assume that there would be no climate impact or change. Five SSP storylines present radically different perspectives on contemporary society and the implied difficulties of addressing climate change mitigation and adaptation (Lehtonen et al., 2021). The scenarios are SSP1: Sustainability (Taking the Green Road), SSP2: Middle of the Road, SSP3: Regional Rivalry (A Rocky Road), SSP4: Inequality (A Road divided), and SSP5: Fossil-fueled Development (Taking the Highway) see Table 11.2 for details.

SSP is divided into two parts: essential and extended. Basic SSPs include nine elements such as demographics, well-being, economic development, environmental and ecological factors, resources, institutions and governance, technological development, and broader societal factors such as lifestyle, attitudes, and policies (O'Neill et al., 2014).

11.5 Technical Approach to Study

The study has three steps, illustrated in Table 11.3. In step 1, the climate effect on cotton is studied. Following step 2, the study demands understanding other variables, if any, to ascertain the sourcing location for cotton. Finally, in the third step, the factor rating method is applied to conclude the results.

Table 11.2 SSP IPCC (2021)

SSP	Scenario	Estimated warming (2041–2060) (°C)	Estimated warming (2081–2100) (°C)	Very likely range in °C (2081–2100)
SSP1-1.9	**Very low GHG emissions:** CO_2 emissions cut to net zero around 2050	1.6	1.4	1.0–1.8
SSP1-2.6	**Low GHG emissions**: CO_2 emissions cut to net zero around 2075	1.7	1.8	1.3–2.4
SSP2-4.5	**Intermediate GHG emissions**: CO_2 emissions will be around current levels until 2050, then falling but not reaching net zero by 2100	2.0	2.7	2.1–3.5
SSP3-7.0	**High GHG emissions**: CO_2 emissions double by 2100	2.1	3.6	2.8–4.6
SSP5-8.5	**Very high GHG emissions:** CO_2 emissions triple by 2075	2.4	4.4	3.3–5.7

Table 11.3 Procedure and sequence of study

Step 1	Step 2	Step 3
Comparing the effect of climate on cotton availability in all supplying countries	*Incorporating other factors necessary in sourcing/location decision*	*Factor rating method*
• Identifying cotton-producing regions in supplying countries • Probing climatic parameters affecting cotton production • Evaluating climate change in cotton-producing regions of each country and scoring • Comparative analysis among supplying countries to find future cotton availability	• Ascertaining other factors essential for the company in their cotton supply chain • Comparing and evaluating those factors in all supplying countries	• Apply the factor rating method to determine feasible countries to continue with

11.5.1 Scoring Criteria

Since the end objective will be to find a suitable country for the manufacturing company to continue and a non-feasible country with which to stop their cotton-sourcing business, this study will use comparative scoring to analyze and apply the factor rating method. Comparative scoring will be based on a Likert Scale, with one being least suitable and five being most appropriate.

11.6 Data Analysis and Discussions

11.6.1 Step 1

- Comparing the effect of climate on cotton availability: In All Supplying Countries Identifying Cotton Producing Regions in Supplying Countries. Within all five of the manufacturing company's cotton-supplying countries, cotton is grown in specific regions, which must be recognized for working over climatic factors and climate change analysis to find the impact on cotton availability.
- Probing climatic parameters affecting cotton production

  The next step is identifying which specific climatic factors affect cotton production and growth. These factors are critical for analyzing future cotton availability in every region.

  1. Cotton yield is significantly affected by the following major climatic parameters, which have specific ranges critical for cotton production. These ranges are found in different published studies and research referenced below.
  2. Secondly, disaster risk is a highly unfavorable and terminal adverse factor for the cotton supply chain (Chari & Ngcamu, 2019). Disasters, including floods, droughts, storms, etc., are lethal for cotton growth, and thus, the risk of such disasters is incorporated in this study.

As suggested in Table 11.3, the climatic factors are brought forward in Table 11.5. Thus, climate analysis will be based on the following four factors:

Evaluating Climate Change in Cotton Producing Regions Each Country and Scoring: The next step is to analyze the situation of the identified climatic factors table in 11.5 ascertained cotton-producing regions in Table 11.4 of supplying countries. Data on current and future projections related to climatic parameters affecting cotton production is majorly taken from the World Bank Group Climate Change Knowledge Portal, IPCC report, and different climatic web portals.

11.6.1.1 Temperature

The average temperature required for cotton growth is 15–32 °C over multiple periods, including germination, vegetative growth, and fruition of the cotton crops (Cetin & Basbag, 2010). However, the extreme temperature range is a minimum of 15 °C and a maximum of 36 °C (Cetin & Basbag, 2010) see Table 11.6 for current and projected temperature of supply sources.

Summarized temperature data of cotton-producing regions of supplying countries is tabulated in Table 11.7 (World Bank, 2024). Also illustrated in Fig. 11.1 is the forecasted temperature in the future by 2100.

Scoring

Table 11.4 Sourcing countries

Sourcing country	Cotton producing region/province	References
Bangladesh	Chittagong	Banglapedia (2021)
China	a. Xinjiang Uyghur b. Jiangxi c. Jiangsu d. Anhui e. Hubei f. Hebei g. Henan h. Shandong	UN (2002)
India	a. Punjab b. Haryana c. Rajasthan d. Gujarat e. Maharashtra f. MP in the Central zone g. Andhra Pradesh h. Karnataka i. Tamil Nadu j. Orissa	CCI (2021)
Pakistan	a. Punjab b. Sindh	Government of Pakistan (2020)
Vietnam	a. Dak Lak b. Dak Nong c. Gia Lai d. Dong Nai e. Binh Thuan f. Binh Phuoc g. Baria—Vung Tau	Robinson (2009)

Table 11.5 Climate factors

Climatic factors with specific ranges critical to cotton production			
Climatic factor	**Ideal range for cotton production**	**Non-productive extremes**	**References**
Temperature	15 °C during germination 21–27 °C during growth 27–32 °C during fruition	< 15 °C > 36 °C	Cetin and Basbag (2010)
Rainfall	550–1050 mm	< 500 mm > 2000 mm	Freeland et al. (2006)
Relative Humidity	60–80%	< 50%	iKisan (2020)
Factor adverse for cotton growth			
Adverse climatic factor		**Reference**	
Disaster risk		Chari and Ngcamu (2019)	

Table 11.6 Current and projected temperature of sourcing countries

Sourcing country	Cotton producing region/province	Current average annual temperature (°C)		Temperature rise (°C)	Projected temperature by 2100 (°C)
Bangladesh	Chittagong	26.31	26.31	4.2	30.5
China	a. Xinjiang Uyghur Autonomous Region	6.78	15.31	5.0	20.31
	b. Jiangxi	18.9			
	c. Jiangsu	16.4			
	d. Anhui	16.98			
	e. Hubei	16.52			
	f. Hebei	16.38			
	g. Henan	15.68			
	h. Shandong	14.8			
India	a. Punjab	23.75	29.10	4.4	33.50
	b. Haryana	29.13			
	c. Rajasthan	29			
	d. Gujarat	29.95			
	e. Maharashtra	28.95			
	f. MP Central zone	28.71			
	g. Andhra Pradesh	30.63			
	h. Karnataka	28.03			
	i. Tamil Nadu	30.82			
	j. Orissa	32			
Pakistan	a. Punjab	31.01	31.55	5.3	36.85
	b. Sindh	32.09			
Vietnam	a. Dak Lak	25.01	26.99	3.3	30.28
	b. Dak Nong	26.98			
	c. Gia Lai	25.76			
	d. Dong Nai	27.95			
	e. Binh Thuan	27.28			
	f. Binh Phuoc	28.14			
	g. Baria—Vung Tau	27.84			

1. As temperature rises, the annual average temperature is approaching the maximum non-productive extreme for cotton production, above 36 °C.
2. The projected temperature by the year 2100 Pakistan is expected to be highest at 36.85 °C, then India at 33.5 °C, Bangladesh at 30.5 °C, Vietnam at 30.28 °C, and China at the lowest at 2.31 °C.

Table 11.7 Current and projected average temperature

Cotton-growing regions of	Current average annual temperature (°C)	Temperature rise (°C)	Projected temperature by 2100 (°C)
Bangladesh	26.31	4.2	30.5
China	15.31	5	20.31
India	29.1	4.4	33.5
Pakistan	31.55	5.3	36.85
Vietnam	26.99	3.3	30.28

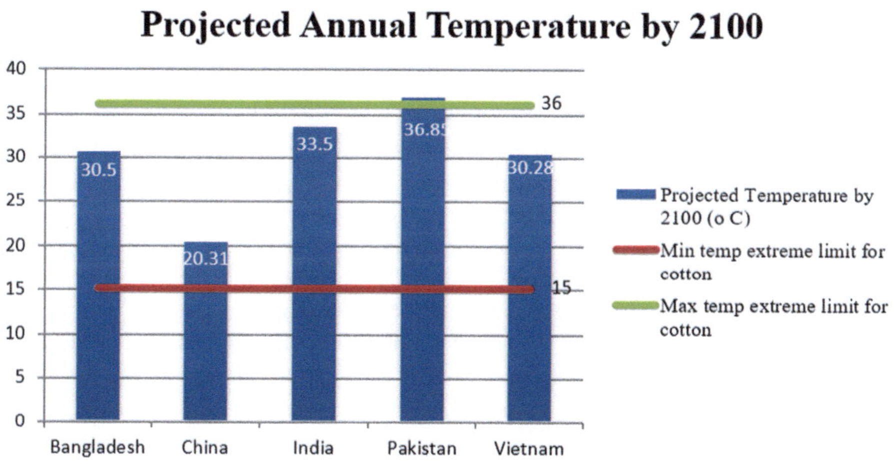

Fig. 11.1 Projected annual temperature by 2100 (World Bank, 2024)

3. This implies Pakistan, being the least suitable, scored 1, India 2, Bangladesh 3, Vietnam 4, and China 5 in terms of temperature. Hence, the results are illustrated in Fig. 11.2;

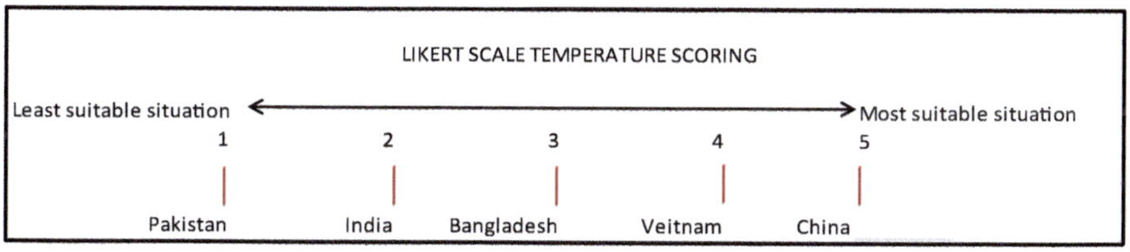

Fig. 11.2 Scoring of countries based on temperature

11.6.1.2 Rainfall

For cotton production, a minimum of 500 mm of rainfall is required. It grows well with annual average rainfall above 550 mm (up to 2000 mm). However, cotton does not grow well in heavy rain (Heuzé & Tran, 2015). Exceptionally high rainfall can result in over-saturation and waterlogging (Pandya et al., 2020). Rainfall data was collected for all five supplying countries and tabulated in Table 11.8.

Summarized rainfall data of cotton-producing regions of supplying countries is tabulated in Table 11.9.

Graphically, rainfall projected data for the end of the century is plotted against minimum and maximum rainfall extremes to analyze the situation and score accordingly in Fig. 11.3 visually.

Scoring (Fig. 11.4)

1. Analysis shows that Pakistan's cotton-growing region, even after the rainfall increase until 2100, will still receive less than the minimum annual rainfall requirement of 500 mm. Thus, drip irrigation will continue to fulfill the deficiency, scoring 1 (least feasible) on the Likert Scale.
2. Vietnam's annual rainfall in cotton-producing regions is expected to surpass the maximum rainfall level, scoring 2. Similarly, Bangladesh's rainfall is expected to cross the upper extreme, scoring three very slightly.
3. The situation of India and China is very suitable, scoring 4 and 5, respectively.

11.6.1.3 Relative Humidity (RH)

For cotton production, relative humidity should be between 60 and 80%, with an ideal 65% + 5% range. Below 50% RH, cotton production is not possible; however, drip irrigation (as used in Pakistan to overcome lower rainfall consequences) can also maintain humidity. RH data were collected for all five supplying countries and tabulated below in Table 11.10.

RH is dropping in all five countries with time except Pakistan. Since Pakistan is currently much below the ideal RH range, rising RH is favorable for Pakistan, although it will still need to lead Pakistan to the perfect range (see Table 11.11). In India, a few cotton-producing regions at higher altitudes will face RH rise (Willett, 2020), while RH will decrease in a few areas. On average, the RH of cotton-producing regions of India is rising to a small extent (see Fig. 11.5).

Scoring (Fig. 11.6)

1. Future projections depict that Pakistan will remain below the suitable relative humidity range; however, it has high moisture relative to temperature, and the rise in humidity is low (moisture-containing capability increases with temperature). Thus, suitable relative humidity will only be reached within the minimum requirement. Hence, Pakistan scored 1 (least suitable) because, at times, it could need drip irrigation to fulfill its needs.

Table 11.8 Current and projected rainfall of sourcing countries (World Bank, 2024)

Sourcing country	Cotton producing region/province	Current average annual rainfall (mm)		Projected annual rainfall by year 2100 (mm)	
Bangladesh	Chittagong	1565	1565	2021	2021
China	a. Xinjiang Uyghur Autonomous Region	255	1074	318	1243
	b. Jiangxi	1744		1916	
	c. Jiangsu	1229		1360	
	d. Anhui	1430		1628	
	e. Hubei	1425		1687	
	f. Hebei	666		862	
	g. Henan	1043		1212	
	h. Shandong	801		962	
India	a. Punjab	649	928	740	1058
	b. Haryana	617		703	
	c. Rajasthan	675		770	
	d. Gujarat	1107		1262	
	e. Maharashtra	901		1027	
	f. MP Central zone	1017		1159	
	g. Andhra Pradesh	1094		1247	
	h. Karnataka	731		833	
	i. Tamil Nadu	998		1138	
	j. Orissa	1489		1697	
Pakistan	a. Punjab	164	122	254	213
	b. Sindh	79		172	
Vietnam	a. Dak Lak	1972	2005	2183	2142
	b. Dak Nong	2108		2176	
	c. Gia Lai	1819		2020	
	d. Dong Nai	2042		2157	
	e. Binh Thuan	2031		2138	
	f. Binh Phuoc	2074		2184	
	g. Baria—Vung Tau	1989		2133	

2. Since the ideal relative humidity value is 65 + 5, the countries near this value scored high, and the countries farther away scored low. Hence, on the Likert scale, Bangladesh scored 2, Vietnam 3, India 4, and China 5 (most suitable).

Table 11.9 Current and projected average rainfall (World Bank, 2024)

Cotton-growing regions of	Current average annual rainfall (mm)	Projected temperature by year 2100 (mm)
Bangladesh	1565	2021
China	1074	1243
India	928	1058
Pakistan	122	213
Vietnam	2005	2142

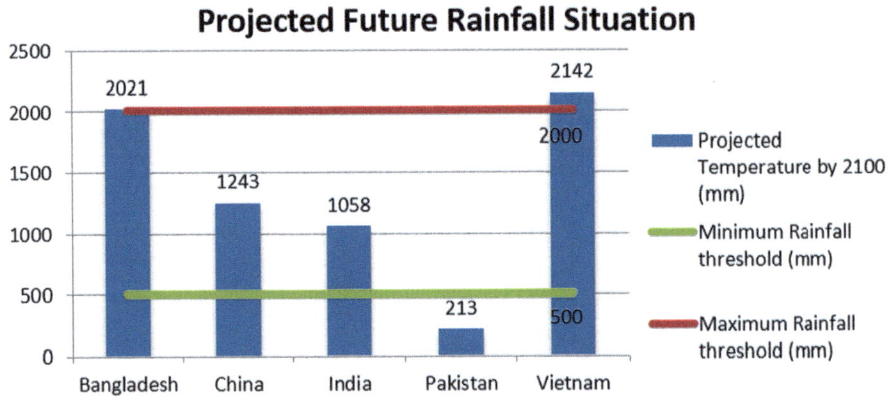

Fig. 11.3 Projected future rainfall (World Bank, 2024)

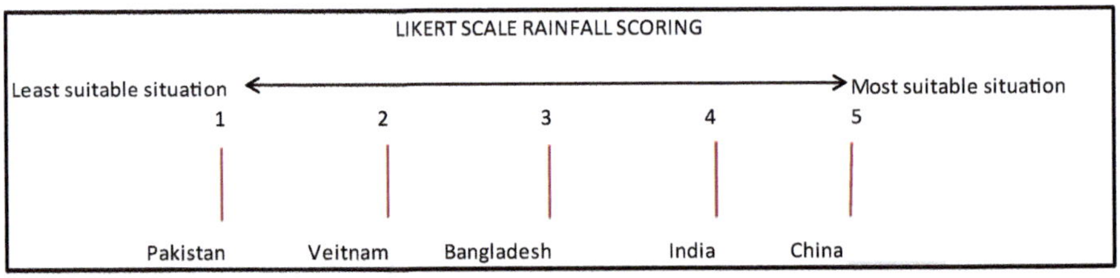

Fig. 11.4 Scoring of countries based on rain fall

11.6.1.4 Disaster Risk

Studies have found that disasters include floods, droughts, storms, and hazards for cotton crops (Chari & Ngcamu, 2019). As per the United Nations Office for Disaster Risk Reduction, developing and developed countries continue to face new and emerging hazards that magnify disaster risks with devastating effects (UNISDR, 2013).

Table 11.10 Current and projected relative humidity (World Bank, 2024)

Sourcing country	Cotton producing region/province	Current average relative humidity (%)		Projected relative humidity by 2100 (%)	
Bangladesh	Chittagong	80.0	80.0	76.9	76.9
China	a. Xinjiang Uyghur Autonomous Region	46.3	64.9	45.7	62.4
	b. Jiangxi	73.5		70.2	
	c. Jiangsu	70.6		68.5	
	d. Anhui	70.2		68.4	
	e. Hubei	73.4		70.8	
	f. Hebei	55.9		52.6	
	g. Henan	65.2		62.5	
	h. Shandong	63.8		60.3	
India	a. Punjab	57.2	61.6	60.4	61.1
	b. Haryana	53.6		57.0	
	c. Rajasthan	46.9		49.6	
	d. Gujarat	61.5		63.4	
	e. Maharashtra	63.0		60.8	
	f. MP Central	57.7		55.2	
	g. Andhra Pradesh	64.9		62.6	
	h. Karnataka	69.6		67.3	
	i. Tamil Nadu	71.3		69.7	
	j. Orissa	70.0		65.3	
Pakistan	a. Punjab	51.1	51.6	53.2	53.8
	b. Sindh	52.2		54.4	
Vietnam	a. Dak Lak	76.8	78.3	72.8	75
	b. Dak Nong	76.7		72.2	
	c. Gia Lai	73.6		68.9	
	d. Dong Nai	80.5		77.8	
	e. Binh Thuan	80.8		78.3	
	f. Binh Phuoc	76.5		72.4	
	g. Baria—Vung Tau	82.9		81.4	

For this purpose, the INFORM Risk Index/Ranking is used worldwide to identify specific risks across a country to support decisions on prevention, preparedness, response, and overall risk management (World Bank, 2024). INFORM risk ranking is a global, open-source risk assessment performed for humanitarian crises and disasters, enabling countries and businesses to make decisions about prevention, preparedness, and response.

Table 11.11 Current and average relative humidity (RH) (World Bank, 2024)

Cotton producing region of	Current average RH (%)	Projected RH by 2100 (%)
Bangladesh	80	76.9
China	64.9	62.4
India	61.6	61.1
Pakistan	51.6	53.8
Vietnam	78.3	75

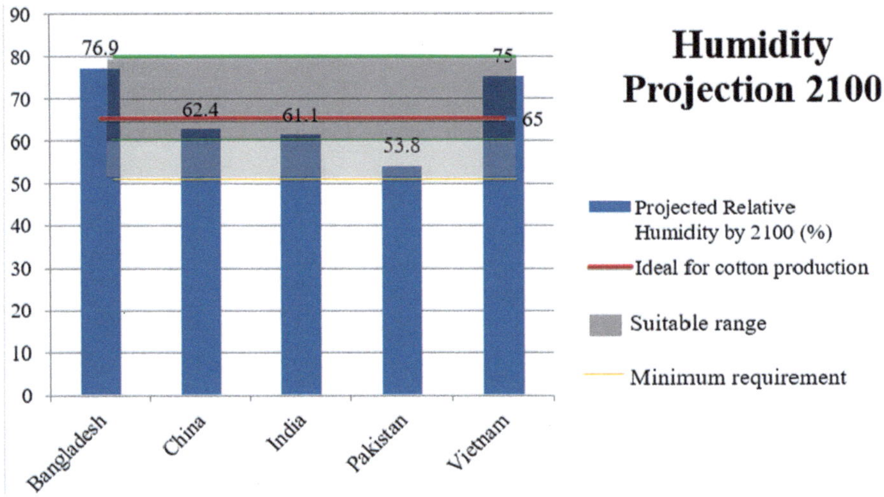

Fig. 11.5 Humidity projection 2100 (World Bank, 2024)

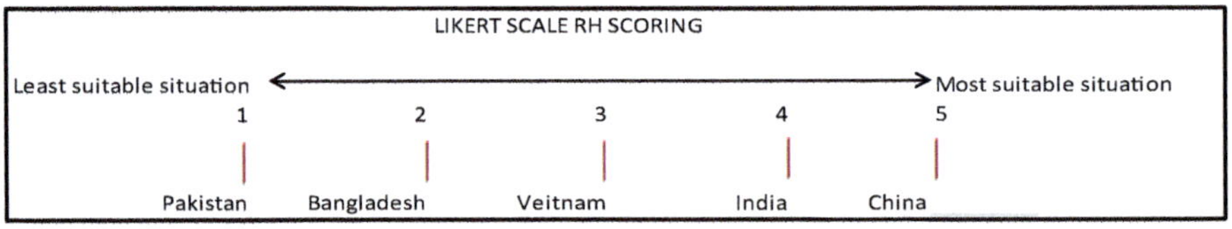

Fig. 11.6 Scoring of countries based on humidity

INFORM Risk ranking caters to floods, cyclones, droughts, and a country's lack of coping capability to find an overall risk level and score a country according to the associated risk. For example, China scored 71st most dangerous out of 191 countries in terms of disaster ranking by INFORM ranking, as shown in Table 11.12.

Similarly, this study uses the INFORM Risk Ranking of all five sourcing countries for comparative analysis (see Table 11.13).

Scoring (Fig. 11.7)

Table 11.12 Inform risk ranking (World Bank, 2024)

Flood (0–10)	Tropical cyclone (0–10)	Drought (0–10)	Vulnerability (0–10)	Lack of coping capacity (0–10)	Overall inform risk level (0–10)	Rank (1–191)
8.4[4.5]	8.1[1.7]	4.6[3.2]	3.1[3.6]	3.6[4.5]	4.3[3.8]	71

Table 11.13 Risk ranking of sourcing countries

Sourcing country	INFORM disaster risk ranking (Out of 191 countries)	Source
Bangladesh	22	World Bank (2024)
China	71	
India	32	
Pakistan	18	
Vietnam	91	

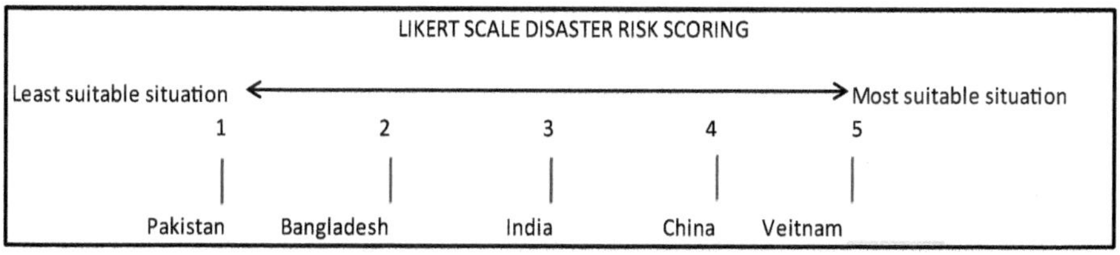

Fig. 11.7 Scoring of countries based on disaster risk

Likert Scale Scoring for Disaster Risk is done by assigning a score of 1 (least suitable) to the country with the highest risk rank among all five countries (i.e., Pakistan with rank 18). Bangladesh scored 2nd and so on as in Fig. 11.7.

11.6.1.5 Comparative Analysis to Identify Cotton Supply Source

All the scores assigned at each stage against all four climatic factors above are consolidated and averaged to find each country's comparative aggregate scoring in Table 11.14.

This shows that Pakistan is expected to face the worst impact on cotton production due to its worst climate change situation, having an aggregate rating of 1 (least suitable) among all five countries. On the other hand, China is in the most appropriate climatic condition, with an average rating of 5 (most relevant). Thus, the impact of climate change will ultimately be quite bearable for cotton crops (see Fig. 11.8).

Conclusively, this depicts that the future cotton availability in Pakistan is most uncertain, whereas it is most promising in China.

Table 11.14 Comparative analysis

Climate factor	Likert scale score				
	Bangladesh	China	India	Pakistan	Vietnam
Temperature	3	5	2	1	4
Rainfall	3	5	4	1	2
Humidity	2	5	4	1	3
Disaster risk	2	4	3	1	5
Average rating (Resource availability—lowest impact of climate change)	2	5	3	1	4

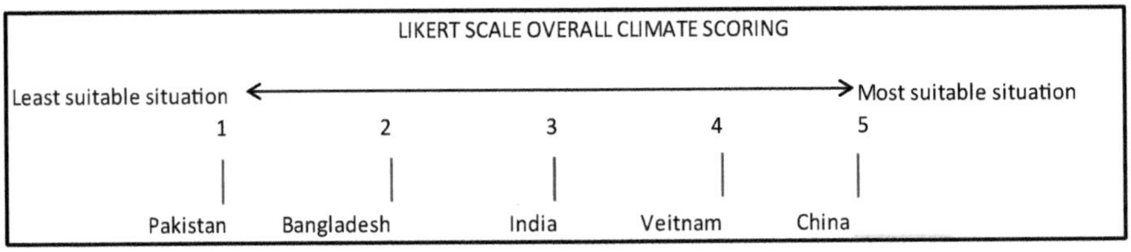

Fig. 11.8 Scoring of countries based on climatic factors

11.6.2 Step 2

Incorporating Other Factors Necessary in Sourcing/Location Decision

Although the impact of climate change on cotton production in all countries has found which countries will be suitable for cotton availability, the manufacturing company cannot decide based on climate change's resulting cotton availability while ignoring other necessary factors. Hence, as declared essential by the company, two other significant factors have been included in this research: price and Distance or Proximity to the Market.

11.6.2.1 Price

1. Since the price of cotton products is directly linked to the price of cotton itself, the current price per kg of cotton in each country has been found.
2. Future prices are projected based on each country's average inflation rate over the last five years.

See Table 11.15 (the prices are extracted from selected country commodity news in 2022).

This shows that Pakistan has the lowest cotton price among all five countries due to the highest average inflation rate in the last five years. However, assuming the inflation rates will remain unchanged, they are expected to increase to the highest

Table 11.15 Current and projected cotton price of sourcing countries

Sourcing country	Cotton price/kg		Inflation rate (%)	Forecasted price/kg (by 2100)*
	(local currency)	(USD)	(2018–2021)	(USD)
Bangladesh	৳ 275.43	2.9	5.5925	191.49
China	¥12.15	1.79	2.0925	8.82
India	₹87.59	1.1	4.855	42.34
Pakistan	PKR 175	0.82	8.725	514.26
Vietnam	₫35,109.73	1.5	2.8475	13.03

*Assuming the same inflation rate as calculated from the average of each country's inflation rate in the last five years

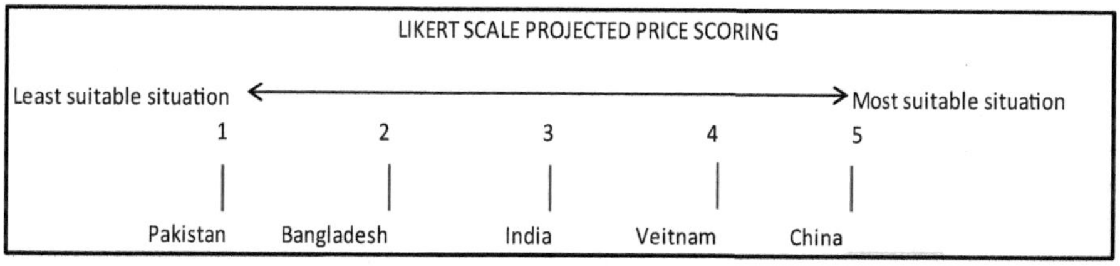

Fig. 11.9 Scoring of countries based on pricing

level by the end of the century. Likert Scoring based on Future Projected Price shown in Fig. 11.9.

11.6.2.2 Proximity to Market/Distance

Proximity to market is figured out in terms of the distance of the supplying country to the destination where cotton has to be sent, i.e., Europe. Distance is observed from the World Map and scores comparatively. Vietnam, being the farthest from Europe, scored 1, while Pakistan, being the nearest, scored 5 (most suitable) in Fig. 11.10.

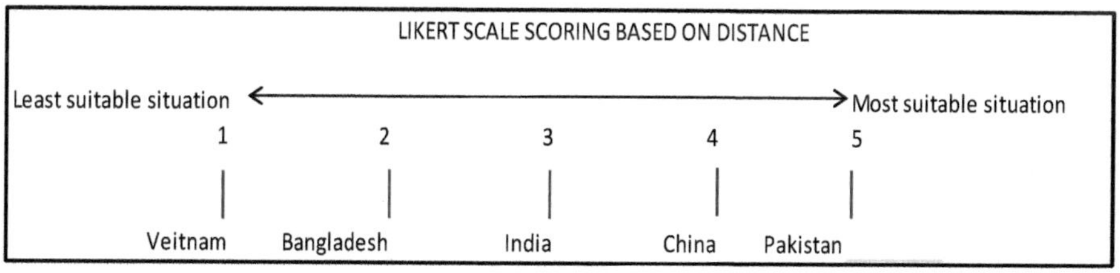

Fig. 11.10 Scoring of countries based on distance

11.6.3 Step 3

Factor Rating Method

The factor rating method is applied to conclusively reach a final decision based on scores calculated on all three factors, including:

1. Availability of cotton (in terms of climate change impact affecting cotton production, calculated in Figs. 11.2, 11.4, 11.6, 11.7, and 11.8).
2. Price (see Fig. 11.9).
3. Distance or Proximity to Market (see Fig. 11.10).

Each factor is assigned weightage based on its importance to the company/ Manufacturing Company Group see Table 11.16.

11.7 Conclusion

This study analyzed and compared the severity of climate change and its impact on cotton production in all the cotton-supplying countries of the manufacturing company Group, including India, China, Vietnam, Pakistan, and Bangladesh. Besides cotton availability projected based on climate change, two other factors include price and proximity to the market. The results demonstrated that China is the most favorable country for the manufacturing company to continue cotton sourcing business with, then Vietnam, followed by India, then Bangladesh, and the least favorable being Pakistan, where the impact of climate change is highest.

Based on Pakistan's worst projected climate conditions among all five countries, it is high time the Pakistan government considered severe actions. Rising temperatures, erratic rainfall, greater intensity of changing seasons, and extreme events—floods and droughts are all characteristics of climate change in Pakistan. Cotton production needs drip irrigation in Pakistan, and it is depicted that climate change may impact irrigation water availability, negatively impacting crop productivity (Raza & Ahmad,

Table 11.16 Factor rating analysis

Factors	Weightage (by the company) (%)	Likert scale scores calculated in previous sections				
		Bangladesh	China	India	Pakistan	Vietnam
Cotton availability (Projected through the impact of climate change)	50	2	5	3	1	4
Price	35	2	5	3	1	4
Proximity to market	15	2	4	3	5	1
Total score		2.0	4.7	3.1	1.6	3.3

2017). Water levels are also rising because of the melting of the Himalayas' glaciers, leaving the country prone to devastating floods (Chaudhary & Clark, 2022).

Climate change significantly impacts industrial ecology, affecting various industrial systems, processes, and resource management. More frequent and severe extreme weather events, such as hurricanes, floods, and droughts, can damage industrial infrastructure, disrupt operations, and lead to production delays. These events can also cause supply chain disruptions by damaging transportation networks and affecting the availability of critical inputs. Climate change increases the risks of disruptions to global supply chains, necessitating greater resilience and adaptation strategies. Industries may diversify suppliers, invest in redundancy measures, and adopt flexible manufacturing processes to mitigate the impacts of climate-related disruptions on their operations. Climate change poses complex challenges for industrial ecology, requiring industries to adopt sustainable practices, enhance resilience, and innovate to mitigate risks and capitalize on emerging opportunities in a changing climate.

A few recommendations have been proposed for the manufacturing company to optimize their sourcing and for Pakistan to combat its critical devastating situation are proposed below:

11.7.1 Study Recommendations

11.7.1.1 Sourcing Company

1. It is recommended that the sourcing company periodically conduct such analysis every 10–20 years to optimize its sourcing business amidst the prevailing climate, supply chain, and global economy.
2. The sourcing company should continue exploring other sourcing countries that may provide them with long-term benefits. Cotton-producing countries like Turkey and Brazil may also be rationally analyzed against prices, geographical suitability, climate change, and availability for cotton product sourcing.

11.7.1.2 Countries Affected by Climate Change

1. Carbon footprint is the primary cause of global warming and the greenhouse effect, leading to ozone depletion and temperature rise. The government and authorities should focus on research and development to improve their industrial ecology and bring sustainability to industrial ecosystems.
2. Another reason for climate change is deforestation and the increased use of chemicals in agriculture. The government should increase plantation drives to combat deforestation, and agricultural authorities should find alternatives to agrarian chemicals (Azeem, 2019).

3. Coal, oil, and gas continue to power much of our electricity and heating. Airplanes and automobiles are also primarily powered by fossil fuels. Reduce your carbon footprint by using less energy at home, switching to a wind or solar energy provider, avoiding long-distance flights, and driving less.
4. All food production, processing, transportation, consumption, and disposal contribute to greenhouse gas emissions. To reduce carbon footprints, buy local and seasonal food and eat more plant-based meals.
5. The agriculture extension department should be responsible for providing timely and on-farm information about climate change.
6. The government should also provide cutting-edge agricultural technology to increase crop yields while remaining cost-effective in the long run (Rashid et al., 2020).
7. Ocean acidification has increased by 30% since the Industrial Revolution, ultimately leading to and negatively impacting crop yield production. Conscious industrial waste, effluent disposal, and necessary dilution are essential (Hafiz et al., 2019).

References

Achparaki, M., Thessalonikeos, E., Tsoukali, H., Mastrogianni, O., Zaggelidou, E., Chatzinikolaou, F., Vasilliades, N., & Raikos, N. (2012). Heavy metals toxicity. *Aristotle University Medical Journal, 39*(1), 29–34.

Azeem, H. M. (2019). Climate change: Causes, outcomes in Pakistan and a way forward. *Daily Times.*

Chaudhary, A., & Clark, A. (2022). *Melting Himalayan glaciers are making Pakistan's floods worse.* Bloomberg.

Banglapedia. (2021). *Cotton.* Banglapedia.

Better Cotton. (2020). *Cotton farming and Covid-19: On the ground in Pakistan.* Better Cotton.

CCI. (2021). *The cotton corporation of India (CCI).* The Cotton Corporation of India Ltd.

Cetin, O., & Basbag, S. (2010). Effects of climatic factors on cotton production in semi-arid regions—A review. *Research on Crops, 11*(3), 785–791.

Chapman, A., Davies, W., Downey, C., & Dookie, D. (2021). ADB climate risk country profile: Micronesia.

Chari, F., & Ngcamu, B. S. (2019). The impact of disaster risks on the economic sustainability of cotton supply chains: Evidence from Chiredzi District, Zimbabwe. *Cogent Social Sciences, 5*(1), 1625742.

Chester, M. V. (2020). Industrial ecology in support of climate change adaptation. *Journal of Industrial Ecology, 24*(2), 271–275. Wiley Online Library.

EPA. (2021). *Value of cotton crops.* EnviroAtlas: Led by the U.S. Environmental Protection Agency.

Freeland Jr, T. B., Pettigrew, B., Thaxton, P., & Andrews, G. L. (2006). *Agrometeorology and cotton production.* World Meteorological Organization.

Government of Pakistan. (2020). *Cotton.* The Pakistan Central Cotton Committee (PCCC).

Gray, E. (2021). *Global climate change impact on crops expected within 10 years,* NASA Study Finds. Global Climate Change.

Habib-ur-Rahman, M., Ahmad, A., Raza, A., Hasnain, M. U., Alharby, H. F., Alzahrani, Y. M., Bamagoos, A. A., Hakeem, K. R., Ahmad, S., & Nasim, W. (2022). Impact of climate change on agricultural production; Issues, challenges, and opportunities in Asia. *Frontiers in Plant Science, 13*, 925548.

Heuzé, V., & Tran, G. (2015). *Cotton (general)*. Feedipedia.

Hughes, K. (2021). *What is next for the cotton sector after Covid-19?*

iKisan. (2020). *Cotton*. IKisan .

Imran, M. A., Ali, A., Ashfaq, M., Hassan, S., Culas, R., & Ma, C. (2018). Impact of Climate Smart Agriculture (CSA) practices on cotton production and livelihood of farmers in Punjab, Pakistan. *Sustainability, 10*(6), 2101.

IPCC. (2021). Summary for policymakers.

Jaffar Iqbal, M. F., Roman, M., Magsi, S., Rauf, H., Arif, Z., Khaliq, T., & Tahir, M. (2023). 15. Climate change effects on cotton planting date and planting density using modelling techniques. *Pure and Applied Biology (PAB), 12*(1), 732–753.

Lehtonen, H. S., Aakkula, J., Fronzek, S., Helin, J., Hildén, M., Huttunen, S., Kaljonen, M., Niemi, J., Palosuo, T., & Pirttioja, N. (2021). Shared socioeconomic pathways for climate change research in Finland: Co-developing extended SSP narratives for agriculture. *Regional Environmental Change, 21*, 1–16.

Malhi, G. S., Kaur, M., & Kaushik, P. (2021). Impact of climate change on agriculture and its mitigation strategies: A review. *Sustainability, 13*(3), 1318.

Morton, J. F. (2007). The impact of climate change on smallholder and subsistence agriculture. *Proceedings of the National Academy of Sciences, 104*(50), 19680–19685.

Moss, R. H., Edmonds, J. A., Hibbard, K. A., Manning, M. R., Rose, S. K., Van Vuuren, D. P., Carter, T. R., Emori, S., Kainuma, M., & Kram, T. (2010). The next generation of scenarios for climate change research and assessment. *Nature, 463*(7282), 747–756.

Nadiruzzaman, M., Rahman, M., Pal, U., Croxton, S., Rashid, M. B., Bahadur, A., & Huq, S. (2021). Impact of climate change on cotton production in Bangladesh. *Sustainability, 13*(2), 574.

O'Neill, B. C., Carter, T. R., Ebi, K., Harrison, P. A., Kemp-Benedict, E., Kok, K., Kriegler, E., Preston, B. L., Riahi, K., & Sillmann, J. (2020). Achievements and needs for the climate change scenario framework. *Nature Climate Change, 10*(12), 1074–1084.

O'Neill, B. C., Kriegler, E., Riahi, K., Ebi, K. L., Hallegatte, S., Carter, T. R., Mathur, R., & Van Vuuren, D. P. (2014). A new scenario framework for climate change research: The concept of shared socioeconomic pathways. *Climatic Change, 122*, 387–400.

Ojeaga, P. I., & Posu, S. M. (2017). Climate change, industrial activity, and economic growth: A cross-regional analysis. *Global Economic Observer, 5*(2), 7–17.

Pandya, P. A., Ghosiya, S. M., Pithiya, V. H., & Dudhatra, S. P. (2020). Effect of rainfall on productivity of cotton. *Emergent Life Sciences Research, 6*, 56–63.

Rashid, M., Husnain, Z., & Shakoor, U. (2020). Impact of climate change on cotton production in Pakistan: An ARDL bound testing approach. *Sarhad Journal of Agriculture, 36*(1).

Raza, A., & Ahmad, M. (2017). Analysing the impact of climate change on cotton productivity in Punjab and Sindh, Pakistan. *Research Journal Social Sciences, 6*(1).

Robinson, J. (2009). Booming textiles make Vietnam hungry for imports.

Sawan, Z. M., Hanna, L. I., El Karim, G. A. G., & McCuistion, W. L. (2002). Relationships between climatic factors and flower and boll production in Egyptian cotton (Gossypium barbadense). *Journal of Arid Environments, 52*(4), 499–516.

Solomon, S. (2007). *Climate change 2007-the physical science basis: Working group I contribution to the fourth assessment report of the IPCC* (Vol. 4). Cambridge University Press.

Song, X., & Ye, C. (2017). Climate change adaptation pathways for residential buildings in southern China. *Energy Procedia, 105*, 3062–3067.

Soviadan, M. K. (2019). Impact of climate change on cotton production: Case of Savannah region, Northern Togo. In M. K. Soviadan, E. M. Koffi-Tessio, A. A. Enete, & N. J. Nweze (Eds.), *Impact of climate change on cotton production: Case of Savannah region, Northern Togo. Agricultural Sciences, 10*, 927–947.

UK Met Office. (2018). *UKCP18 guidance: Representative concentration pathways.*

UN. (2002). *Integrated assessment of trade liberalization and trade-related policies a country study on the cotton sector in China.*

UNISDR. (2013). *Disaster risk reduction in the United Nations.*

Volik, B. G., Mitelman, M. G., & Durnev, V. N. (2004). *On the concept of industrial safety.* Abstracts of the Pacific Basin Nuclear Conference, 193.

Wadanambi, R. T., Wandana, L. S., Chathumini, K., Dassanayake, N. P., Preethika, D. D. P., & Arachchige, U. S. (2020). The effects of industrialization on climate change. *Journal of Research Technology and Engineering, 1*(4), 86–94.

Waddle, B. A. (1984). Crop growing practices. *Cotton, 24*, 233–263.

Willett, K. Dr. (2020). *Investigating climate change's 'humidity paradox.'* World Economic Forum.

World Bank. (2024). Climate Change Knowledge Portal. https://climateknowledgeportal.World Bank.org/country/pakistan/trends-variability-projections#country-map

Chapter 12
Industrial Ecology in Food Supply Chain

Industrial ecology in the food supply chain is a holistic approach that aims to optimize resource use and minimize waste through sustainable practices. It involves analyzing and improving the flow of materials and energy through food products' production, processing, distribution, consumption, and disposal systems. The concept integrates principles from industrial ecology with the food supply chain to create a more sustainable and efficient system. This includes practices like waste minimization, Industrial symbiosis, and life cycle assessment. Based on the starting statement, the chapter examines the fundamental factors contributing to inefficiencies in the food supply chain in developing countries, intending to devise approaches to establish a more resilient and sustainable system. This investigation offers insights into the fundamental challenges that impact the food supply chain. An exhaustive literature review is done before conducting the study to establish a theoretical foundation. By employing a qualitative research approach, this study identifies barriers to a sustainable food supply through unstructured interview questions with critical stakeholders and rice processing mills in Pakistan.

Furthermore, a proposed structure for the food supply chain procedure offers pragmatic resolutions to augment circularity. The barriers to attaining a closed-loop food supply chain are assessed in rice processing mills. By utilizing a decision-making approach, conclusions are deduced about the obstacles encountered, focusing on the food industry. The study presented in this chapter is a comprehensive examination of the complexities inherent in the food supply chain, elucidating the origins and consequences of bottlenecks while furnishing a strategic blueprint for establishing a circular and sustainable food economy in developing countries in the context of industrial ecology.

A. Shah et al., *Industrial Ecology*, Industrial Ecology,
https://doi.org/10.1007/978-981-97-3619-5_12

12.1 Circular Economy

The entire amount of trash created includes the wasteful disposal of edible items across the supply chain. According to estimates, one-third of the world's edible food supply (about 1.3 billion tons) is wasted or lost around the globe. Growing food waste is therefore becoming a concern for environmental governance and global food security, with consequential ecological, monetary, and social effects (Farooque et al., 2019).

In response to these pressing waste management and environmental concerns, the government has implemented several legislative, regulatory, and financial steps to bolster its circular economy (CE) initiative. Restorative and regenerative design concepts under the domain of CE make it an industrial system that is more resilient than the conventional, primarily linear economic model, which helps us in terms of make, use, and disposal (Baig et al., 2020).

In addition to transportation, information, and money, the supply chain also involves the movement of product-based knowledge. It is important to note that suppliers engage in more than just production and manufacturing activities. They also engage in various forms of knowledge transfer and exchange with their clients. Knowledge management's ability to boost the digital performance of the supply chain is a vital part of the industry's ongoing digital transformation in light of today's rapid digital development (Leigh & Li, 2015).

In today's more competitive and global corporate climate, it is getting harder to rein in the adverse effects of wasteful and ecologically destructive consumption and manufacturing habits. The concept of a CE emerged in response to the rising need to reduce the link between economic growth and negative impacts on the environment and scarce natural resources. CE is supposed to improve the environment and the economy since it re-purposes materials or product components and reduces waste by prolonging the life of things (Kayikci et al., 2022).

The CE is frequently called a closed-loop supply chain that prioritizes repair and renewal. For the industrial system to embrace the concept of "end-of-life" with restoration, to get rid of hazardous materials, and to cut down on waste across the supply chain, it must explicitly use circular practices such as eco-product design and circular procurement. As a result, the core mission of the CE is to improve resource use and contribute to more positive environmental outcomes throughout the whole supply chain (Gebhardt et al., 2022).

12.2 Circular Food Supply Chain

Managing the waste produced by society through the use of a circular food business model, where a guide regarding how the waste is reused and recycled to generate value-added products to meet the current demands of society, has been identified as a crucial and potentially lucrative step toward achieving sustainable development by several studies (Papargyropoulou et al., 2014). The concept behind the zero-waste regeneration system known as the CE is that a firm's garbage may be recycled and used as a resource by another enterprise. By prolonging the lifespan of resources and items already in circulation, the CE decreases the need to mine for raw materials and the accumulation of trash in landfills. The CE approach has drawn much interest from businesses and government officials due to its foundation on promises of conserving the environment and raising GDP (Batista et al., 2017).

According to proponents of the green product movement, these goods have incrementally positive impacts on the long-term survival of firms and play a part in the more significant endeavor to create sustainable competitive advantages in the market (Mehmood et al., 2021). Environmentally friendly design and operations can help prolong a business's viability by lessening its adverse effects on the planet. Eco-design has been shown to have a direct and significant relationship to social environmental sustainability after an economic analysis of several aspects of environmental performance (Farooque et al., 2019). Meanwhile, waste creation should be planned out of the supply chain system. However, a collecting system must be in place to recover resources from the trash that is inevitably produced at each step in the supply chain, from agricultural production to post-harvest handling and storage to processing to distribution to consumption (Baig et al., 2020).

The research aims to understand Pakistan's obstacles in pursuing a sustainable and circular food supply chain, among other things. The study intends to discover and examine the challenges to realizing a sustainable model by exploring the complex network of factors impacting the nation's food supply chain. More than just an analysis, the study hopes to set a precedent for creating a food system that can withstand the test of time. Also, the research aims to support the field by building a thorough framework based on different organizational theories. By illuminating the problems and reasons for the failed integration of food supply chain management, this framework hopes to help Pakistan move towards a more sustainable and circular food supply chain by developing strategic interventions.

12.3 Industrial Ecology in Food Supply Chain

All the steps in a supply chain can harm the environment, but Sustainability Supply Chain Management (SSCM) works to mitigate such effects. When a central company works with its suppliers and consumers to boost the ecological performance of its goods, processes, services, and supply chains, that's the first step toward creating ecologically sustainable supply chains (Leigh & Li, 2015; Lutfi et al., 2022). The term "reverse logistics" refers to the procedure through which customers return items to the central business. A company's disposal responsibilities have traditionally been limited to unsold inventory. The goal of reverse logistics is for the central business to recoup or responsibly dispose of the returned goods while creating value where feasible (McLachlan, 2021).

When a buyer sends back an item they no longer desire, the reverse logistics process might lead to its resale or usage as a warranty replacement. Expanding on the idea of supply chains in place, reverse logistics has the potential to boost economic output. To be successful, reverse logistics must have buy-in from customers and upper management, clear and consistent communication, reasonable prices, a formal structure, strategic scheduling of activities, and an eye toward environmental impact (Baig et al., 2020; Farooque et al., 2019).

People generally believe in and work hard to implement the sustainable development agenda. This study examines and evaluates these four notions, highlighting their similarities, distinguishing characteristics, and connections. Before going over the four concepts—material flows, the temporal domain, related system limits, the actor-network, and the analysis and comparison—a few observations on sustainable development and a quick evaluation of pertinent research criteria will be given (Leigh & Li, 2015).

Ensuring no food goes to waste at any point in the supply chain, even at the factory, is a pressing concern for businesses today. To solve this problem, the modern food business needs dependable and powerful instruments to track food from the farm to the processing facility and the client, ensuring complete transparency along the whole supply chain.

12.4 Barriers to Circular Food Supply Chain

As per the UN Food and Agriculture Department, around 1/3rd of the world's food supply is either misplaced or thrown away, and this is well recognized as a significant source of landfill trash. A more sustainable food supply chain and proper CE are necessary to eliminate these waste problems (Baig et al., 2020). Several obstacles prevent supply chain management from fully embracing the CE. CE implementation is hindered by many factors, such as inadequate funding, a lack of necessary skills and knowledge, and inadequate infrastructure, especially in a nation like Pakistan (Yadav et al., 2020).

Weak environmental legislation and enforcement, a lack of economies of scale, poor organizational culture, and leadership all contribute to skepticism about the positive effects of a CE. These issues contribute significantly to Pakistan's circular food supply chain, facing several obstacles (Farooque et al., 2019). Integrating the CE into the SC and building a circular food supply chain can help countries worldwide overcome these obstacles and take strong measures to minimize losses of this kind (Gebhardt et al., 2022).

12.5 Research Approach

This section discussed the research strategy and methods that were employed here. This study is qualitative and exploratory in nature. With the knowledge gained from this study, it will be better able to advise food processing companies on minimizing food waste by eliminating the bottlenecks in their supply chain.

Several companies in Pakistan use the circular food supply chain. In this respect, the authors contacted each company's single point of contact. Subsequently, they interviewed that individual in the presence of additional department heads and officials from procurement, operations, marketing, and sales. Interviews are done with the highest-ranking executives, managers, supply-chain chiefs, and business owners.

However, our prioritized company, the food processing company, has given authentic sources of information related to the food supply chain and causes of barriers that influence their operation, which increase their lead time plus the cost, although the supply chain motive is to reduce the cost. For this case, this study uses the inductive nature of assessing the root causes of the obstacles the Food Processing Company has identified, and it will employ a qualitative methodology.

Purposive sampling was used to interview the production manager for this investigation, along with five of his colleagues. The same activity was performed for the other three companies' representatives.

Some of the food processing firms in Pakistan have been contacted to gather data that can be used to determine the specific challenges each firm has as it works to develop a circular food supply chain. Four companies were contacted for this study to learn more about food supply chain procedures and the obstacles faced while applying Circular Supply chain management.

As adopting reverse logistics is a crucial step toward a successful GSCM implementation, this study is primarily concerned with the Recycling aspect of Circular supply chain management. Rising consumer expectations and regulatory mandates for reverse logistics have created several challenges for businesses throughout the recycling process. Complexity in the design is one of the few issues, and cost is another. A questionnaire for the interview was designed to collect data and capture evaluators' views.

Following is the Interview questionnaire illustrated in Table 12.1 which was asked during the focused group interview session.

12.6 Research Analysis

This study offers several novel insights. It advances the theoretical knowledge of CSC, a new frontier in sustainability that goes beyond conventional supply chain sustainability paradigms, and this adds to the literature in the first place. Secondly, the study creates a theoretical framework to pinpoint obstacles to CE integration in SCM. This study offers a multi-stakeholder analysis of the challenges to circular

Table 12.1 Interview questions

Interview questions based on literature review	
1.	What are the main functions of your company?
2.	What do you know about the circular food supply chain?
3.	What do you think? How supportive are the government policies/rules and regulations related to CSC?
4.	Do you consider circular supply chain management when setting organizational goals and strategies?
5.	How much is your top management committed to and willing to implement CSC?
6.	What are the reasons for not investing in the implementation of CSC?
7.	How much are your customers aware of the circular supply chain and the importance of system sustainability?
8.	How much do your key partners, like vendors, suppliers, and distributors, have awareness and commitment towards the implementation and corporation of the circular supply chain?
9.	In your opinion, what are the barriers to managing key partners to implement CSC?
10.	What do you think is how supportive the stakeholders will be in implementing CSC?

food supply chains, moving beyond the dominant focal firm viewpoint and adopting a dyadic supply chain perspective that considers the members of the downstream supply chain who receive the least attention, such as customers and retailers.

In pursuance of the research objective, the following barriers are identified through the extensive literature review and information and knowledge extracted through the interview sessions of the professionals working in Pakistan's food industry (see Table 12.2). Professionals contacted for the interview are associated with different departments to ensure the overall clarity of the system and the issues faced by the organization.

Table 12.2 Barriers of the food processing company

S. No.	Barriers
1	Lack of government support towards Environmentally friendly policies
2	Market competition and uncertainty
3	Complexity of design to reuse/recycle used products
4	Lack of technical expertise
5	Lack of new technology, materials, and processes
6	Lack of an environmental partnership with suppliers
7	Poor supplier commitment/unwillingness to exchange information
8	Lack of support and guidance from regulatory authorities
9	Lack of training of the employees within an organization

Subsequently, through the detailed analysis of the knowledge and information acquired from the literature review and interview session of food industry professionals in Pakistan, the above-listed barriers are categorized into the following themes.

12.6.1 Industry Incentives in the Adoption of Sustainable Process

The environmental costs of industry are proliferating, and there is little hope for an economic recovery. If businesses are going to invest in green and circular ideas, they need financial incentives to do so. Developing sustainable/regenerative products is challenging since there are no governmental incentives to encourage using a circular supply chain. An insufficient amount of available funds prevents this because developing countries like Pakistan are so far behind in adopting green and circular models in terms of cutting-edge technology and manufacturing transformation, which presents a significant challenge for the food sectors in these countries. Among the obstacles cited is a lack of official backing for greener initiatives. The identified barrier is:

- Insufficient support from the government for eco-friendly initiatives.

12.6.2 Lack of Customer Awareness and Participation in CSC Activities

Improvements in consumer purchasing behavior and the uptake of eco-friendly products and services rely heavily on raising awareness of and supporting personal responsibility among consumers. A lack of familiarity with circular models might give the impression, from an institutional point of view, that "public perception and views are not engaged," which could slow the spread of such models across the supply chain. To name a few of the obstacles:

- Market competition and unpredictability.

Inadequacy in Knowledge and Awareness of CSC Initiatives

Applying CSC principles, however, requires high-level scientific competence that is currently lacking within the framework of a business supply chain. Because of this lack of understanding, organizations cannot fully embrace CSCM to enhance circular product and network design and promote greater reuse, remanufacturing, recycling, and repair.

- An inability to perform technical tasks.

Lack of coordination and collaboration among SC members

Organizations that rely on their suppliers, retailers, and distributors must work together and coordinate their efforts to maximize profits and maintain a positive reputation in the market. Collaborating and working together is essential since no one business can implement methods for recycling and remanufacturing all waste goods.

- Suppliers not working together on environmental issues.
- Suppliers are not committed to or prepared to share knowledge.

Lack of support and participation of stakeholders

To successfully implement CSCM concepts like efficient management of natural resources and fair resource allocation, managers must secure the involvement and cooperation of all relevant stakeholders in the system's development. It is challenging to adopt any process or technical innovation and to streamline the efforts of stakeholders to implement CSCM without a suitable level of support and participation from the stakeholders, ignoring advice and assistance from regulators.

Lack of management commitment to CSC adoption

A determined managerial approach to sustainable development drives transformations and environmental improvements. CSCM concepts can only be put into practice with the backing of management, a dedicated mindset, a solid basis in CE, and a feasible implementation strategy. In reality, management is unreliable and never follows through. Thus, it is thought that a lack of managerial commitment is one of the biggest hurdles to adopting CSCM.

- Investments are high with little return.
- Itemized costs include the price tag of eco-friendly packaging,
- The money is spent on recycling and reusing old items and shipping. Concern Over Potential Defeat.

The discussed themes are the factors preventing the widespread adoption of a circular food supply chain in Pakistan's food industry. This information will determine the relative importance of each barrier and the order in which it should be addressed to facilitate the widespread adoption of such a system.

The outcomes of integrating CE into the food supply chain in Pakistan are identified through the development of a theoretical framework in light of the preceding discussion. The identified obstacles to the combination of CE and SCM are depicted in Fig. 12.1 as a framework for barriers in supply chain integration.

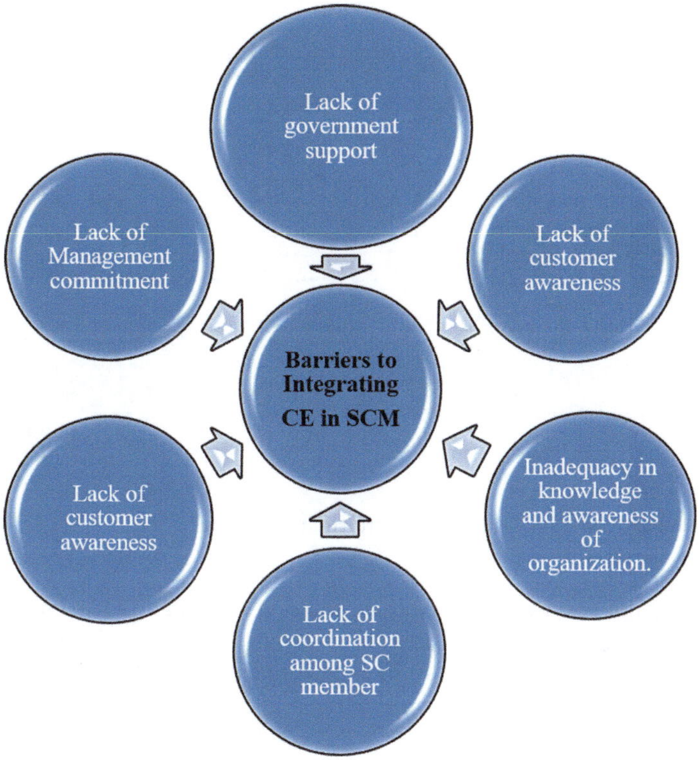

Fig. 12.1 Framework on obstacles in the food supply chain's integration of CE

12.7 Findings

Unlike the linear model, a circular food supply chain aims towards zero waste, which sends trash to landfills at various points. Implementing circularity into food product design, packaging, and supply chain activities is essential. On the one hand, waste creation should ideally be planned out of the supply chain system. However, there must be a mechanism to collect the trash that will inevitably be created at each step of the SC, from initial agricultural production through post-harvest storage, processing, distribution, and consumption (see Fig. 12.2). CE can facilitate the shift toward circular supply chains since it replicates natural ecosystems by regenerating trash into valuable feedstock. Although uncommon, there is a rising interest in circular food supply chains and a bright future for such applications. The creation and execution of genuinely circular food supply networks face formidable obstacles. This work's theoretical contributions will help us understand supply chain sustainability in the food industry. In the first place, it cements the CSC as a promising new area for studying and improving the sustainability of supply chains. To achieve circularity in materials and realize the zero-waste objective of the CSC, a regenerative and

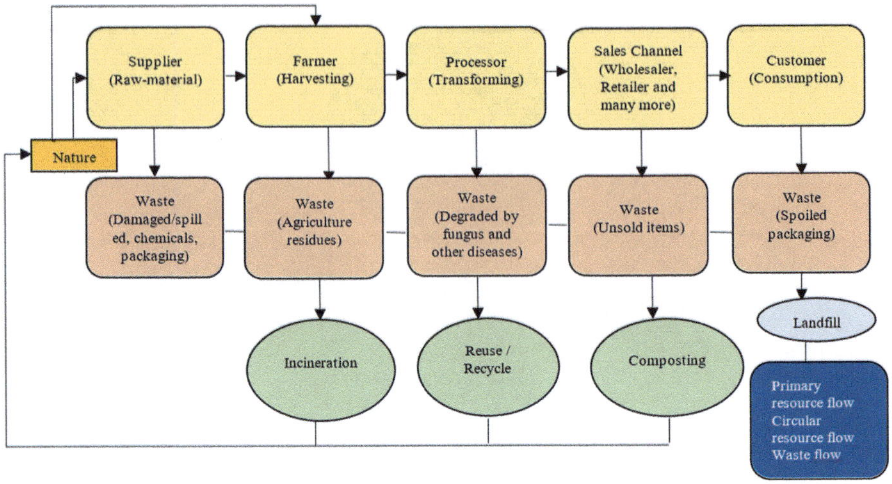

Fig. 12.2 Circular food supply chain flow chart (rice processing)

restorative design must be employed in addition to system-wide innovations. In addition to being grounded in literature, the framework has also been modified and verified with the help of feedback from experts in the food industry.

12.8 Conclusion

The rapid use of international resources and solid waste creation poses a significant danger to global sustainability in developing countries. Over the past decade, the government of Pakistan has spent extensively in adopting CE to address the problems caused by the recycling of waste products. However, progress has been slow because of challenges in establishing networks of companies working together to recycle their outputs into new inputs. Concerning Pakistan food supply chains, this research examines and evaluates the unique challenges associated with incorporating CE into SCM. Food supply chains have significant adverse effects on the environment; however, studies on their sustainability policies are lacking. The findings point to a lack of market demand or pressure and lax environmental legislation and enforcement as significant cause constraints. In addition, the most noticeable challenge is the supply chain's lack of cooperation and assistance.

The study discusses the practical and theoretical implications for addressing and evading the obstacles identified. It adds to the existing body of knowledge that can inform future studies of the sustainability of supply chains.

- Over the past ten years, the Pakistani government has made significant investments in implementing CE to address the issues related to the recirculation of waste materials.

- This study examines the obstacles to incorporating CE into SCM within Pakistani food supply chains.
- This study creates a theoretical framework based on various organizational theories to pinpoint obstacles to incorporating CE in SCM.
- The findings indicate that the primary causes of barriers are inadequate environmental regulations and implementation and a lack of market preference or pressure.
- The most notable barrier is the lack of cooperation or support from supply chain actors.

12.9 Summary

This chapter examines the factors contributing to Pakistan's food supply chain inefficiencies. The ultimate objective is to develop approaches to establish a more resilient and sustainable system. By utilizing qualitative research approaches and conducting unstructured interviews with key stakeholders, this project seeks to identify and evaluate barriers that hinder the development of a sustainable food supply. The study makes Additional contributions in the form of a framework for the food supply chain process designed to promote Pakistan's circular economy. Significantly, the assessment of obstacles to attaining a closed-loop food supply chain incorporates knowledge acquired from a dialogue with the Production Manager of a food processing organization. By employing a decision-making approach, the research derives conclusions about the particular obstacles encountered within the food industry. In addition to illuminating the causes and consequences of bottlenecks in the food supply chain, the research offers a nuanced comprehension of the matters under consideration. To tackle the critical issue of waste management in the food industry, this chapter integrates multiple theories of organizational structure into a unified framework. It provides practical recommendations that can assist the sector in efficiently harnessing its waste resources.

References

Baig, S. A., Muhammad, A., Aysha, B., Muhammad, H., & Rizwan, S. (2020). Barriers to the adoption of sustainable supply chain management practices: Moderating role of firm size. *Cogent Business and Management, 7*(1).

Batista, L., Davis-Poynter, S., Ng, I., & Maull, R. (2017). Servitization through outcome-based contract—A systems perspective from the defence industry. *International Journal of Production Economics, 192*, 133–143.

Farooque, M., Zhang, A., & Liu, Y. (2019). Barriers to circular food supply chains in China. *Supply Chain Management, 24*(5), 677–696.

Gebhardt, M., Alexander, S., & Hendrik, B. (2022). The future of the circular economy and its effect on supply chain dependencies: Empirical evidence from a Delphi study. *Transportation Research Part E: Logistics and Transportation Review, 157*(October 2021).

Kayikci, Y., Subramanian, N., Dora, M., & Bhatia, M. S. (2022). Food supply chain in the era of industry 4.0: Blockchain technology implementation opportunities and impediments from the perspective of people, process, performance, and technology. *Production Planning and Control, 33*(2–3), 301–321.

Leigh, M., & Li, X. (2015). Industrial ecology, industrial symbiosis and supply chain environmental sustainability: A case study of a large UK distributor. *Journal of Cleaner Production, 106*, 632–643.

Lutfi, A., Alsyouf, A., Almaiah, M. A., Alrawad, M., Abdo, A. A. K., Al-Khasawneh, A. L., Ibrahim, N., & Saad, M. (2022). Factors influencing the adoption of big data analytics in the digital transformation era: Case study of Jordanian SMEs. *Sustainability (Switzerland), 14*(3), 1–17.

McLachlan, C. J. (2021). Developing a framework for responsible downsizing through best fit: The importance of regulatory, procedural, communication and employment responsibilities. *International Journal of Human Resource Management* 1–29.

Mehmood, A., Ahmed, S., Viza, E., Bogush, A., & Ayyub, R. M. (2021). Drivers and barriers towards circular economy in agri-food supply chain: A review. *Business Strategy and Development, 4*(4), 465–481.

Papargyropoulou, E., Rodrigo, L., Julia, K.S., Nigel, W., & Zaini, B. U. (2014) The food waste hierarchy as a framework. *Journal of Cleaner Production 76*, 106–115.

Yadav, V. S., Singh, A. R., Raut, R. D., & Govindarajan, U. H. (2020) Blockchain technology adoption barriers in the Indian agricultural supply chain: An integrated approach. *Resources, Conservation and Recycling 161*(November 2019), 104877.

Chapter 13
Impact of COVID-19 on Industrial Ecology Advancement: An Example from Automotive Industry

Numerous industries and geographical areas have had positive and harmful effects from the COVID-19 pandemic on industrial ecology. Global supply chain disruption is one of COVID-19's most notable and immediate effects on industrial ecology. Production delays and shortages have resulted from disruptions in the supply of materials and components caused by lockdown procedures, travel restrictions, and facility closures. The pandemic has also culminated in consumer demand patterns and behavior modifications. Demand has significantly decreased in industries like travel, hotel, and entertainment, while it has surged in e-commerce, healthcare, and telephony. The distribution of resources and industrial processes have had to be adjusted due to these changes. The pandemic has highlighted the importance of these traits to industrial systems. Businesses are now compelled to review their supply chains, diversify their sources, and spend money on technology that improves adaptability and agility.

13.1 Introduction

In the closing months 2019, healthcare practitioners in Wuhan, located in Hubei Province, China, encountered a succession of pneumonia cases that exhibited suspicious origins. The pathogenic agent responsible for the disease was ultimately recognized as a newly detected coronavirus, designated explicitly as SARS-CoV-2, and the resulting illness was classified as COVID-19. The advent of this novel virus has triggered a worldwide health crisis, attracting the immediate attention of international health authorities. Upon receiving notification on December 31, 2019, the World Health Organisation (WHO) commenced a sequence of investigations, ultimately resulting in the proclamation of a Public Health Emergency of International Concern on January 30, 2020. In 2020, the virus rapidly spread across international borders, facilitated by modern global travel. In the following weeks, there was a significant

A. Shah et al., *Industrial Ecology*, Industrial Ecology,
https://doi.org/10.1007/978-981-97-3619-5_13

increase in cases, which quickly turned into a pandemic—officially declared by the World Health Organisation on March 11, 2020.

This event signified the commencement of an unparalleled worldwide health crisis, exerting significant strain on social, economic, and healthcare systems to their utmost capacity. The rapid transmission of COVID-19, driven by asymptomatic cases and delayed reactions in certain areas, posed unique problems to global public health systems. Faced with the pressing nature of the problem, countries implemented several techniques to reduce the transmission of the virus. The policies varied from strict lockdowns to adopting advanced track-and-trace systems, intending to mitigate the public health consequences. Global cooperation facilitated the rapid exchange of genomic information regarding the virus, sparking competition for the creation of vaccines. Concurrently, the public health narrative emphasized the importance of essential preventive measures, including hand cleanliness, social distancing, and face-concealing. Simultaneously, scientists and policymakers were confronted with emerging epidemiological data to develop a cohesive response.

13.2 COVID-19 and Sustainable Development

The onset of the COVID-19 pandemic has profoundly impacted the progress of the Sustainable Development Goals (SDGs), posing a considerable challenge to the global commitment to promoting a more sustainable and improved future for all individuals. The global health crisis shed light on inherent weaknesses within systems while offering prospects for revitalizing development from a sustainability perspective.

Disruptions to the SDGs Landscape: The extensive dissemination of the COVID-19 virus resulted in a complex situation that threatened crucial goals across the range of Sustainable Development Goals (SDGs). As economies weakened, the revival of food insecurity and rising economic inequality threatened the imperatives of SDG 1 (No Poverty) and SDG 2 (Zero Hunger). The disruption of educational systems significantly impacts SDG 4, which pertains to Quality Education, resulting in a disruption of learning for a particular cohort of children. Concurrently, the worldwide burden on healthcare systems has highlighted the goals of Sustainable Development Goal 3 (Good Health and Well-being) and the significant disparities in healthcare accessibility, particularly in settings with limited resources. On the other hand, the lockdown measures implemented during the pandemic unintentionally yielded a temporary advantage for the environment by diminishing emissions and pollutants while offering a look into the possibility of attaining the targets outlined in SDG 13 (Climate Action). The pause in environmental activities prompted a conversation on preserving ecological benefits while revitalizing the economy.

The "Building Back Better" Imperative: The post-COVID-19 recovery period has adopted the philosophy of "building back better," emphasizing the importance of inclusive growth, resilience, and sustainability as fundamental pillars. This concept promotes the implementation of reconstruction initiatives that aim to bring about not only rehabilitation but also significant change by integrating economic revitalization with the overarching goals of the Sustainable Development Goals (SDGs), particularly SDG 8 (Decent Work and Economic Growth) and SDG 9 (Industry, Innovation, and Infrastructure). Moreover, it strengthens the demand for cooperative efforts following SDG 17 (Partnerships for the Goals), emphasizing the necessity of synchronized global reactions to emergencies.

Policy Shifts Toward Green Recovery: Green recovery strategies have been advocated by both national and international organizations as strategic approaches to maintain economic growth while simultaneously enhancing environmental stewardship. The policies mentioned serve as the foundation for investments in renewable energy to support SDG 7 (Affordable and Clean Energy) and promote sustainable farming practices as a component of SDG 12 (Responsible Consumption and Production). These policies are fundamental pillars of this movement. This approach is exemplified by prominent plans, such as the European Union's Next Generation EU plan and projects resembling the United States' Green New Deal.

Enhancing Health Resilience and Achieving Educational Equity: The imperative to strengthen healthcare systems has become a pressing concern, acknowledging the inherent connection between public solid health capabilities and maintaining economic and social equilibrium. The pandemic's effect on education necessitated inventive measures to narrow the digital gap, guaranteeing that the advancement towards SDG 4 is not permanently hindered while also tackling wider disparities, as emphasized by SDG 10 (Reduced Inequalities).

Industrialization, Employment, and the Emerging Economic Model: The labor market saw significant transformations, highlighting the need to reassess and readjust the strategy towards Sustainable Development Goal 8. The global health crisis has acted as a catalyst for rediscovering the "essential work" concept and reassessing labor rights and industry norms. This has resulted in a shift towards a sustainable economic framework incorporating innovation and infrastructure, as highlighted in Sustainable Development Goal 9.

Achieving Equilibrium in the Future: The epidemic has highlighted the imperative for a development approach that is robust, equitable, and sustainable. The unforeseen environmental respite has sparked a conversation on the importance of conserving and managing resources sustainably, presenting a challenge to maintain these achievements. As nations traverse the aftermath of the pandemic, the Sustainable Development Goals (SDGs) provide a comprehensive framework for reconstructing societies with enhanced resilience and improved readiness to address global challenges. These goals prioritize sustainability as a fundamental principle.

In conclusion, the dynamic relationship between the COVID-19 pandemic and sustainable development has revealed the intricate nature of our globally interdependent society. The situation has emphasized the necessity of a comprehensive and all-encompassing recovery strategy that focuses on urgent actions and tackles long-term viability. The Sustainable Development Goals (SDGs) serve as a guiding principle in this undertaking, outlining a trajectory for a future that aligns economic revitalization with the well-being of our planet and its inhabitants.

13.3 Integrated Long-Term Solutions for Multiple Problems

The concept of the Circular Economy and Sustainable Production: In the post-Covid era, a circular economy offers a pathway to sustainability by separating economic growth from the consumption of resources. This concept facilitates implementing a closed-loop system, wherein resources are repurposed, recycled, or reused, reducing trash generation. The Netherlands' plan to achieve a circular economy by 2050 sets a model that successfully reduces the consumption of raw materials and emissions. Philips and Interface have implemented business models that exemplify practical applications, explicitly emphasizing product-as-a-service and closed-loop recycling. Governments can promote this transition by providing incentives for adopting circular business processes and allocating funds toward research on sustainable materials.

Renewable Energy and Economic Resilience: The transition to renewable energy is crucial for achieving sustainable recovery since it tackles issues related to energy security and the environment while promoting economic growth. The Energiewende in Germany offers a comprehensive framework for mitigating reliance on fossil fuels and cultivating a prosperous renewable industry. Policy instruments such as feed-in tariffs, renewable portfolio standards, and the gradual elimination of fossil fuel subsidies are crucial in expediting the transition toward renewable energy sources.

Resilient and Smart Infrastructure: Establishing robust infrastructure is crucial in mitigating susceptibility to forthcoming catastrophes. Japan's seismic-resistant infrastructure and Singapore's Smart Nation plan serve as prime examples of resilience and flexibility in both physical and digital domains. Policies should set down resilience criteria and provide incentives for modifying current infrastructure, incorporating intelligent technologies to enhance efficiency and facilitate efficient catastrophe response.

Social Equity and Inclusion: A broad recovery must also tackle the social disparities that the pandemic has intensified. Implementing inclusive policies can improve healthcare accessibility, as exemplified by Rwanda's community-based health insurance model, and assist marginalized communities. To achieve fair solutions, it is imperative to implement universal healthcare, education, and training programs and engage in participatory policy formation.

Innovation and Technology for Sustainability: Technological advancement catalyzes the promotion of sustainable development. Carbon capture and storage (CCS) methods provide prospects for decarbonizing heavy industries. Collaboration among governmental entities, private enterprises, and academic institutions has the potential to cultivate innovation, as evidenced by the implementation of legislation that facilitates green technology research and the adoption of clean technologies.

Sustainable Agriculture for Food Security: The emergence of COVID-19 has revealed deficiencies in food supply systems, underscoring the imperative for sustainable agriculture. Regenerative strategies that restore soil health and biodiversity ensure resilient food systems. Policies have the potential to facilitate the advancement of organic farming, local food networks, and the promotion of plant-based diets, thereby augmenting both food security and environmental well-being.

Urban Planning for Health and Sustainability: Urban planning substantially impacts health and environmental results. The "15-min city" notion advocates for developing inclusive and diverse urban environments that prioritize green spaces and active transportation, fostering sustainable urban lifestyles. Policy initiatives such as zoning reforms and incentives for green buildings can influence urban development, promoting improved health and sustainability.

Leveraging Financial Instruments: Green bonds and sustainability-linked loans are financial tools that can be used to finance sustainable projects, thereby encouraging responsibility for environmental impact. Policies can support sustainable development financing by implementing well-defined norms and procedures for green finance and offering tax incentives for green investments.

Enhancing Community Involvement to Achieve Optimal Solutions: The effectiveness of long-term sustainability solutions is contingent upon the active involvement of the local community. Including communities in urban planning and policymaking can yield more effective and universally acceptable outcomes. Public consultation, community-led initiatives, and sustainability education are essential to effective policies.

Therefore, for a post-Covid recovery, it is crucial to incorporate circular economy concepts, renewable energy, resilient infrastructure, and social inclusion. This should be backed by innovative technologies, sustainable agriculture, strategic urban design, practical financial instruments, and community participation. These components establish a unified policy structure to achieve a sustainable and resilient future. It is imperative to strengthen these integrated methods to build a new standard where sustainable development is not merely an ideal but a practical reality for all industrial and social endeavors.

13.4 Sustainability and Ethical Considerations

The ethical components of sustainability have been emphasized in the post-Covid period, prompting global and local actors to reassess the moral foundations of their recovery initiatives. The intersection of ethical considerations with sustainable development raises significant inquiries regarding equity, responsibility, and the enduring consequences of our collective decisions.

Ensuring Fairness in the Allocation of Vaccines: The fair allocation of COVID-19 vaccinations is a compelling exemplification of ethical considerations in global health. Although measures like COVAX have been implemented worldwide to provide equitable access, the distribution of vaccines has revealed significant discrepancies. Affluent nations obtained substantial vaccination supplies, frequently at the cost of less wealthy countries, intensifying worldwide disparities. Morally, this prompts inquiries on the ideals of fairness and unity in worldwide healthcare, underscoring the necessity for structures that guarantee universal access to life-saving vaccines and treatments, irrespective of wealth disparities.

The Responsibility of Developed Nations: Due to its history and ongoing role in environmental degradation and climate change, the developed world bears a disproportionate burden in spearheading global sustainability initiatives. This obligation includes assisting developing countries through the transfer of technology, provision of financial aid, and implementation of capacity-building projects. From an ethical standpoint, this is based on reparative justice and acknowledging a collective yet distinct worldwide need to address environmental issues and promote sustainable development.

Use of Resources and Efforts for Recovery: The ethical considerations surrounding the utilization of resources in recovery endeavors are crucial. As nations allocate resources toward revitalizing their economies, it is ethically necessary to consider the enduring ecological consequences of these investments. The decisions made in the present will have enduring implications for subsequent generations, prompting inquiries over intergenerational fairness and the ethical utilization of natural resources. Recovery plans must encompass the immediate restoration of economic activity and consider these endeavors' long-term viability and ecological impact, guaranteeing their alignment with the overarching objectives of environmental conservation and societal welfare.

The Intersection of Social Justice and Sustainability: Sustainability projects should also encompass social justice considerations, guaranteeing that sustainable development's advantages are distributed fairly and that no community bears an inequitable share of the negative consequences of environmental degradation or recovery endeavors. This entails taking into account the consequences for indigenous populations, marginalized communities, and the global South, who frequently experience the most significant burden of environmental exploitation and climate change, although having made the least tremendous contributions to it.

The Intersection of Ethical Consumerism and Corporate Responsibility: In the aftermath of the COVID-19 pandemic, there has been an increasing examination of the moral implications around consumption and corporate accountability. Ethical consumerism promotes the adoption of options that encompass both environmental sustainability and social equity, taking into account the entirety of the supply chain, spanning from production to disposal. Similarly, there is a growing demand for corporate entities to demonstrate accountability for their sustainability practices. This demand encompasses not only the reduction of environmental damage but also the active promotion of societal well-being. This expectation shift reflects a more comprehensive comprehension of ethical responsibility within business.

International Unity and Moral Administration: The global epidemic has brought attention to the interdependence of our international society, emphasizing the importance of ethical governance that emphasizes unity and collaboration rather than nationalistic strategies. The post-Covid era should prioritize the establishment of global alliances that uphold the dignity and rights of all individuals while also advocating for inclusive, transparent, and equitable policies within the ethical framework of sustainability.

13.5 Impact of Covid 19 on Industrial Ecology

The COVID-19 epidemic has significantly changed industrial ecology, impacting supply chains, industry, and environmental goals. Shaping a more sustainable and resilient future will depend heavily on industrial systems' resilience, adaptability, and innovation as the globe navigates the epidemic and its aftermath. Governments, organizations, and individuals worldwide have implemented various methods and activities to reduce the impacts of COVID-19 on industrial ecology. Among the crucial actions are:

Green Recovery Investments: Several nations have unveiled stimulus packages and recovery plans prioritizing investing in sustainable technologies, renewable energy sources, and green infrastructure. These investments aim to quicken the shift to a more robust and sustainable economy while promoting economic growth.

Promotion of the Circular Economy: The circular economy's tenets, which emphasize minimizing waste and maximizing the reuse, recycling, and repurposing of materials and resources, are gaining attention from corporations and governments. Programs, including expanded producer responsibility schemes, product redesign, and waste reduction targets, are being implemented to encourage a more circular approach to production and consumption.

Support for Sustainable Industries: Governments are offering financial assistance and other incentives to sectors of the economy that help the environment, like eco-friendly manufacturing, clean transportation, and renewable energy. Subsidies, tax breaks, and legislative initiatives are examples of how to promote the adoption of sustainable technologies and practices.

Improved Resilience Planning: Industrial systems need resilience, as the pandemic has shown. Businesses and governments are spending money on initiatives like supply chain diversification, bolstering domestic manufacturing capacity, and implementing risk management plans to increase resilience against shocks and disruptions in the future.

Digitalization and Innovation: The pandemic has created new avenues for innovation in industrial ecology because of the rapid uptake of digital technologies.

Companies and governments invest in digitalization projects to increase productivity, adaptability, and sustainability, including digital supply chain management, smart manufacturing, and predictive maintenance.

International Cooperation: Collaboration is needed to address COVID-19 and industrial environmental concerns. Governments, organizations, and stakeholders collaborate to exchange best practices, plan responses, and advance international collaboration on vaccine distribution, supply chain resilience, and climate change mitigation.

Public Awareness and Education: An essential part of reducing the effects of COVID-19 on industrial ecology is educating the public about the value of sustainability and environmental stewardship. Governments, non-governmental organizations, and corporations are launching public awareness campaigns, educational efforts, and awareness-raising activities to encourage sustainable behavior and consumption patterns.

In broad terms, governments, corporations, civil society organizations, and individual citizens must work together to adopt a multifaceted and cooperative approach to mitigate the effects of COVID-19 on industrial ecology. In the post-pandemic period, the world can recover more effectively and forge a more sustainable and just future by prioritizing sustainability, resilience, and innovation.

Financial losses because of COVID-19 preventing the implementation of industrial ecology

Due to COVID-19, there may be significant and varied financial losses if industrial ecology measures are not implemented. The following are some possible locations for monetary losses to happen:

(a) **Inefficiency of Resources**: Businesses need to use appropriate industrial ecological practices to keep using resources intensively, which would increase production costs and material waste. Increased energy, raw materials, and waste disposal costs may follow.
(b) **Supply Chain Disruptions**: As demonstrated by the COVID-19 pandemic, supply chains are more susceptible to disruptions when industrial ecological precautions are not implemented. Reliance on remote suppliers and a weak supply chain can expose businesses to more significant risks of shortages, delays, and cost changes.
(c) **Regulatory Compliance Costs**: Governments increasingly enforce regulations and policies to promote sustainability and reduce adverse environmental effects. Businesses that do not use industrial ecology techniques may find it more expensive to comply with the increasingly stringent environmental laws and norms.
(d) **Reputation and Brand Damage**: Investors and customers increasingly examine environmental policies in today's socially conscious industry. If industrial ecology measures are not implemented, sales and market share may decline due to harm to the company's reputation and loss of customer trust.

(e) **Missed Innovation Opportunities**: By pushing businesses to develop innovative ways to reduce waste, use fewer resources, and adopt sustainable practices, industrial ecology promotes innovation. Businesses that invest in industrial ecology gain chances to create cutting-edge goods, procedures, and business plans that can boost productivity and competitiveness.
(f) **Greater Susceptibility to Upcoming Shocks**: The COVID-19 epidemic has highlighted how crucial resilience is for industrial systems. Businesses that do not use industrial ecology techniques can be more susceptible to shocks and disruptions in the future, like supply chain breakdowns, pandemics, and natural disasters, which can cause considerable financial losses.

Businesses' profitability, competitiveness, and long-term sustainability may be negatively impacted by the significant and far-reaching financial losses that can arise from failing to adopt industrial ecological principles. Businesses may cut expenses, lower risks, and take advantage of growth and innovation opportunities in a world economy that is changing quickly by using the concepts of industrial ecology.

13.6 Effects of COVID-19 on Automotive Industry

To understand the effects of COVID-19, a study was conducted on the automotive industry in Pakistan to ascertain the pandemic impact on the advancement of industrial ecology. The selected manufacturing sector collaborates with foreign assemblers and brand names from developed countries, thus having access to updated knowledge on industrial ecology. There are currently 11 such car manufacturers/assemblers in Pakistan, out of which we have selected three top brand names to study and understand the implementation of industrial ecology. To evaluate the effects of COVID-19 on industrial ecology policy in the manufacturing sector, a benchmark has been selected in the shape of UN SDGs. The UN SDGs selection provides an excellent framework for an apple-to-apple comparison as each organization is dynamically different and serves a different target market.

13.6.1 Industrial Ecology Implementation in the Automotive Industry

Multinational corporations can give their host-country affiliates instructions on how to cut back on the excessive usage of natural resources. To decrease the quantity of virgin natural resources needed to produce goods, host-country subsidiaries may want to upgrade their operations, technologies, and procedures throughout their subsidiary operational sites. Industries can also help their suppliers and subsidiaries in host countries to lower their carbon dioxide emissions at operating sites by increasing renewable energy consumption and improving energy efficiency. Companies can help improve resource management practices by identifying subsidiary and supplier operational sites at risk from resource depletion and environmental disasters and assessing their resource management strategies in the context of the local population. Multinationals may benefit from these internal investments by paying less to operate and comply with environmental regulations, ensuring a reduction in their ecological footprint. Impact of COVID-19 on Sustainability. The whole study was based on the research objectives:

- The study aims to gauge the impact of COVID-19 on the automotive sector of Pakistan concerning industrial ecology with UN SDGs as a benchmark.
- The SDGs provide financial strategies, targets, global guidelines, and implementation. This study will help map the impact of COVID-19 on UN SGDS in Pakistan by conducting a systematic review of previously published research, articles from various databases, and company performance.

13.6.2 Evaluation Quantifiable Measures

To thoroughly understand the study themes of the last five years, we did a structured literature search on journal articles and print media. We evaluated by citation frequency and tried evenly distributing our literature research throughout the observed time. We created a list of important search terms and potential links between them to acquire high-quality results.

Below are the goals selected:

- Goal 7: Ensure access to affordable, reliable, sustainable, and modern energy for all.
- Goal 9: Build resilient infrastructure, promote inclusive and sustainable industrialization, and foster innovation.
- Goal 12: Ensure sustainable consumption and production patterns.

Interview questions were extracted from the main focus of our research based on UN SDG's 03 selected goals and their indicators, which are essential to achieving these goals. Questions Q1-Q3 were extracted from UN SDG 7, Q4-Q6 were derived from Goal 9, and Q7-Q12 were from Goal 12. Also, the questions were put in front of industry professionals who are the most knowledgeable about sustainability and reporting for vetting.

13.6.3 Research Participants

Interviews were conducted with experts of OEMs who are well versed in the topics that cover the firm's performance, strategy, and practices, such as operations or manufacturing managers or someone in a comparable position. To ensure that operative-level knowledge is guaranteed, we selected experts who either had a product development background or belonged to engineering. The reason for interviewing those is to understand how they look after sustainability at an operational level.

13.6.4 Deep Dive into Goals

As mentioned, participating member countries will achieve the United Nations Sustainable Development Goals by 2030. Detailed description of the goals relevant to our project are:

SDG Goal 7

The UN's seventh Sustainable Development Goal (SDG) guarantees everyone access to modern, affordable, dependable, and sustainable energy. Goal 7 and industrial ecology interact in many ways, which are as follows:

1. Conservation of Energy and Resource Efficiency: Industrial ecology encourages the economical use of resources, such as energy. Industrial ecology promotes sustainability and energy saving by streamlining industrial operations, cutting waste, and encouraging recycling and reuse. This is in line with the goals of Goal 7, which are to increase energy efficiency and guarantee access to renewable energy sources.
2. Renewable Energy Integration: Industrial ecology encourages the use of renewable energy sources in industrial processes, such as hydroelectric, solar, and wind power. By incorporating renewable energy into their industrial processes, businesses may lessen their dependency on fossil fuels, cut greenhouse gas emissions, and aid in the shift to sustainable energy systems, a crucial element of Goal 7.
3. Decarbonization of the Industrial Sector: A significant portion of the world's greenhouse gas emissions comes from the industrial sector. The goal of Goal 7 is to reduce the carbon intensity of energy systems and mitigate climate change. Industrial ecology methods, such as carbon capture and storage, energy efficiency improvements, and adopting low-carbon technologies, can assist in decarbonizing industrial activity.
4. Access to Clean Energy Technology: Industrial ecology encourages the creation and application of clean energy technology that can enhance access to energy, especially in developing nations and underprivileged areas.
5. Achieving Goal 7's objective of universal access to electricity and clean cooking facilities requires increasing access to clean and sustainable energy sources, which industrial ecology promotes through innovation and knowledge transfer.

6. Synergies with Other Goals: Because industrial ecology places a strong emphasis on sustainable production and consumption patterns, pollution control, and resource efficiency, it has connections to other SDGs, particularly Goal 7. Several sustainable development goals, such as those about poverty alleviation, climate action, and sustainable development, are supported by industrial ecology, which promotes circular economy principles, lessens environmental consequences, and increases the resilience of industrial systems.

In conclusion, industrial ecology is essential to achieving Goal 7 of the Sustainable Development Goals because it fosters resource efficiency, integrates renewable energy sources, decarbonizes the industrial sector, makes clean energy technologies accessible, and works with other SDGs. In addition to tackling more general sustainability issues, nations and industries may help ensure everyone can access modern, inexpensive, sustainable, and reliable energy by implementing industrial ecology concepts and practices.

SDG Goal 9

Building resilient infrastructure, advancing equitable and sustainable industrialization, and encouraging innovation are the three objectives of Goal 9 of the Sustainable Development Goals (SDGs) of the United Nations. Goal 9 and industrial ecology interact in several significant ways; a few are mentioned as follows:

1. Sustainable Industrial Practices: Industrial ecology encourages industries to use ecologically responsible and sustainable practices. It helps create resilient and sustainable industrial infrastructure by maximizing resource utilization, decreasing waste production, and limiting environmental effects. This is in line with Goal 9's promotion of sustainable industrialization.
2. Circular Economy Principles: The reuse, recycling, and repurposing of materials and resources are given priority in the circular economy, which is emphasized by industrial ecology. Industries may assist Goal 9's objective of constructing resilient infrastructure using circular economy techniques, which will lessen their reliance on virgin resources, minimize waste generation, and create more resilient and sustainable production processes.
3. Innovation and Technology Transfer: Industrial ecology promotes innovation and technology transfer within industrial sectors. Industrial ecology supports Goal 9's fundamental components of innovation and competitiveness by encouraging the creation and uptake of cleaner technology, sustainable manufacturing methods, and resource-efficient practices.
4. Infrastructure Development: Industrial ecology plans, designs, and oversees industrial infrastructure to minimize environmental effects and optimize resource efficiency. To create resilient and sustainable infrastructure—which is necessary to meet the goals of Goal 9—industrial ecology incorporates environmental considerations into infrastructure development initiatives.
5. Resilience and Adaptability: Industrial ecology strongly emphasizes the importance of these traits to industrial systems.

Industrial ecology helps to strengthen the resilience of industrial infrastructure by encouraging supply chain diversification, flexible production methods, and risk management techniques. These actions help to lessen the effects of shocks and disruptions like pandemics and natural disasters.

6. Encouragement of Inclusive Development: Social, economic, and environmental aspects of inclusive development are emphasized by industrial ecology. Goal 9's primary objective of inclusive growth is supported by industrial ecology, which promotes sustainable industrialization methods that benefit local communities, improve livelihoods, and generate employment opportunities.

In conclusion, industrial ecology plays a critical role in accomplishing Goal 9 of the Sustainable Development Goals by encouraging sustainable industrial practices, circular economy concepts, innovation and technology transfer, resilient infrastructure development, resilience and adaptability, and inclusive development. By incorporating these ideas and practices into their industrial activities, countries and industries can help construct resilient, inclusive, and sustainable infrastructure, promoting economic growth and prosperity while reducing environmental impacts.

SDG Goal 12

Ensuring sustainable patterns of consumption and production is the emphasis of Goal 12 of the Sustainable Development Goals (SDGs) of the United Nations. Goal 12 and industrial ecology are closely related because industrial ecology explicitly tackles the need for more sustainable practices in industrial systems and processes. This is how Goal 12 and industrial ecology are related:

1. Resource Efficiency and Circular Economy: Industrial ecology encourages the circular economy's tenets, which optimize resource and material reuse, recycling, and repurposing while minimizing waste. Goal 12's objectives are aligned with industrial ecology's contribution to more sustainable production and consumption patterns through waste reduction and resource optimization.
2. Life Cycle Thinking: Industrial ecology promotes a life cycle perspective on goods and procedures, considering the environmental effects at each step, from raw material extraction to recycling or disposal at the end of a product's useful life. Goal 12's discussion of the shift to more sustainable production and consumption habits is supported by industrial ecology, which evaluates and reduces the environmental impact of products and processes throughout their life cycle.
3. Strategies for Preventing Pollution and Reducing Emissions: Industrial ecology advocates for reducing emissions and pollutants caused by industrial operations. Goal 12's objectives to reduce pollution and minimize the release of hazardous chemicals are supported by industrial ecology, which employs cleaner production methods, pollution control strategies, and emissions reduction technologies to help minimize environmental effects.
4. Collaborative Ways and Partnerships: To encourage sustainable consumption and production, industrial ecology places a strong emphasis on partnerships and collaborative ways amongst stakeholders, including enterprises, governments, academic institutions, and civil society. Industrial ecology supports the creation

of creative solutions to sustainability problems and the adoption of best practices by encouraging communication, knowledge exchange, and collaboration. This helps fulfill Goal 12's aim of supporting sustainable production and consumption partnerships.

5. Education and Awareness: Industry stakeholders and the public are encouraged to learn about sustainable production and consumption practices through industrial ecology. Industrial ecology supports attempts to shift towards more sustainable lifestyles and behaviors, in line with Goal 12's purpose of fostering sustainable development, by increasing understanding of the environmental and social implications of consumption and production patterns and encouraging sustainable alternatives.
6. Policy Support and Implementation: Industrial ecology is a foundation for creating and applying laws and rules that support environmentally friendly production and consumption. Goal 12 aims can be achieved by establishing effective policies and initiatives that promote sustainable practices with the help of industrial ecology, which offers insights into the economic and environmental implications of various production and consumption patterns.

Overall, through encouraging resource efficiency, circular economy concepts, pollution prevention, cooperative approaches, education and awareness, and policy support for sustainable consumption and production, industrial ecology plays a critical role in achieving Goal 12 of the Sustainable Development Goals. By incorporating these ideas and practices into their industrial operations, countries and industries may help create more resilient and sustainable economies and societies.

13.6.5 Results and Discussion

Based on the interview questions extracted from the UN SDG, the answers were noted to understand the current situation of Pakistani companies concerning three goals that were initially identified. Furthermore, after assessing the local progression, we compared the findings with the parent company's global sustainability report to understand the achievement of the UN SDG and the commitment of both companies to achieve these targets.

UN SDG 07 Comparative Analysis

The goal is related to adopting reliable and modern energy incorporation in the company to reduce carbon emissions globally. After reviewing the literature on sustainability reports published by the company, it was found that the parent company is actively looking to reduce carbon emissions by incorporating reliable and modern energy solutions such as solar power and electric battery-related solutions. By modifying the distribution nodes globally, around 25% of global CO_2 reduction was reported through hybrid technologies, better efficient engines, and optimized transportation routes.

The local counterpart is trying to incorporate solar power into the grid to maximize the share of renewable energy; currently, around 5% of the current power distribution is renewable. Furthermore, steps are being taken to replace the diesel-operated fork truck fleet with battery-operated ones completely. However, due to the pandemic, the sales volume drastically dropped, resulting in financial troubles for the company halting all the upgrade works in the facility. It was discussed that there is a further plan to incorporate more solar power into the current distribution system to minimize fossil fuel consumption.

UN SDG 9 Comparative Analysis

Increasing a country's GDP and taking the developing countries to self-sustainability is the main focus of this goal. Relating to improvements in local companies, this goal is designed to ensure that the population of the country benefits from the activities that these multinationals are performing, for example, any upgradation project that increases the employment opportunities or incorporates locally manufactured items in the product to strengthen small local businesses which result in overall impact on the country's GDP. Decreasing the use of natural resources ensures that the local environment remains healthy. Focusing on tree plantation will negate the effect of CO_2 emissions that the plant is currently producing.

The parent company has launched multiple projects all over the globe where their manufacturing sites are present, encouraging the local businesses to take part and manufacture those products that meet their standards so that locals of those countries can benefit from it, simultaneously increasing employment opportunities in those countries.

Due to the pandemic, the local company has not yet taken such initiatives in light of UN SDG, resulting in decreased sales volume. The pandemic also affected the localization process of the product as visits to vendors were prohibited, and they had to test the product virtually, ultimately resulting in quality losses. The business model is such that the host company does not invest, and the local company has to make every change and every upgrade from its profit and loss.

UN SDG 12 Comparative Analysis

This goal embarks on the various factors for sustainability improvement, including efficient use of natural resources, waste reduction by recycling and reusing techniques, controlling those activities that are acting on the environment and humans negatively, sustainable public procurement methods, and ensuring those patterns of consumption and production which are sustainable.

Promotion of environmental conservation employing promotion on a global level to encourage the use of those materials that generate comparatively less Volatile Organic Compounds, which will help improve the environment; this can be done by having those type of vehicle interiors which has meager amounts of VOCs, also to implement such processes and practices which results in reduce amount VOC emissions during the paint job. This can be achieved through the improvement of thinners. The local counterpart in Pakistan had also worked to reduce volatile organic compounds in their processes by incorporating bonding agents or materials in interior parts with less VOC emissions, reducing VOC in the painting area through thinner improvements.

13.7 Conclusion

The COVID-19 pandemic has brought attention to the complex relationship between human health, economic stability, and environmental sustainability, emphasizing the pressing requirement for comprehensive approaches that tackle these interrelated issues. To recover from the pandemic, we must prioritize sustainability and ethical considerations. Our goal should not only be to reconstruct what was lost but also to establish systems that are more robust, equitable, and sustainable.

The debate surrounding sustainability in the aftermath of the COVID-19 pandemic is intricately linked to ethical deliberations on fairness, impartiality, and accountability principles. The equitable distribution of vaccines worldwide has brought attention to the wider matter of resource allocation and the obligation of rich governments to assist disadvantaged populations in developing nations. This extends beyond healthcare to include economic assistance, the transfer of technology, and the development of skills and abilities, guaranteeing that all countries may participate in and gain advantages from sustainable development.

Moreover, the ongoing pandemic has stimulated a reassessment of our consumption behaviors and utilization of resources, necessitating a contemplation of the enduring consequences of our recuperation approaches on the environment and forthcoming cohorts. Adopting a circular economy, utilizing renewable energy sources, and developing resilient infrastructure signifies a dedication to mitigating our ecological impact and fostering a sustainable equilibrium between human actions and natural systems.

The COVID-19 epidemic has provided valuable insights that should shape our approach to sustainability. It directs us toward a future where ethical considerations play a crucial role in decision-making processes. By adopting these principles, we can strive to attain a global state in which progress is not solely sustainable but also characterized by fairness and justice, guaranteeing a harmonious cohabitation between mankind and the natural world for future generations.

Chapter 14
Challenges of Industrial Ecology in the Manufacturing Sector

Industrial ecology is a comprehensive approach to promoting sustainable industrial growth. It emphasizes the efficient use of materials and energy within industrial processes. The objective is to emulate industrial systems based on natural ecosystems, in which waste from one process is transformed into input for another, resulting in minimal waste and optimal resource utilization. Although industrial ecology can substantially contribute to sustainability, it encounters several obstacles. This analysis delves into the difficulties, supported by theoretical and practical perspectives.

14.1 Complexity of Industrial Systems

Industrial systems are naturally intricate, encompassing many processes, technology, and stakeholders. To effectively use industrial ecology ideas, one must profoundly comprehend the intricacy and interconnections inside industrial systems. The book "Industrial Ecology" by Graedel and Allenby, published in 1995, offers fundamental insights into analyzing and optimizing intricate systems. It emphasizes the importance of multidisciplinary approaches in tackling this task. This problem encompasses technological and organizational aspects, necessitating collaboration across several sectors and specialties. The influential book "Industrial Ecology" by Graedel and Allenby underscores the necessity of an interdisciplinary methodology that integrates engineering, economics, and environmental science perspectives to manage and optimize intricate systems efficiently.

Implementing industrial ecology in large industrial systems is a formidable task because of the intricate interconnection of processes, materials, energy flows, and stakeholder relationships. These systems exhibit a wide range of inputs and outputs, operate at different sizes, and have varied levels of interconnection across different industrial operations. To achieve sustainability in these systems, it is crucial to thoroughly understand their intricacies and devise novel strategies to manage them

A. Shah et al., *Industrial Ecology*, Industrial Ecology,
https://doi.org/10.1007/978-981-97-3619-5_14

successfully. Here is a more detailed analysis of the intricacy of industrial systems within the framework of industrial ecology:

- Interrelation of Processes: Industrial systems are interconnected components within a broader framework of processes that engage in mutual interactions with one another and the surrounding environment. The difficulty is comprehending these interactions and using them to establish closed-loop systems where waste from one operation is transformed into input for another. Ehrenfeld and Gertler (1997) present a case study in Kalundborg, Denmark, titled "Industrial Ecology in Practice: The Evolution of Interdependence at Kalundborg," which showcases an early instance of a system where different industrial organizations collaborate and share resources like energy, water, and materials. This case study highlights the potential benefits of industrial symbiosis. Industrial systems exhibit a broad spectrum of operations, varying in size from local to global and including diverse activities, each with distinct material and energy needs. Enhancing the sustainability of these activities requires technological advancements and modifications in organizational practices and policies. In their publication, "Modelling and Mapping Industrial Systems to Analyze and Manage Material Flows and Resource Use," Despeisse et al. (2012) explore several strategies for mapping and modeling industrial systems to enhance the management of material flows and resource utilization across different scales.
- Technological Complexity and Innovation: The technological foundation of industrial systems is continuously advancing, with new procedures, substances, and energy sources developing often. Incorporating these innovations into current systems in a manner that adheres to industrial ecological principles might pose difficulties. In their publication "Industrial Ecology: The Role of Manufactured Capital in Sustainability" (2013), Lifset et al. examine the impact of technical innovation on the objectives of industrial ecology, highlighting the importance of research and development endeavors that prioritize sustainability. Effective implementation of industrial ecology necessitates the engagement and collaboration of many stakeholders, encompassing enterprises, governments, communities, and consumers. Collaboration is crucial yet challenging due to each stakeholder group's differing interests, aims, and influence. In their study titled "Levels of Institutional Capacity and Actor Expectations about Industrial Symbiosis: Evidence from the Dutch Stimulation Programme 1999–2004," Boons and Spekkink (2012) investigate the significance of developing institutional capacity to promote stakeholder involvement and cooperation in industrial symbiosis projects.
- Regulatory and policy frameworks: The regulatory and policy framework can either support or impede the implementation of industrial ecology. Creating and enforcing rules that promote the efficient use of resources, minimize waste, and safeguard the environment is essential. In their article "Reflections on implementing industrial ecology through eco-industrial park development," Gibbs and Deutz (2007) examine the importance of policy in promoting the establishment of eco-industrial parks. They emphasize the necessity of regulatory frameworks

that facilitate the adoption of industrial ecology methods. Industrial systems require the capacity to adjust and recover from changes in market circumstances, resource availability, and environmental limitations, demonstrating adaptability and resilience. Incorporating resilience into these systems is crucial for guaranteeing their long-term viability. In their article "Organising self-organizing systems: Towards a theory of industrial symbiosis," published in 2012, Chertow and Ehrenfeld examine the idea of self-organizing systems in industrial ecology. They highlight the significance of adaptation and resilience in pursuing sustainable industrial systems.

A comprehensive strategy is needed to tackle the intricacies of industrial systems while implementing industrial ecology principles. This strategy should encompass technical advancements, engagement with stakeholders, the implementation of supporting regulations, and continuous research and teaching. By addressing these difficulties, industrial ecology may substantially contribute to advancing sustainable, efficient, and resilient industrial systems.

14.2 Economic and Financial Barriers

Adopting methods supported by industrial ecology generally necessitates significant initial investment in new technology or process redesigns. Many firms are greatly concerned about the economic viability of such initiatives. In "Industrial Ecology: Paradigm Shift or Normal Science?", Ehrenfeld (2000) explores the alignment of monetary incentives and financial processes with the aims of industrial ecology to overcome the obstacles at hand. This includes tax incentives, financial support for environmentally friendly technology, and pricing systems that account for environmental expenses.

Economic and budgetary restrictions substantially affect the deploying industrial ecology concepts inside organizations and across sectors. These hurdles include many factors, such as the upfront expenses of implementing new technology, the financial uncertainty linked to novel business models, and the economic viability of sustainability efforts in competitive marketplaces. It is essential to overcome these obstacles to promote the progress of industrial ecological practices that contribute to sustainable development. Here is a more detailed analysis of these challenges.

Firstly, the significant initial investment costs are associated with implementing technologies and processes that align with industrial ecology. This encompasses expenses for acquiring new equipment, modifying current facilities, and instructing personnel. The perceived risk associated with such expenditures, particularly for small and medium-sized firms (SMEs) with few resources, can be a substantial disincentive. The study by Ormazabal et al. (2016) examines the difficulties that small and medium-sized enterprises (SMEs) encounter while adopting environmental management systems. It emphasizes that financial limitations are a significant obstacle to implementing sustainable practices.

Secondly, companies may exhibit reluctance to participate in industrial ecology efforts due to the ambiguity around the potential return on investment (ROI). Although sustainable methods can result in long-term cost savings by improving efficiency and minimizing waste, it cannot be easy to measure these advantages in advance accurately. Schaltegger and Csutora (2012) examine the notion of "eco-efficiency" and the challenges associated with quantifying the economic effects of environmental endeavors. They emphasize the necessity for more precise measurements and valuation techniques to evaluate return on investment (ROI).

Thirdly, Insufficient access to funding poses a notable obstacle for sustainability projects. Conventional financial institutions may hesitate to finance initiatives that prioritize the environment, as they view them as having a greater level of risk and emerging as potential alternatives are novel finance mechanisms such as green bonds, sustainability-linked loans, and crowdsourcing for green initiatives. The World Bank's, 2018 report on green finance emphasizes the increasing interest in funding methods that promote environmental sustainability, such as industrial ecology initiatives.

Fourthly, firms prioritize immediate financial performance in fiercely competitive marketplaces above long-term sustainability objectives. Focusing on short-term goals might discourage investments in industrial ecology, which may not result in immediate economic gains. In their influential study "Towards a New Conception of the Environment-Competitiveness Relationship," Porter and van der Linde (1995) contend that incorporating environmental sustainability into fundamental corporate strategy may bolster competitiveness by promoting innovation and efficiency.

Fifthly, The economic feasibility of recycling and waste reduction, which are essential aspects of industrial ecology, can be influenced by market pricing for raw resources and recyclable products. When new resource prices are low, the financial benefits of recycling or waste reduction decrease. Gregson et al. (2015) analyze the obstacles in attaining a circular economy, such as the economic variables that impact recycling and resource recovery initiatives, and advocate for governmental interventions to create a fair and equitable environment.

Finally, unfavorable legislative and regulatory frameworks might worsen the economic and financial obstacles to adopting industrial ecology. Implementing policies that internalize environmental externalities, such as implementing carbon pricing and enacting legislation that requires recycling and waste reduction, can provide economic incentives for adopting industrial ecology practices. Korhonen et al. (2018a, 2018b) examine the impact of policy on advancing industrial symbiosis and emphasize specific policy tools that might aid the shift towards more environmentally friendly industrial systems.

A comprehensive strategy is needed to address industrial ecology's economic and financial obstacles. This strategy should involve creating new and creative funding methods, establishing more precise measures for evaluating the economic advantages of sustainability projects, and implementing policies and regulations that support these initiatives. By tackling these obstacles, corporations and sectors may enhance the incorporation of industrial ecology concepts into their activities, facilitating the shift towards sustainable development.

14.3 Lack of Regulatory Support

Successful implementation of industrial ecology often requires supportive legislative and regulatory frameworks that promote or require recycling, waste reduction, and energy efficiency. The lack of such frameworks can impede the implementation of industrial ecological techniques. In their publication "Eco-Industrial Strategies: Unleashing Synergy Between Economic Development and the Environment" (Cohen-Rosenthal & Musnikow, 2007), the authors explain how policy reforms might facilitate industrial symbiosis, a crucial aspect of industrial ecology. The lack of regulatory support in promoting the concepts of industrial ecology is a significant barrier, as effective laws and regulations are essential for fostering the adoption of sustainable industrial practices. Regulatory frameworks may enforce recycling, incentivize the optimal utilization of resources, and advocate for energy preservation, thereby aiding a transition towards more sustainable industrial practices. Nevertheless, the rate at which rules adjust to environmental requirements and adopting inventive industrial ecology methods can frequently be sluggish, resulting in a disparity between the prospective accomplishments in sustainability and the current state of industrial practices.

Recent talks and research have emphasized the necessity for regulatory settings that are more adaptable and encouraging to address this problem. The "Circular Economy Package" implemented by the European Union is a legislative strategy to promote recycling and minimize waste. The European Commission (2018) incorporates many techniques to encourage the design of products that can be easily recycled, implement expanded producer responsibility, and establish ambitious recycling goals for member states. Regulatory frameworks are vital in offering explicit guidance and assistance to enterprises and sectors seeking to embrace industrial ecological concepts. Moreover, a study conducted by Tseng et al. (2018a, 2018b) in the field of "Sustainable Industrial Production: Practices and Circular Economy" highlights the significance of incorporating circular economy ideas into regulatory frameworks to improve industrial sustainability. The study supports the implementation of legislation that requires waste reduction and resource efficiency and promotes innovation in sustainable product design and industrial symbiosis. The United Nations Environment Programme (UNEP) has highlighted the significance of policy in endorsing industrial ecology by advocating for green public procurement (GPP) laws. The GPP regulations promote the procurement of sustainable and recyclable goods by government agencies, therefore establishing a demand for items that align with the concepts of industrial ecology (UNEP, 2017).

Nevertheless, there are still obstacles to overcome to ensure that rules can keep up with the rapid progress of technology and the ever-changing concept of sustainability. Korhonen et al. (2018a, 2018b) emphasize in their publication "Circular Economy: The Concept and its Limitations" the necessity of policy frameworks that endorse existing industrial ecological practices and accommodate future advancements and alterations in the sector. A collaborative endeavor, including politicians, industry stakeholders, and the academic community, is needed to tackle the absence of regulatory support. This encompasses formulating policies that encourage sustainable practices, implementing well-defined regulatory benchmarks for recycling and waste reduction, and establishing channels for continuous communication between regulators and the industrial sector to ensure that regulations remain pertinent and conducive to achieving industrial ecology objectives.

14.4 Technological Challenges

To advance industrial ecology, new technologies that facilitate the effective recycling of materials, the recovery of energy, and the integration of processes must be developed and implemented. Nevertheless, technical advancement is sometimes sluggish, and current technologies may not entirely align with industrial ecology concepts. In their paper "Industrial Ecology: Goals and Definitions," Lifset and Graedel (2002) discuss the importance of ongoing technological innovation and adaptation within industrial ecology.

To tackle the technological difficulties in implementing industrial ecology, we must overcome barriers associated with creating, adopting, and combining new and existing technologies that support environmentally friendly industrial operations. The significance of these difficulties lies in the essential role of technical innovation in facilitating resource efficiency, energy recovery, and the establishment of closed-loop systems that imitate natural ecosystems. A critical technological obstacle in industrial ecology is the advancement of technology that can effectively collect and recycle materials, particularly those now deemed non-recyclable. Advancements in this field have the potential to diminish waste and greatly enhance resource utilization efficiency. Recent breakthroughs in chemical recycling methods, such as pyrolysis and depolymerization, provide innovative solutions for previously challenging polymers. These methods disassemble plastic waste into its molecular constituents, which may be utilized to manufacture new polymers or other goods.

Nevertheless, there are still obstacles to overcome when it comes to expanding the use of these technologies, decreasing energy usage, and guaranteeing economic feasibility. The study conducted by Vollmer et al. (2020) titled "Technological, environmental and Economic Aspects of Asphalt Recycling for Road Construction" examines the possibilities and constraints of recycling technologies in the specific context of asphalt recycling. The findings of this study offer valuable insights that can be applied to broader initiatives aimed at recovering materials.

Although Enhancing energy efficiency in industrial operations and using renewable energy sources are vital elements of industrial ecology. Nevertheless, notable technological obstacles must be overcome, such as the requirement for substantial upfront capital, the fluctuating nature of renewable energy sources, and the incorporation of these sources into current energy networks. The study by Zheng and Luo (2020) focuses on integrating renewable energy and energy efficiency in data centers. "Strategies and Sustainability Assessment" analyses the methods of incorporating renewable energy into data centers, emphasizing the technological and economic obstacles and suggesting adaptable solutions for industrial systems. Meanwhile, industrial symbiosis is a process in which several industrial facilities share resources such as materials, water, and energy to improve the system's efficiency.

The technological issue at hand involves the development of the necessary infrastructure and mechanisms to enable this sharing. This encompasses sophisticated logistics, information and communication technologies (ICT), and systems for handling and transporting goods and energy. The case study by Chertow and Park (2016) titled "Scholarship and Practice in industrial symbiosis: 1989–2014" examines the advancements and technological obstacles in industrial symbiosis. It highlights the importance of ongoing innovation in integrating systems and managing logistics. Using digital technologies, such as the Internet of Things (IoT), big data analytics, and artificial intelligence (AI), brings out both advantageous prospects and obstacles for industrial ecology.

These technologies have the potential to significantly improve the surveillance, examination, and fine-tuning of material and energy movements in industrial systems. Nevertheless, it is vital to tackle the obstacles associated with data security, privacy, the intricacy of data analysis, and incorporating digital technologies into current industrial procedures. The paper by Li et al. (2021) titled "Big data in smart farming: A review" demonstrates the immense possibilities of using big data analytics in the agricultural sector, which is progressively included in larger industrial ecological frameworks. The paper emphasizes the technological obstacles and options for enhancing resource utilization using data analytics.

A comprehensive strategy is needed to successfully address the technological obstacles in implementing industrial ecology. This strategy should encompass research and development, investment in innovative technologies, and establishing supportive policies and frameworks. These recent references indicate that advancements are being achieved in several areas, including materials recycling, energy efficiency, and digitization. Nevertheless, to fully harness the potential of industrial ecology to contribute to a sustainable industrial future, it is crucial to maintain a constant stream of new ideas, foster cooperation among all parties involved, and expand the implementation of successful technologies.

14.5 Cultural and Organizational Resistance

Incorporating industrial ecological ideas may require substantially modifying the business culture and organizational structure. Resistance to change is a prevalent obstacle since corporations prioritize immediate profits over long-term sustainability objectives. Gibbs and Deutz (2007) analyze the cultural and organizational changes necessary to create a favorable environment for implementing industrial ecology techniques by developing eco-industrial parks. Implementing industrial ecological ideas into organizations frequently faces cultural and organizational opposition. The opposition arises from deeply ingrained habits, perceived economic uncertainties, and difficulty reconciling sustainability objectives with current business structures.

To overcome these obstacles, it is necessary to make a focused and collaborative attempt to encourage a culture shift, promote knowledge acquisition inside the organization, and showcase the lasting advantages of using industrial ecological methods. One of the main challenges in discussing entrenched habits and perceptions is the resistance to change from established operational routines. Additionally, there is a perception that sustainable practices, like those promoted by industrial ecology, may clash with goals related to productivity or profitability. In their analysis of environmental behaviors titled "Mind the Gap: Why do people act environmentally and what are the barriers to pro-environmental behavior?", Kollmuss and Agyeman (2002) emphasize that deeply ingrained habits and established corporate cultures frequently prioritize immediate benefits over long-term sustainability. This poses a significant obstacle to the implementation of industrial ecology practices.

Nevertheless, regarding economic and financial matters, Organisations may be hesitant to embrace industrial ecology concepts owing to apprehensions over the financial consequences, such as the expenses associated with implementing new technology and processes, the uncertainty around the return on investment, and the belief that environmental rules are burdensome. Schaltegger and Wagner (2011) explore the integration of sustainability into the fundamental structure of businesses to address environmental concerns and transform them into economic prospects, as discussed in their study "Sustainable Entrepreneurship and Sustainability Innovation: categories and Interactions." Insufficient knowledge and awareness regarding the concept and implementation of industrial ecology might result in resistance. Education and training are essential for overcoming this obstacle, as they can clarify the concept of industrial ecology for personnel at all hierarchical levels and showcase its tangible advantages. In "The State of Sustainability Reporting in Universities," Lozano (2013) highlights the significance of sustainability education in cultivating a culture of sustainability inside organizations, including industrial ecology concepts.

The conventional compartmentalized framework of numerous organizations might impede the interdepartmental cooperation required to implement industrial ecological methods efficiently. It is crucial to dismantle these isolated groups and encourage collaboration across different fields of study to tackle intricate sustainability issues. In their article "Strategic Perspectives of Corporate Sustainability Management to Develop a Sustainable Organization," Baumgartner and Rauter (2017) explore methods for reorganizing companies to include sustainability principles related to industrial ecology effectively.

Moreover, the significance of leadership in overcoming cultural and organizational opposition cannot be emphasized enough. Influential leaders dedicated to sustainability and with a distinct vision for how industrial ecology may positively impact the organization are essential for facilitating transformative transformation. They can inspire and encourage people, direct resources towards sustainability projects, and incorporate industrial ecology ideas into the company's strategic objectives. Visser and Crane (2010) emphasize the significance of leadership in promoting corporate sustainability initiatives in their article "Corporate Sustainability and the Individual: Understanding what drives sustainability professionals as change agents." Identical By implementing incentive and reward systems that acknowledge and encourage sustainable behaviors, resistance may be overcome if individual and departmental objectives are aligned with those of industrial ecology. In their article "How to Become a Sustainable Company," Eccles et al. (2012) explore the use of incentives linked to sustainability measures to motivate employees to embrace and endorse industrial ecology projects.

14.6 Knowledge and Information Sharing

The efficacy of industrial ecology frequently relies on the efficient dissemination of knowledge and optimal methodologies within corporations, sectors, and across industries. Nevertheless, the transmission of information may be restricted due to competitive considerations and concerns around intellectual property. In "Uncovering' Industrial Symbiosis," Chertow (2007) emphasizes the significance of promoting information sharing to facilitate the recognition and execution of symbiotic possibilities that play a crucial role in industrial ecology.

Effective dissemination and exchange of knowledge and information are essential for effectively implementing industrial ecology. However, there are substantial obstacles that might hinder advancements in this domain. The problems encompass various concerns, including those linked to gaining a competitive advantage and protecting intellectual property and the dispersion of information across several fields. Tackling these obstacles is crucial for cultivating a cooperative atmosphere that encourages sustainability and optimal utilization of resources. Here is a detailed analysis of the difficulties encountered while exchanging knowledge and information within the field of industrial ecology, for example. Topics of interest include competitive advantage, intellectual property and open innovation, interdisciplinary collaboration, digital technologies for information sharing, government and institutional roles, and cultural and organizational barriers.

Companies frequently see their sustainability strategies and technology as proprietary to get a competitive advantage in the market. This viewpoint can result in a hesitancy to provide knowledge that could benefit broader industrial ecological initiatives. To tackle this issue, it is crucial to implement tactics that involve the development of collaborative platforms that facilitate the exchange of information while safeguarding competing interests. Tseng et al. (2018a, 2018b) examine the combination of Industry 4.0 and circular economy principles, proposing that using big data and collaborative platforms can enable the sharing of anonymous or aggregated data. This sharing can help overcome competitive obstacles without sacrificing individual competitive advantages. Intellectual Property and Open Innovation: Intellectual property rights may have positive and negative effects, as they safeguard inventors' rights while possibly impeding the unrestricted sharing of knowledge. Chesbrough (2003) introduced the notion of open innovation, which offers a framework for addressing these difficulties. It suggests approaches that enable enterprises to exchange and gain from external ideas and technology, leading to faster and more sustainable breakthroughs in industrial ecology. This strategy necessitates meticulous handling of intellectual property to foster cooperation while preserving exclusive benefits.

The multidisciplinary character of industrial ecology, which involves integrating several subjects such as engineering, environmental science, economics, and sociology, presents difficulties in effectively exchanging knowledge. Every field of study has unique methods, terms, and viewpoints, which might hinder comprehension and collaboration. Fam et al. (2017) propose that more funding should be allocated to multidisciplinary research and education to close these gaps. This would help create a collaborative culture necessary for tackling intricate sustainability issues. Digital technologies provide exceptional potential for exchanging information but pose problems safeguarding data, protecting privacy, and addressing the digital divide. It is essential to prioritize the security, accessibility, and user-friendliness of digital platforms to enable efficient information sharing. The study conducted by Rajput and Singh (2019) emphasizes the significance of digital platforms in facilitating sustainable supply chains. It underscores the need for a robust IT infrastructure and rules that provide fair access to information.

The significance of governments and organizations in fostering information exchange cannot be underestimated. Nevertheless, the lack of policy coherence, regulation obstacles, and insufficient financial support for sustainability projects can hinder information transmission and collaborative efforts. Organizations like the United Nations Industrial Development Organisation (UNIDO) have fostered global industrial symbiosis. However, there is still a need for more comprehensive policy frameworks that promote information exchange at both national and international levels. Cultural and organizational barriers can substantially impact the willingness and capacity of people and departments to exchange information. The culture and structure of the organization influence these barriers. Departments operating in isolation, without motivating factors for knowledge sharing, and a culture that does not prioritize cooperation can impede the flow of information. In their study, Zeng et al. (2017) emphasize the significance of organizational culture in environmental management systems. They propose that cultivating a culture that incentivizes knowledge exchange and promotes cooperation is crucial for successfully implementing sustainable practices.

To overcome these challenges, adopting a comprehensive strategy that involves cultivating a collaborative culture, implementing intellectual property management practices that promote sharing, allocating resources to digital technologies and infrastructure, encouraging interdisciplinary research and education, and advocating for policies and frameworks that provide support. To expedite the achievement of sustainability and resource efficiency objectives, the area of industrial ecology can make significant progress by overcoming these obstacles.

14.7 The Challenges of Industrial Ecology and Industrial Symbiosis

Industrial ecology and symbiosis are innovative strategies that strive to convert conventional industrial processes into sustainable, circular systems that imitate natural ecosystems. Although these concepts show potential for decreasing environmental effects and improving resource efficiency, they encounter many inherent obstacles that might impede their general adoption and implementation. It is essential to tackle these problems to fully harness the benefits of industrial ecology and symbiosis in advancing sustainable industrial growth.

Interdisciplinary Complexity

The inherent transdisciplinary complexity of industrial ecology is one of the significant issues. Optimizing industrial systems necessitates integrating information from engineering, environmental science, economics, and social sciences. Managing a multidisciplinary approach can be challenging due to divergent terminology, approaches, and views across several professions. To overcome this difficulty, it is essential to promote robust collaboration between different disciplines and establish shared frameworks that enhance comprehension and enable coordinated efforts across several domains (Despeisse et al., 2017).

Economic and Financial Barriers

The shift towards industrial ecology and symbiosis typically entails substantial initial expenses for implementing novel technology, process reconfiguration, and system incorporation. Small and medium-sized firms (SMEs) with restricted capital access may find these financial obstacles especially intimidating. In addition, unpredictable fluctuations in market pricing for commodities can influence the economic feasibility of recycling and waste reduction initiatives, which are essential aspects of industrial symbiosis (Gregson et al., 2015). To overcome these obstacles, it is crucial to create new financial models and legislative incentives that decrease the economic risks related to sustainability projects.

Technological Innovation and Adoption

Technological innovation plays a crucial role in effectively implementing industrial ecological concepts. This includes developing efficient recycling processes, integrating renewable energy, and using waste-to-resource technology. Nevertheless, the progress of technology can sometimes be sluggish, and current infrastructures may not be conducive to implementing new sustainable methods. To effectively tackle these difficulties, it is essential to expedite research and development in critical technological domains while cultivating an environment of innovation across various businesses (Ghisellini et al., 2016).

Policy and Regulatory Frameworks

Creating a supportive policy and regulatory framework is crucial in promoting the implementation of industrial ecology and symbiosis activities. However, current legislation may not wholly support or encourage the cooperative, multi-sector methods essential to industrial symbiosis. Supporting and formulating policies that endorse resource sharing, waste reduction, and circular economy concepts might facilitate implementing these activities (Chertow, 2007).

Stakeholder Engagement and Collaboration

Efficient involvement and cooperation with relevant stakeholders are crucial for recognizing and achieving the potential benefits of industrial symbiosis. This process involves sharing resources, energy, and by-products across many organizations. To cultivate mutually beneficial partnerships, it is essential to develop trust and build effective communication channels between enterprises, government organizations, and communities (Lombardi & Laybourn, 2012).

Cultural and Organizational Change

The reluctance of cultures and organizations to adapt might hinder the implementation of industrial ecology and symbiosis. Organizations may be required to undertake substantial cultural transformations to prioritize long-term sustainability above short-term economic benefits. Promoting a culture that prioritizes sustainability while offering education and training on the advantages and methods of industrial ecology can help assist this shift (Korhonen et al. 2018a, 2018b).

Figure 14.1 illustrates the challenges of industrial ecology and industrial symbiosis. The infographic illustrates the difficulties associated with implementing industrial ecology and industrial symbiosis. The design aims to effectively convey the various challenges, such as the complexity of different disciplines, economic and financial obstacles, technological advancements and acceptance, policy and regulatory structures, involvement and cooperation of stakeholders, and cultural and organizational transformation. Each of these challenges is represented using user-friendly icons and visual elements. This visualization facilitates comprehension of the extensive range of factors required to implement sustainable industrial practices successfully.

Concerning interdisciplinary complexity, This section will likely include images such as linked gears or a web, which symbolize the intricate interplay between several areas necessary for industrial ecology. The text highlights the importance of interdisciplinary collaboration across engineering, environmental science, and economics in successfully establishing sustainable industrial systems. For economic and financial barriers: This infographic section focuses on the difficulties associated with the expenses involved in adopting sustainable practices.

The statement highlights the necessity of significant investments in emerging technology and the potential lack of rapid financial gains, which may discourage enterprises from embracing industrial ecological concepts. The section for technological innovation and adoption: Here, gears, light bulbs, or computer screens might

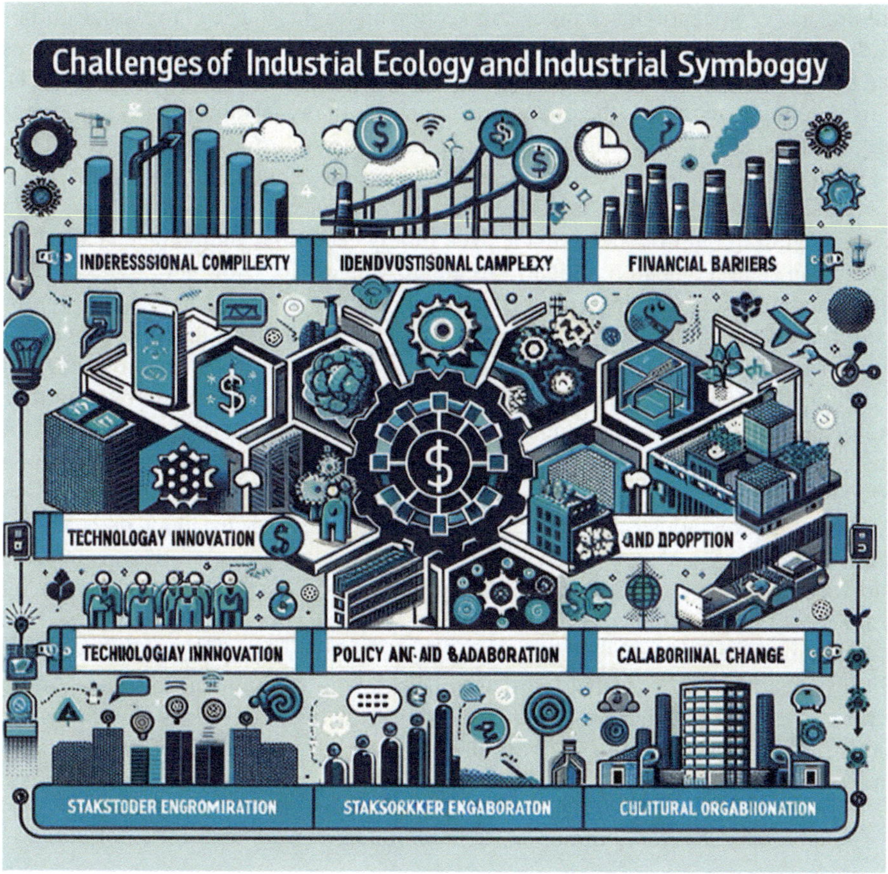

Fig. 14.1 Challenges of industrial ecology and industrial symbiosis

be employed to symbolize the significance of technical progress in attaining industrial sustainability. This section highlights the difficulties of developing, expanding, and incorporating novel technologies that effectively utilize resources and eliminate waste for section policy and regulatory frameworks. This portion may include icons such as gavels, policy documents, or legislative buildings to illustrate how supportive government policies and laws enable industrial ecology. It indicates that having clear and favorable regulatory settings is crucial for promoting sustainable industrial practices and overcoming bureaucratic obstacles. The stakeholder engagement and collaboration section is represented via symbols such as handshakes, network diagrams, or community symbols; this section highlights the importance of cooperation across different parties, including enterprises, governments, and communities.

The statement emphasizes the difficulty of establishing collaborations and trusts to exchange resources and mutually beneficial interactions. The last section is related to cultural and organizational change. This section highlights the need to change corporate culture and organizational structures to prioritize sustainability. These changes can be represented through human figures, organizational charts, or cultural symbols. This statement recognizes the opposition to change in organizations and emphasizes the significance of cultivating a culture that promotes industrial ecology projects.

Implementing industrial ecology and symbiosis poses significant problems, yet it is crucial for establishing sustainable industrial practices. The challenges associated with integrating multiple disciplines, overcoming economic and budgetary hurdles, innovating technologically, navigating regulatory landscapes, promoting stakeholder engagement, and catalyzing cultural and organizational transformation are significant. Nevertheless, these difficulties emphasize the significance and potential influence of industrial ecology in developing systems replicating the effectiveness and durability of natural ecosystems. Industrial ecology has the potential to revolutionize industrial operations by overcoming these challenges. This can lead to a decrease in environmental effects and an improvement in economic sustainability and resource effectiveness.

The future of industrial practices relies on our capacity to adopt the ideas of industrial ecology and symbiosis, shifting away from linear models of consumption and waste towards circular, linked systems. Achieving success in this undertaking will necessitate coordinated endeavors by enterprises, governments, researchers, and communities to cultivate a favorable milieu for innovation, cooperation, and sustainable expansion. As we face these difficulties, achieving a sustainable industrial future is no longer just a goal but a realistic possibility. This future envisions a world where industrial operations are in sync with the natural environment, benefiting the earth and its inhabitants.

14.8 Green Supply Chain

Green Supply Chain Management (GSCM) is an innovative strategy that combines environmental and sustainable practices with existing supply chain management frameworks. The primary objective of GSCM is to reduce the environmental impact of a product's whole life cycle, encompassing the extraction of raw materials, production, distribution, and end-of-life processes such as disposal, recycling, or reuse. This strategy includes optimizing internal processes for sustainability and extends to managing environmental impacts throughout the whole supply chain network. GSCM strives to harmonize ecological, economic, and social results by prioritizing environmentally friendly materials, minimizing waste, and adopting energy-efficient procedures. This approach ensures company operations align with environmental stewardship and sustainable development objectives. The adoption of GSCM is motivated by an increasing acknowledgment of the pressing necessity to tackle environmental issues such as climate change, resource depletion, and pollution. Companies are increasingly motivated to embrace green supply chain processes due to growing regulatory demands and customer demand for sustainable products. Implementing

Green Supply Chain Management (GSCM) not only aids in minimizing the environmental footprint of organizations but also provides a competitive edge by optimizing operational efficiency, elevating brand reputation, and satisfying the demands of environmentally concerned consumers. Furthermore, GSCM offers a structure for enterprises to actively address environmental rules and reduce risks related to limited resources and unpredictable commodity prices, guaranteeing long-term economic durability and sustainability.

Nevertheless, shifting towards an environmentally friendly supply chain has distinct difficulties. These include the requirement for a substantial initial investment in sustainable technologies and practices, the challenge of navigating the intricacies of managing supply chain sustainability across diverse regions and regulatory frameworks and establishing the requisite knowledge and cooperation among stakeholders. Although there are difficulties, the advantages of GSCM are convincing, providing not just ecological enhancements but also significant financial savings by enhancing operational effectiveness and optimizing resources. As organizations and society place greater importance on sustainability, implementing green supply chain practices becomes vital to achieving a more sustainable future. This highlights the significance of the supply chain in promoting environmental and social progress. The next part will cover the essential components of Green Supply Chain Management (GSCM), the advantages of adopting GSCM, the difficulties encountered in its implementation, and the solutions that may be employed.

14.9 Key Components of GSCM

14.9.1 Sustainable Sourcing

Sustainable sourcing, a fundamental aspect of Green Supply Chain Management (GSCM), entails acquiring goods and services that reduce environmental harm and consider social and ethical considerations. Methods encompass utilizing sustainable resources, collaborating with suppliers that use environmentally friendly methods, and considering the overall environmental effect of obtained materials (Seuring & Müller, 2008). Additionally, it prioritizes the reduction of the carbon footprint and environmental degradation linked to sourcing operations, as well as ensuring that suppliers comply with principles of social responsibility. Sustainable sourcing seeks to not only fulfill the immediate requirements of enterprises but also to guarantee the long-term viability and sustainability of supply resources, as well as the welfare of communities engaged in the supply chain. Here is a more in-depth analysis of sustainable sourcing.

Renewable and Recycled Materials: Companies emphasize utilizing renewable resources rather than non-renewable ones in their supply networks. IKEA aims to exclusively use renewable or recycled materials in its products by 2030. This goal requires a change in sourcing techniques to prioritize using sustainable resources.

Patagonia's dedication to procuring 100% sustainable cotton significantly reduces environmental harm in the textile sector (Pedersen & Gwozdz, 2014).

Supplier Environmental Performance: Sustainable sourcing entails assessing suppliers' environmental performance and sustainability strategies. Apple's Supplier Responsibility Report delineates its supplier standards, encompassing utilizing renewable energy and mitigating carbon emissions. Apple collaborates closely with its suppliers to enhance environmental performance, exemplifying how firms may exert influence on their supply chains to achieve higher levels of sustainability (Apple Inc., 2020a, 2020b).

Ethical Sourcing and Fair Trade: Besides environmental factors, sustainable sourcing also includes social and moral dimensions, such as labor rights, equitable salaries, and the influence on local communities. Ben and Jerry's and similar companies procure essential ingredients such as sugar, cocoa, and vanilla from suppliers who have obtained Fairtrade certification. This guarantees that farmers are remunerated relatively and operate following ethical standards. This method promotes social sustainability and enhances supply networks' robustness and resilience (Nicholls & Opal, 2005).

Local Sourcing and Reduced Transportation: Minimising the distance that goods travel may significantly reduce a supply chain's carbon footprint. Many firms use local sourcing to reduce transportation emissions and bolster local economies. As shown by Martinez et al. (2010), the farm-to-table trend in the restaurant business prioritizes obtaining products from nearby farmers. This approach not only reduces emissions associated with transportation but also promotes the availability of fresher and more sustainable food choices.

Certification and Standards: Another component of sustainable procurement is the strict adherence to established sustainability standards and certifications. Certifications such as the Forest Stewardship Council (FSC) for wood and paper goods and the Marine Stewardship Council (MSC) for seafood establish standards for sustainable practices and give customers confidence in the items they buy. Companies that use these certifications within their sourcing strategy actively contribute to more comprehensive environmental conservation initiatives (Auld et al., 2008).

Sustainable sourcing is a strategic method to reduce a company's supply chain's adverse environmental and social effects. By meticulously choosing resources, collaborating with environmentally conscious suppliers, prioritizing ethical procedures, and adhering to sustainability certifications, businesses may substantially contribute to achieving global sustainability objectives. With the rise in consumer awareness and regulatory restrictions, firms are finding it essential to engage in sustainable sourcing to stay competitive in a market that is becoming more environmentally sensitive.

14.9.2 Eco-friendly Manufacturing

Eco-friendly manufacturing is a crucial component of Green Supply Chain Management (GSCM) that focuses on minimizing environmental effects during production. This method aims to reduce waste, decrease emissions, save energy and resources, and employ sustainable materials. The objective of eco-friendly manufacturing is to reduce the environmental impact of industrial processes while simultaneously promoting innovation and efficiency, which can result in long-term economic advantages for organizations. To implement these practices, one must combine technical innovation, process improvement, and a dedication to sustainable ideals. Here, we explore the topic of environmentally sustainable manufacturing in further detail;

Energy Efficiency and Renewable Energy Use: An essential approach in environmentally sustainable manufacturing is maximizing energy efficiency and promoting renewable energy sources utilization. Siemens AG has achieved notable progress in decreasing its carbon footprint via investments in energy-efficient technology and a commitment to achieving carbon neutrality by 2030. Siemens (2020) implements strategies to improve energy efficiency in their industrial processes and relies on renewable energy sources like wind and solar power.

Waste Minimization and Circular Economy Practices: Minimising waste production and adopting circular economy ideas are fundamental to environmentally conscious manufacturing. Toyota's "Zero Waste" concept embodies this strategy, intending to reduce waste in every manufacturing phase. Toyota employs comprehensive recycling initiatives, optimized material utilization, and methods for repurposing by-products throughout its manufacturing facilities, resulting in a substantial waste reduction disposed of in landfills (Toyota Motor Corporation, 2020).

Sustainable Materials and Product Design: Incorporating sustainable materials and creating products prioritizing durability, repairability, and recyclability are essential for environmentally conscious production. Adidas collaborates with Parley for the Oceans to create footwear and clothing using recovered plastic from the ocean. This program not only decreases the use of new plastics but also tackles the problem of ocean pollution, demonstrating how the acquisition of sustainable materials and the design of products may help preserve the environment.

Pollution Control and Reduction: Incorporating strategies to regulate and diminish pollution at every stage of the production process is another crucial element of environmentally conscious manufacturing. For instance, BASF SE, a chemical company, has used sophisticated pollution control technology to address its notable environmental footprint. The "Verbund" idea developed by BASF connects manufacturing facilities, energy distribution, and waste management systems, reducing emissions and enhancing resource efficiency (BASF SE, 2020).

Adoption of Advanced Manufacturing Technologies: Utilising sophisticated manufacturing technologies such as additive manufacturing (3D printing), automation, and the Industrial Internet of Things (IIoT) helps improve environmentally friendly production operations. These technologies provide greater accuracy in material utilization, decreased energy usage, and reduced waste generation. General Electric's use of additive manufacturing for aviation engine components exemplifies how these technologies may substantially minimize material and energy consumption and decrease emissions (General Electric, 2020).

Eco-friendly manufacturing is a progressive strategy that combines environmental sustainability with economic efficiency and innovation. Companies may significantly reduce their ecological effect and perhaps discover new growth and competitive possibilities by implementing energy-efficient procedures, minimizing waste, using sustainable materials, decreasing pollution, and using sophisticated technology. With the growing demand for sustainable products and increasing regulatory demands, firms must adopt eco-friendly production techniques to succeed in the contemporary economy.

14.9.3 Logistics and Distribution Optimization

Logistics and Distribution Optimisation in the Green Supply Chain Management (GSCM) framework aims to improve transportation and delivery activities' effectiveness and ecological impact. This field seeks to minimize the release of carbon emissions, decrease energy usage, and minimize waste production that occurs during the transportation of commodities from producers to consumers. Strategies encompass the optimization of routes, the transition to more sustainable transportation choices, improvements in packaging, and the adoption of new technology. Here is a more in-depth analysis of logistics and the optimization of distribution:

Route Optimization and Efficient Vehicle Loading: Maximising transportation routes and enhancing vehicle loading efficiency are essential for minimizing fuel usage and emissions. UPS and other companies have used sophisticated route optimization software that minimizes needless travel, decreases fuel consumption, and cuts emissions. UPS's ORION (On-Road Integrated Optimisation and Navigation) system exemplifies the utilization of technology to accomplish these objectives, showcasing substantial decreases in mileage and fuel usage throughout their delivery network (UPS, 2020).

Modal Shift and Intermodal Transport: Transitioning goods delivery from high-emission modes, such as road and air, to lower-emission alternatives like rail or water, may significantly diminish the environmental consequences of logistics. Intermodal transportation, a method that utilizes many modes of transportation for a single shipment, provides adaptability and can potentially reduce expenses and emissions. Maersk Line, a prominent global shipping firm, has prioritized enhancing its environmental performance by optimizing its fleet's energy efficiency and expanding multimodal transportation options (Maersk Line, 2020).

Sustainable Packaging Solutions: Minimising packaging waste and using sustainable packaging materials are crucial components of environmentally conscious logistics. Companies are investigating biodegradable, repurposed, and recyclable packaging alternatives to reduce environmental impact. Amazon introduced the "Frustration-Free Packaging" initiative to minimize waste by utilizing entirely recyclable packaging and optimizing package dimensions to enhance transportation efficiency.

Electric and Alternative Fuel Vehicles: Integrating electric vehicles (EVs) and vehicles fueled by alternative sources such as biofuels and natural gas into logistical operations may significantly decrease greenhouse gas emissions and air pollution. DHL, a prominent logistics corporation, has pledged to accomplish carbon–neutral logistics by 2050. To achieve this goal, DHL is allocating resources towards acquiring electric delivery vehicles, bicycles, and drones, aiming to enhance the sustainability of last-mile deliveries (DHL, 2020).

Digitalization and Advanced Technologies: digital technologies such as the Internet of Things (IoT), artificial intelligence (AI), and blockchain revolutionize logistics and distribution. These technologies have the potential to enhance tracking capabilities, optimize route planning, improve fleet management, and increase transparency in the supply chain. The TradeLens platform, developed by IBM and Maersk, utilizes blockchain technology to enhance efficiency and sustainability in global commerce. It achieves this by boosting transparency and decreasing the need for paperwork (IBM, 2020).

Collaborative Distribution Networks: Engaging in partnerships with other companies to exchange logistics and distribution resources can result in more optimal transportation capacity utilization, hence decreasing the number of cars on the road and the emissions that come with them. The "Physical Internet" idea advocates for a globally accessible logistics system that facilitates the sharing of transportation resources, aiming to optimize and minimize the environmental consequences of cargo transportation (Montreuil, 2011).

Implementing optimization solutions in logistics and distribution is crucial for reducing the environmental effects of supply chains. Through technology, implementation of sustainable practices, and promotion of collaboration, organizations can significantly improve the effectiveness and sustainability of their logistical operations, making a valuable contribution to more significant environmental and societal objectives.

14.9.4 Consumer Engagement and Product Stewardship

Consumer Engagement and Product Stewardship are integral to Green Supply Chain Management (GSCM). They highlight the role of businesses in directing customers towards sustainable consumption and assuming accountability for the complete lifespan of their goods. This strategy includes creating items that are long-lasting, easy to fix, and can be recycled. It also involves involving customers in sustainability efforts and managing the disposal of products at the end of their life cycle through recycling or take-back programs. Here, we delve deeper into these principles, providing examples to illustrate them.

Consumer Engagement in Sustainability: Including customers in sustainability practices is essential to enhance the effectiveness of green supply chain activities. Informing customers about the ecological advantages of sustainable products and activities can impact their purchase choices and promote more conscientious consumption behaviors. Patagonia's "Worn Wear" program supports repairing, sharing, and recycling their items. Additionally, it educates consumers on the significance of extending the lifespan of their clothes to minimize the environmental effects. To promote sustainability and engage customers, Patagonia offers repair workshops and online tutorials to empower individuals to mend their apparel.

Product Stewardship and Extended Producer Responsibility: Product stewardship and extended producer responsibility (EPR) are approaches that expand the accountability of manufacturers to include the period after consumers have used a product. These techniques strive to reduce items' ecological footprint by guaranteeing effective disposal management at the end of their life cycle. As an illustration, Apple has established an extensive recycling initiative, providing complimentary recycling services for all its products and striving to incorporate recycled components into its new products. Apple's disassembly robot, Daisy, can dismantle 15 distinct iPhone models to retrieve precious parts for recycling, demonstrating a solid dedication to product stewardship (Apple Inc., 2020a, 2020b).

Circular Economy and Sustainable Product Design: The concepts of the circular economy are essential for involving consumers and ensuring responsible management of products. These principles prioritize the design of products with a focus on durability, reusability, and recyclability—Adidas, under its Futurecraft. The loop program has created a running shoe that is entirely recyclable. Upon reaching the end of its lifespan, the shoe may be sent back to Adidas, where it will be disassembled and used to manufacture new high-performance footwear. This process aligns with the ideas of the circular economy since it ensures that materials are continuously utilized and prevents them from being disposed of in landfills (Adidas, 2020).

Collaborative Efforts for Sustainability: Effective product stewardship requires collaboration among firms, consumers, and waste management agencies. The "New Plastics Economy" program of the Ellen MacArthur Foundation aims to collaborate with businesses, governments, and other organizations to reconsider and revamp the future of plastics, with a particular focus on packaging. This effort seeks to establish a circular economy for plastics by enhancing recycling rates, promoting recycled content utilization, and the active involvement of consumers in circular practices (Ellen MacArthur Foundation, 2020a, 2020b).

Consumer involvement and product stewardship are crucial approaches for mitigating the environmental footprint of products and advocating for sustainable consumption habits. Companies may contribute substantially to the shift towards a more sustainable and circular economy by considering the whole lifespan of their goods, including customers in sustainability initiatives, and accepting responsibility for managing the items at the end of their lives. Implementing these practices positively affects the environment and allows businesses to cultivate brand loyalty, distinguish themselves in the market, and meet the increasing customer demand for sustainable products.

14.10 Challenges in Implementing GSCM

Implementing Green Supply Chain Management (GSCM) involves strategically incorporating environmental and ethical factors into the core of supply chain operations. The introduction of this paradigm shift poses various problems that organizations must skillfully manage to achieve sustainable goals. GSCM aims to revolutionize conventional supply chains by including environmentally friendly procurement, production, distribution, and end-of-life methods. Nevertheless, achieving a completely environmentally friendly supply chain is challenging because of economic, technological, legal, and behavioral obstacles. Tackling these difficulties is not just about improving the reputation of companies. However, it is crucial for enterprises' sustainable success and environmental responsibility in a more environmentally aware global market.

Economic and financial obstacles frequently hinder the implementation process of GSCM. Implementing sustainable supply chain methods sometimes requires substantial initial expenditures in environmentally friendly technology, sustainable materials, and process reconfigurations. These early expenses provide formidable obstacles for some organizations, particularly small and medium-sized firms. Furthermore, the economic justification for these investments may be obscured by uncertainty around the return on investment, especially when the advantages, albeit significant, are realized over extended periods. Therefore, firms have a substantial problem connecting economic imperatives with environmental goals along their Green Supply Chain Management (GSCM) journey. Technological issues are crucial in determining the practicality and efficiency of GSCM initiatives. Technological developments can potentially improve sustainability, but the speed of invention, rates of adoption, and integration into current systems can vary greatly. Organizations frequently struggle with finding, acquiring, and efficiently implementing the appropriate technology that might enhance productivity, decrease inefficiencies, and mitigate environmental effects. The absence of technical proficiency and infrastructure assistance exacerbates this difficulty, rendering the transition to more environmentally friendly supply chains daunting.

Moreover, the adoption of GSCM is significantly impacted by regulatory frameworks and the culture inside an organization. Organizations operating globally may struggle to navigate different jurisdictions' intricate and ever-changing environmental rules. Furthermore, cultivating a culture of sustainability inside organizations and their supply chain partners necessitates a deliberate endeavor to synchronize beliefs, practices, and ambitions to achieve common environmental goals. To successfully address the organizational and cultural problems related to GSCM, it is crucial to overcome resistance to change, dismantle silos, and promote collaborative approaches to sustainability. To summarise, the transition to Green Supply Chain Management is vital for environmental sustainability and has the potential to bring economic advantages. However, this process is filled with difficulties and obstacles. To effectively apply GSCM, it is necessary to overcome financial limitations, technological challenges, regulatory intricacies, and restrictions related to organizational culture. Organizations dedicated to this shift must embrace a comprehensive strategy that utilizes cutting-edge ideas, promotes collaborations, and nurtures a robust commitment to sustainability to overcome these obstacles and accomplish their green supply chain goals.

14.10.1 High Initial Costs

High initial expenses hinder the introduction of Green Supply Chain Management (GSCM) methods. Adopting more environmentally friendly supply chain operations requires significant investments in new technology, sustainable materials, and process redesign to mitigate environmental effects. These expenditures can be intimidating, particularly for small and medium-sized firms (SMEs) with limited financial resources compared to more prominent organizations. Notwithstanding these difficulties, the enduring advantages of GSCM, such as decreased operational expenses, adherence to regulatory norms, and improved business standing, might compensate for the original financial investment. This is a detailed analysis of the significant upfront costs involved in implementing Green Supply Chain Management (GSCM):

Costs Associated with Sustainable Materials: Acquiring sustainable resources typically involves higher costs than traditional materials. This is a result of the limited supply, increased costs of production, and the emphasis placed on certifying and verifying sustainability qualifications. As an illustration, the shift to organic cotton in the textile sector, pursued by firms such as Patagonia, incurs increased expenses due to adopting more labor-intensive farming methods and the resulting decrease in crop yields. Nevertheless, using organic cotton diminishes the application of pesticides and promotes the development of more robust ecosystems, hence supporting overarching sustainability objectives (Patagonia, 2020a, 2020b).

Process Redesign and Training: Implementing Green Supply Chain Management (GSCM) sometimes requires restructuring current operations to enhance productivity and minimize ecological footprints. This might entail the restructuring of production lines, the use of novel manufacturing techniques, and the education of personnel in sustainable methodologies. Process redesign and personnel training incur substantial expenses. However, they are necessary for GSCM to attain efficiency improvements and environmental advantages. For instance, Dell Technologies has made investments in revamping its packaging procedures by using bamboo, mushrooms, and recycled plastics to provide packaging solutions that are more environmentally friendly. Dell's packaging approach decreased its environmental effect and generated long-term economic benefits (Dell Technologies, 2020).

Economic Incentives and Support Mechanisms: Several economic incentives and assistance systems might be utilized to address the issue of expensive upfront expenses. Governments and international organizations provide grants, low-interest loans, and tax incentives to promote enterprises adopting sustainable practices. The European Union's Horizon 2020 program offers financial support for research and innovation projects that involve sustainability and green supply chain efforts. This financing helps cover the initial expenses for enterprises participating in these projects (European Commission, 2020a, 2020b, 2020c).

To summarise, although the significant upfront expenses associated with adopting Green Supply Chain Management (GSCM) may be challenging, the deliberate allocation of resources towards sustainable technologies, materials, and process redesign may result in significant advantages in the environment, economy, and reputation. Companies that successfully overcome these early financial obstacles generally discover that long-term cost reductions, increased productivity, and improved brand reputation justify the investment. To ensure the long-term viability and resilience of the business, it is crucial to see the adoption of GSCM as a strategic investment. This involves utilizing existing support systems to alleviate the financial load.

14.10.2 Supply Chain Complexity

The intricate nature of supply chains is a considerable obstacle to implementing Green Supply Chain Management (GSCM). Contemporary supply chains are complex networks that extend across different nations, encompass various stakeholders, and are subject to diverse rules and norms. The intricate nature of the supply chain might pose challenges in attaining the desired level of transparency and coordination necessary for the successful implementation of sustainable practices throughout the supply chain. The difficulties associated with overseeing such intricacy involve guaranteeing supplier adherence to environmental regulations, controlling the ecological consequences of logistics and transportation, and adjusting to diverse regulatory frameworks. The following is an in-depth analysis of the difficulties linked to supply chain intricacy in GSCM.

Ensuring Supplier Compliance with Environmental Standards: An essential obstacle in overseeing intricate supply chains is guaranteeing that all suppliers comply with environmental and sustainability criteria. Managing suppliers in nations with lax ecological rules can pose significant difficulties. Apple Inc. has effectively tackled this difficulty by mandating that its suppliers adhere to its Supplier Code of Conduct, which encompasses extensive environmental criteria. Apple often performs audits and offers training to assist suppliers in meeting these criteria, showcasing a proactive strategy for managing supplier compliance within an intricate global supply chain (Apple Inc., 2020a, 2020b).

Managing the Environmental Impact of Logistics and Transportation: The logistics and transportation aspect of the supply chain is an additional area where complexity might impede the implementation of Green Supply Chain Management (GSCM). Efficiently planning and coordinating transportation routes and modes is necessary to minimize carbon emissions. DHL, an international logistics corporation, has tackled this obstacle by introducing its GoGreen initiative, which encompasses strategies such as employing electric cars for deliveries, optimizing shipping routes, and providing consumers with the opportunity to offset carbon emissions. The program showcases how enterprises may effectively handle the ecological consequences of logistics in intricate supply networks (DHL, 2020).

Adapting to Varying Regulatory Environments: Supply chains that extend across numerous nations must manage a varied and complex array of environmental legislation and requirements. Various regulations might make establishing consistent Green Supply Chain Management procedures difficult across all activities. H&M, a multinational fashion retailer, has addressed this difficulty by collaborating closely with suppliers and local governments to comprehend and adhere to local environmental legislation, all while attempting to uphold rigorous global sustainability standards. H&M's endeavors to employ sustainable materials and minimize water and energy consumption in its supply chain are guided by a comprehensive comprehension of legislative mandates in various countries (H&M, 2020).

Leveraging Technology to Overcome Complexity: Advanced technologies like blockchain, the Internet of Things (IoT), and big data analytics provide effective methods for handling supply chain complexity in GSCM. IBM's blockchain-based Food Trust network allows food merchants, suppliers, and producers to track the source and safety of food items. This technology improves the clarity and responsibility in the supply chain, making it easier to execute GSCM procedures more efficiently (IBM, 2020).

Collaborative Efforts for Sustainability: Effective collaboration among all stakeholders in the supply chain is crucial for successfully navigating the intricacies of Global Supply Chain Management (GSCM). The Sustainable Garment Coalition, including several brands, retailers, and manufacturers, has created the Higg Index, a collection of instruments for assessing and enhancing the environmental and social sustainability of garment and footwear supply chains. The collaborative approach enables member firms to exchange best practices and tackle the problems of implementing GSCM in an intricate global supply chain (Sustainable Apparel Coalition, 2020a, 2020b).

Ultimately, although the intricate nature of supply chains poses considerable obstacles to adopting GSCM, these obstacles may be overcome. By effectively managing suppliers, using cutting-edge technology, adjusting to regulatory frameworks, and promoting cooperation, businesses may successfully negotiate the complexities of their supply chains to accomplish their sustainability objectives. The firms Apple, DHL, H&M, IBM, and the Sustainable Apparel Coalition provide examples of how companies tackle the intricacies of managing environmentally friendly supply chains.

14.10.3 Lack of Awareness and Expertise

Implementing Green Supply Chain Management (GSCM) necessitates dedication to environmental sustainability and a comprehensive comprehension of intricate supply chain dynamics and the particular strategies that might mitigate environmental effects. A significant obstacle to the successful implementation of Green Supply Chain Management (GSCM) is stakeholders' limited knowledge and skills at different levels of the supply chain. This challenge involves many concerns, from basic comprehension of sustainability concepts to specialized expertise in environmentally friendly materials, energy-efficient manufacturing methods, and sustainable logistics. Addressing this lack of awareness and experience is essential to promote the wider adoption of GSCM procedures. Here, we go into this difficulty further, discussing methods for conquering it, backed by examples.

The Awareness Gap: An inherent obstacle in implementing GSCM is the insufficient knowledge among enterprises, suppliers, and consumers on the significance of sustainability in supply chain operations and the possible environmental consequences of their actions. Inadequate comprehension of these matters may result in stakeholders lacking the will to modify existing practices or allocate resources toward sustainable alternatives. An illustrative instance is provided by Zhu and Sarkis (2004a, 2004b), who emphasize the significance of environmental training and education in enhancing suppliers' environmental performance. They propose that heightened awareness can result in adopting more sustainable practices throughout the supply chain.

The Expertise Challenge: Despite stakeholders' awareness of the significance of sustainability, they may lack the skills to apply Green Supply Chain Management (GSCM) techniques effectively. This encompasses understanding sustainable sourcing alternatives, implementing methods to minimize waste and emissions throughout production, and devising environmentally friendly shipping and distribution strategies. The intricacy of contemporary supply chains, which frequently extend across numerous nations and regulatory contexts, exacerbates the issue of acquiring knowledge. Corporations such as Unilever have significantly invested in training and capacity-building initiatives for their suppliers to tackle this issue. The objective is to improve their suppliers' capability to adhere to sustainability criteria (Unilever, 2020a, 2020b).

Strategies for Building Awareness and Expertise: An excellent approach to addressing the lack of knowledge and skill is to create and distribute instructional materials about GSCM. This encompasses many educational formats such as seminars, webinars, online courses, and best practice guidelines that are easily accessible to diverse individuals involved in the subject matter. The World Wildlife Fund (WWF) provides training programs and materials to encourage businesses and suppliers to adopt sustainable supply chain practices (WWF, 2020). Another strategy is utilizing partnerships and collaborative networks to exchange knowledge and implement best practices in GSCM. The Sustainable Supply Chain Foundation (SSCF) offers a forum for enterprises to work together to tackle sustainability issues in their supply chains. This platform enables the sharing knowledge and creative solutions (SSCF, 2020).

Examples of Addressing the Challenge: Multiple corporations and organizations have implemented programs to improve knowledge and proficiency in Green Supply Chain Management (GSCM). For instance, Nike has created the Materials Sustainability Index (MSI), a publicly accessible tool that offers information about the ecological effects of different materials utilized in their goods. Nike's objective in providing this tool to designers and product developers is to promote the adoption of environmentally friendly materials in the garment industry (Nike, 2020a, 2020b). The Circular Economy 100 (CE100) program of the Ellen MacArthur Foundation aims to unite enterprises, governments, and academic institutions to enhance their ability to implement circular economy principles, such as GSCM. The program provides seminars, courses, and collaborative projects to improve participants' comprehension and application of circular economy ideas (Ellen MacArthur Foundation, 2020a, 2020b).

Ultimately, it is crucial to tackle the issue of insufficient knowledge and competence to carry out Green Supply Chain Management (GSCM) effectively. Employing education, training, and cooperative endeavors, businesses and their supply chain partners may acquire the knowledge and expertise necessary to shift towards more environmentally friendly operations. These programs promote environmental sustainability and provide opportunities for innovation and a competitive edge in the global economy.

14.10.4 Measurement and Reporting

Measurement and reporting are essential elements of adopting Green Supply Chain Management (GSCM), as they allow for evaluating the efficiency of sustainability programs, identifying areas that need improvement, and communicating progress to stakeholders. Nevertheless, creating a thorough and dependable system for assessing and documenting environmental performance throughout the supply chain poses considerable difficulties. These problems encompass the creation of standardized measures, the gathering of precise and complete data, and the guarantee of openness and reliability in reporting. Resolving these problems is crucial for organizations dedicated to GSCM, as it empowers them to make well-informed choices, establish credibility with stakeholders, and contribute to global sustainability initiatives. Here, we explore the intricacies of measuring and reporting in GSCM:

Developing Standardized Metrics: A significant obstacle in measuring and reporting for GSCM is the absence of standardized measures for evaluating environmental performance. Without globally recognized metrics, organizations may have difficulty assessing the effectiveness of their sustainability initiatives or making comparisons with industry standards. The Global Reporting Initiative (GRI) and the Carbon Disclosure Project (CDP) have made substantial progress in tackling this matter by creating standardized frameworks for sustainability reporting. GRI's Sustainability Reporting Standards provide instructions for reporting several sustainability factors, such as environmental, social, and governance (ESG) performance. These principles promote comparability and transparency among organizations and industries (GRI, 2020).

Collecting Accurate and Comprehensive Data: To assess the environmental consequences of supply chain operations precisely, it is necessary to gather data from several sources, such as suppliers, logistics providers, and manufacturers. Managing complicated global supply chains with several stakeholders may be extremely tough. Technology is essential in addressing this dilemma since it provides technologies like blockchain and the Internet of Things (IoT) that enable the recording and verifying of sustainability data along the supply chain. IBM's Blockchain Transparent Supply solution allows firms to track the origin and lifespan of products in real time, improving the precision and dependability of sustainability data (IBM, 2020).

Ensuring Transparency and Credibility in Reporting: Transparency and credibility are crucial elements in ensuring the effectiveness of sustainability reporting since they establish confidence among customers, investors, and other stakeholders. Nevertheless, apprehensions over greenwashing, which refers to firms overstating their environmental accomplishments, can erode trust in sustainability reports. Third-party verification and certification are essential in resolving these issues since they offer independent proof of a company's sustainability claims. The Science-Based-Target-Initiative (SBTI) provides a rigorous procedure for firms to establish and verify their objectives for reducing greenhouse gas emissions following the most up-to-date climate knowledge. This ensures the legitimacy of their sustainability endeavors (SBTi, 2020).

Examples of Effective Measurement and Reporting: Several companies have shown exemplary leadership in Green Supply Chain Management (GSCM) by effectively measuring and reporting their efforts. For example, Schneider Electric releases a yearly sustainability report that adheres to GRI criteria. This report thoroughly summarizes its environmental performance, encompassing energy usage, greenhouse gas emissions, and waste handling. The report furthermore emphasizes advancements made towards the company's sustainability objectives, showcasing a dedication to openness and responsibility (Schneider Electric, 2020a, 2020b).

Unilever's Sustainable Living Plan sets ambitious goals for promoting health and well-being, minimizing environmental harm, and strengthening livelihoods. The firm offers extensive progress reports on its website, which include information on greenhouse gas emissions, water use, and sustainable sourcing procedures. These reports demonstrate how good measurement and reporting may contribute to successfully implementing Green Supply Chain Management (GSCM) (Unilever, 2020a, 2020b).

Measurement and reporting are essential for the successful deployment of GSCM. Companies may successfully monitor their environmental impact, make informed decisions, and demonstrate their commitment to sustainability by creating standardized measures, utilizing technology for data collecting, and guaranteeing transparent and credible reporting. As the subject of Global Supply Chain Management (GSCM) progresses, further innovation in measurement and reporting methods will be crucial for promoting global sustainability objectives.

14.11 Strategies for Implementation

Implementing Green Supply Chain Management (GSCM) necessitates a comprehensive methodology that includes all facets of an organization's operations, including procurement, manufacturing, logistics, and customer involvement. The core of GSCM is integrating sustainability into every aspect of the supply chain, intending to reduce environmental effects while preserving or improving economic performance. This transformative undertaking requires meticulous strategizing, synchronization, and dedication at every level of the organization and its network of suppliers and partners. Successful implementation of Green Supply Chain Management (GSCM) involves adopting sustainable sourcing practices, investing in environmentally friendly technologies and processes, optimizing logistics and distribution to minimize carbon footprints, and promoting product stewardship and circular economy principles. By incorporating these approaches, firms may support global environmental sustainability objectives and attain operational efficiency, bolster brand reputation, and fulfill the growing customer need for sustainable products.

Nevertheless, achieving complete integration of GSCM techniques comes with obstacles, such as the requirement to surmount substantial initial expenses, negotiate the intricacies of worldwide supply chains, establish proficiency and consciousness, and establish strong systems for measurement and reporting. To overcome these difficulties, adopting a comprehensive strategy that utilizes technology advancements, encourages cooperation and the exchange of information among relevant parties, and conforms to favorable policies and regulatory structures is necessary. Moreover, fostering a culture of sustainability inside the organization and its supply chain is essential for promoting enduring dedication and transformation. As organizations worldwide work to address environmental concerns and meet stakeholders' expectations, the strategic adoption of Green Supply Chain Management (GSCM) is a crucial strategy to achieve sustainable development and gain a competitive advantage in the changing global market.

14.11.1 Stakeholder Collaboration

Stakeholder cooperation is crucial in implementing Green Supply Chain Management (GSCM) effectively. It entails actively involving all participants, whether directly or indirectly, in the supply chain, such as suppliers, manufacturers, customers, regulatory agencies, and non-governmental organizations (NGOs). This collaborative strategy aims to synchronize endeavors toward sustainability, distribute responsibilities, and utilize collective skills to tackle environmental concerns. Efficient engagement with stakeholders may result in higher ecological performance, advancements in sustainable practices, and increased transparency and trust among all involved parties. This is an in-depth analysis of the importance of stakeholder participation in Green Supply Chain Management (GSCM).

Enhancing Supplier Engagement: Close collaboration with suppliers is essential for effectively incorporating sustainability into the supply chain. Unilever and similar companies have implemented supplier engagement programs that prioritize sustainability training, capacity development, and the exchange of best practices. Unilever's Partner with Purpose program collaborates with suppliers to enhance their environmental performance, focusing on decreasing greenhouse gas emissions, practicing water conservation, and managing waste. Unilever cooperates with suppliers to ensure its supply chain supports its sustainability objectives (Unilever, 2020a, 2020b).

Collaborative Platforms and Initiatives: Global supply networks are frequently so intricate that they need coordinated action that surpasses the efforts of individual enterprises. Collaborative platforms and multi-stakeholder initiatives offer a structure for exchanging information, resources, and optimal methods. The Sustainable Clothing Coalition (SAC) demonstrates this strategy by uniting clothing and footwear firms, suppliers, NGOs, and academic institutions to jointly propel the sector towards sustainability. The Sustainable Apparel Coalition's Higg Index is a standardized instrument that quantifies and evaluates the sustainability performance of a firm or product. It enables comparison and enhancement of sustainability practices across the sector.

Customer and Consumer Engagement: Including customers and consumers in sustainability initiatives is crucial to stimulate the desire for sustainable products and encourage responsible consumption. Initiatives that promote recycling, product take-back, and the conscientious use of items can significantly diminish environmental effects. Patagonia's Worn Wear program encourages repairing and reusing its apparel while involving customers in the sustainability story. This fosters a community of ecologically aware consumers who prioritize the durability of products above constant consumption (Patagonia, 2020a, 2020b).

Regulatory Bodies and Policy Advocacy: Engaging in policy advocacy and collaborating with regulatory agencies are essential techniques for promoting the advancement of GSCM. Regulatory frameworks have the power to encourage or require sustainable activities, making it vital for firms to actively engage in defining these policies. The World Business Council for Sustainable Development (WBCSD) collaborates with governments and international organizations to promote policies that endorse sustainability in business, specifically focusing on Green Supply Chain Management (GSCM) practices. The WBCSD collaborates with prominent tire manufacturers through its Tyre Industry Project to tackle the environmental consequences of tire and rubber production and use (WBCSD, 2020).

Public–Private Partnerships: Public–private partnerships (PPPs) can gather resources, experience, and innovation to support sustainable supply chain activities. The New Plastics Economy program, led by the Ellen MacArthur Foundation, is a public–private partnership (PPP) to establish a circular economy for plastics. The initiative aims to convene corporations, governments, and other relevant parties to reassess and revamp the plastics industry, specifically focusing on enhancing recycling rates, minimizing plastic waste, and promoting advancements in packaging (Ellen MacArthur Foundation 2020a, 2020b).

Ultimately, the cooperation of stakeholders is crucial for the successful execution of Green Supply Chain Management (GSCM). Companies may utilize collaborative relationships throughout the supply chain and other areas to harness combined capabilities to tackle environmental issues, develop sustainable methods, and facilitate the shift towards a more sustainable economy. The case studies of Unilever, the Sustainable Apparel Coalition, Patagonia, the WBCSD, and the Ellen MacArthur Foundation illustrate that collaboration across many stakeholders is crucial for attaining comprehensive and significant sustainability results.

14.11.2 Technology and Innovation

Technology and innovation are essential in implementing Green Supply Chain Management (GSCM), providing revolutionary solutions that improve environmental sustainability across the supply chain. Innovation empowers organizations to diminish their environmental impact, enhance efficiency, and fulfill the increasing need for sustainable products by utilizing improved materials, production methods, logistics, and recycling technology. In the following discussion, we explore the fundamental role of technology and innovation in supporting the success of GSCM activities.

Advanced Materials and Manufacturing Technologies: Innovation in materials science and manufacturing technology enables corporations to manufacture items more sustainably. In collaboration with Parley for the Oceans, Adidas utilizes repurposed ocean plastic to produce top-quality athletic apparel, showcasing the potential of cutting-edge materials in addressing environmental issues such as ocean pollution (Adidas & Parley, 2020). Nike's Flyknit technology precisely knits the top section of shoes, resulting in a 60% reduction in material waste compared to traditional cut-and-sew processes. This demonstrates how technical developments in production may effectively minimize waste (Nike, 2020a, 2020b).

Energy Efficiency and Renewable Energy: Technological innovations are leading to substantial enhancements in energy efficiency and integrating renewable energy sources in industrial and supply chain processes. Tesla's Gigafactory 1, engineered explicitly for manufacturing batteries for electric vehicles and energy storage goods, is exclusively powered by sustainable energy sources such as solar, wind, and geothermal. The factory's objective is to attain a state of net-zero energy consumption while minimizing its carbon emissions. This demonstrates the feasibility of incorporating renewable energy technology into industrial processes (Tesla, 2020).

Logistics and Transportation Innovations: Technology and innovation play a crucial role in the logistics and transportation industry by improving route optimization, boosting vehicle economy, and facilitating the shift towards more environmentally friendly transportation methods with reduced emissions. UPS's ORION (On-Road Integrated Optimisation and Navigation) routing software employs sophisticated algorithms to calculate the most optimal delivery routes, resulting in substantial fuel savings of millions of gallons per year and a considerable reduction in CO_2 emissions (UPS, 2020). Moreover, the logistics sector is poised to transform by introducing electric and hydrogen fuel cell cars from businesses such as Nikola and Rivian. These vehicles provide sustainable alternatives to traditional diesel-powered trucks.

Circular Economy and Recycling Technologies: Technology and innovation expedite the shift towards a circular economy by enhancing the effectiveness of resource utilization and recycling. The Circular Economy 100 program of the Ellen MacArthur Foundation facilitates collaboration among enterprises, entrepreneurs, and regions to work on circular economy projects. These efforts focus on developing novel business models and technology that enable the recycling and reuse of materials (Ellen MacArthur Foundation 2020a, 2020b). HP Inc. has pledged to enhance the amount of recycled material in its goods by employing sophisticated recycling techniques to convert used plastic cartridges and other plastics from consumers into new HP products HP Inc (2020). This initiative aims to decrease the reliance on new plastic materials.

Digital Technologies for Supply Chain Transparency: Emerging digital technologies such as blockchain, the Internet of Things (IoT), and big data analytics are transforming the level of transparency and traceability in supply chains. These tools allow firms to continuously monitor and control their supply chains, ensuring that sustainability criteria are consistently satisfied. IBM's Food Trust utilizes blockchain technology to provide a transparent and traceable food supply chain. This enables all parties involved, including farmers and merchants, to obtain detailed product information, guaranteeing food safety and sustainability (IBM Food Trust, 2020).

Ultimately, technology and innovation are essential for the effective execution of GSCM. By utilizing progress in materials, manufacturing, renewable energy, logistics, recycling, and digital technologies, firms can overcome environmental obstacles, enhance the sustainability of their supply chains, and contribute positively to a more sustainable future. The cases of Adidas, Nike, Tesla, UPS, HP, and IBM provide evidence that adopting innovation is crucial for enhancing environmental performance and promoting sustainability in supply chain management.

14.11.3 Education and Training

Education and training are essential for effectively implementing Green Supply Chain Management (GSCM) by addressing a significant challenge: the limited knowledge and competence about sustainable practices in organizations and their supply chains. Companies may cultivate the essential knowledge and skills required for sustainable supply chain operations by allocating resources to education and training programs for their workers, suppliers, and partners. This investment promotes a culture of sustainability and enables stakeholders to recognize and adopt measures that decrease environmental effects, enhance efficiency, and adhere to legal requirements. The following is an in-depth analysis of the significance of education and training in Green Supply Chain Management (GSCM).

Building Internal Capacity: Companies are becoming more aware of the need to develop internal capability for sustainability by implementing specialized training programs for their workers. For example, Schneider Electric provides comprehensive training programs on energy efficiency, renewable energy, and sustainability solutions. These programs are offered through Energy University. These programs aim to improve workers' comprehension of sustainability issues and solutions, empowering them to contribute to the company's sustainability objectives and fostering innovation within their responsibilities (Schneider Electric, 2020a, 2020b).

Supplier Development Programs: Enhancing suppliers' sustainability capacities is another crucial element of education and training in GSCM. Several organizations implement supplier development programs to enhance suppliers' environmental performance through training and resource provision. Walmart's Project Gigaton seeks to motivate suppliers to eradicate one gigaton of greenhouse gas emissions from Walmart's worldwide value chain by 2030. As a component of this endeavor, Walmart provides tools, resources, and training to assist suppliers in establishing and attaining their objectives for reducing emissions (Walmart, 2020).

Collaborative Learning Platforms: Collaborative learning platforms and industry-wide initiatives provide chances for exchanging information and implementing best practices in GSCM. The Higg Index, developed by the Sustainable Garment Coalition (SAC), is both a measurement tool for sustainability performance and a forum for exchanging knowledge and adopting industry best practices in the garment sector. The SAC provides its members with opportunities to engage in workshops, webinars, and collaborative projects that support ongoing learning and enhancement of sustainability practices (Sustainable Apparel Coalition, 2020a, 2020b).

Academic and Professional Education: Academic institutions and professional bodies have a crucial role in promoting education and training in GSCM. Universities provide specialized courses and educational programs centered around sustainability and supply chain management, equipping future business leaders with the necessary skills to address sustainability issues. In addition, professional certifications, such as the Certified Supply Chain Professional (CSCP) designation offered by APICS (now part of the Association for Supply Chain Management, ASCM), now include sustainability topics. This ensures supply chain professionals have the necessary knowledge and skills to implement Green Supply Chain Management (GSCM) (ASCM, 2020).

Leveraging Digital Technologies: The advent of digital technology has dramatically increased the scope and availability of educational and training programs. Coursera and edX, online platforms, provide sustainability and supply chain management courses from prestigious institutions and organizations. This enables people and corporations to acquire specialised expertise in GSCM conveniently. These platforms support continuous learning and empower professionals to remain informed about the most recent developments and methodologies in sustainability (Coursera, 2020; edX, 2020).

Ultimately, education and training are essential for integrating sustainability into the core of supply chain operations. Companies may expedite the implementation of Green Supply Chain Management (GSCM) practices, foster sustainability innovation, and accomplish their environmental and social goals by investing in the growth of their workers, suppliers, and partners. The examples mentioned above illustrate that a comprehensive strategy for education and training, including internal programs, collaborative efforts, academic collaborations, and digital learning, may develop the necessary expertise and capability to implement GSCM successfully.

14.11.4 Policy and Incentives

Policy and incentives are essential for promoting the deployment of Green Supply Chain Management (GSCM) techniques. The adoption of sustainable supply chain methods may be influenced by governments, regulatory organizations, and industry groupings by implementing rules and providing incentives. Strategic policies and attractive incentives can motivate corporations to allocate resources toward green technology, embrace sustainable practices, and actively seek advancements in environmental management. On the other hand, the lack of supportive regulations or incentives might create substantial obstacles to the broad implementation of Green Supply Chain Management (GSCM). The following discussion explores the significance of policy and incentives in Green Supply Chain Management (GSCM).

Regulatory Policies and Compliance Requirements: Regulatory policies and requirements establish the lowest acceptable benchmarks for environmental performance in supply chains. Governments can enforce the adoption of Green Supply Chain Management (GSCM) techniques by implementing explicit ecological rules. The Waste Electrical and Electronic Equipment (WEEE) Directive and the Restriction of Hazardous Substances (RoHS) Directive, implemented by the European Union, have substantially impacted electronics manufacturers and suppliers. These directives have prompted the adoption of environmentally friendly practices such as recycling and reducing hazardous substances during production (European Commission, 2020a, 2020b, 2020c).

Economic Incentives: Economic incentives such as tax breaks, subsidies, and grants can significantly reduce financial barriers to implementing Green Supply Chain Management (GSCM) techniques. These incentives enhance the economic feasibility of enterprises' investments in energy-efficient technology, renewable energy sources, and sustainable materials. The U.S. Department of Energy's Loan Programs Office offers financial support to promote implementing renewable energy and energy efficiency projects. This allows enterprises from many industries to invest in environmentally friendly technology U.S. Department of Energy (2020).

Voluntary Programs and Partnerships: Collaboration, information exchange, and recognition may be facilitated via voluntary programs and partnerships between government agencies and the commercial sector, therefore promoting the adoption of Green Supply Chain Management (GSCM) techniques. The SmartWay Transport Partnership, initiated by the U.S. Environmental Protection Agency (EPA), is a commendable voluntary program that effectively assists enterprises in enhancing the sustainability of their logistics operations. SmartWay offers tools and resources to measure, compare, and decrease emissions from goods transportation, encouraging organizations to improve their logistics and supply chain processes (U.S. Environmental Protection Agency, 2020).

Carbon Pricing Mechanisms: Carbon pricing methods, such as carbon taxes or cap-and-trade systems, motivate corporations to decrease their carbon emissions by imposing a financial burden on the release of carbon dioxide. These mechanisms incentivize corporations to allocate resources toward implementing carbon reduction initiatives throughout their supply chains to save expenses. The European Union Emissions Trading System (EU ETS) is the world's most extensive carbon trading system. It is a prime illustration of how carbon pricing may incentivize enterprises across different sectors to embrace environmentally friendly practices and technology to decrease emissions (European Commission, 2020a, 2020b, 2020c).

Green Procurement Policies: Public sector firms can be influenced by governments to adopt green supply chain management (GSCM) practices through the implementation of green procurement rules. These policies mandate the purchase of goods and services that adhere to specific environmental criteria. These rules can stimulate the market for sustainable products, incentivizing providers to embrace more environmentally friendly techniques. The Green Purchasing Network (GPN), established by the government of Japan, aims to encourage the acquisition of ecologically friendly products and services. This initiative is a model for private sector enterprises to emulate (Green Purchasing Network, 2020).

Ultimately, policies and incentives serve as influential mechanisms for encouraging GSCM techniques. Governments and regulatory bodies can influence companies to incorporate sustainability into their supply chain operations by establishing a regulatory framework that promotes sustainability, providing economic incentives for green investments, facilitating voluntary collaboration programs, implementing carbon pricing to incentivize emissions reductions, and adopting green procurement policies. The examples mentioned earlier illustrate that employing a comprehensive strategy involving policies and incentives may successfully promote the extensive adoption of Green Supply Chain Management (GSCM), contributing to larger environmental sustainability objectives.

14.12 Case Study

IKEA, the largest global furniture store, has established a reputation for its unwavering dedication to sustainability and responsible management of the environment. IKEA's green supply chain management (GSCM) methods are essential to its corporate identity and operational efficiency. Their business strategy covers the entire supply chain, from procuring raw materials to managing the end-of-life of their products. This case study examines IKEA's strategy for implementing Green Supply Chain Management (GSCM), specifically emphasizing sustainable sourcing, environmentally friendly production, logistics optimization, and involving consumers in sustainability efforts IKEA (2020).

Sustainable Sourcing: IKEA's dedication to environmentally friendly supply chain management starts with the sustainable sourcing of its raw materials. The firm has established ambitious objectives to guarantee that all of its wood and paper originate from more sustainable sources, defined explicitly as either recycled or certified by the Forest Stewardship Council (FSC). In 2020, IKEA disclosed that 98.5% of its wood and 100% of its cotton complied with sustainability criteria. This accomplishment results from a strong partnership with suppliers and substantial investments in sustainable forestry and cotton growing methods.

IKEA is committed to reducing its environmental footprint by implementing Eco-friendly Manufacturing. The corporation has adopted strategies to diminish energy usage and mitigate the release of greenhouse gas emissions in its production facilities. As an illustration, IKEA has made financial commitments towards sustainable energy by implementing photovoltaic systems on the rooftops of several retail outlets and storage facilities. The corporation prioritizes waste reduction and recycling, aiming to achieve a "climate-positive" status by 2030.

Logistics and Distribution Optimization: IKEA's shipping systems are meticulously engineered to maximize efficiency and sustainability. The firm has enhanced its transportation network to minimize fuel use and carbon emissions. This involves utilizing electric cars for home deliveries in specific markets and creating items that can be packaged flat, optimizing shipping efficiency and minimizing transportation emissions. IKEA promotes environmental responsibility by providing services like bicycle delivery and collaborating with electric car charging stations, encouraging customers to reduce their ecological footprint.

Consumer Engagement and Product Stewardship: IKEA actively involves customers in its sustainability efforts through numerous projects, promoting consumer engagement and product stewardship. The firm offers comprehensive information on its goods' environmental characteristics and advocates for using sustainable resources. IKEA's "Buy Back & Resell" program, which allows consumers to sell their used IKEA furniture for resale or recycling, demonstrates its dedication to the circular economy and responsible management of its products. In addition, IKEA's commitment to innovation is evident in its investment in developing sustainable materials, such as bio-based polymers, showcasing its progressive approach to product design and user involvement.

Impact and Challenges: IKEA's GSCM policies have resulted in substantial environmental advantages, such as decreased carbon emissions, heightened utilization of renewable energy, and enhanced resource efficiency. Nevertheless, the corporation recognizes that there are still obstacles to overcome, especially in attaining its ultimate objective of becoming climate-positive and circular. IKEA is actively striving to enhance the sustainability of its supply chain through a combination of innovation, engagement with stakeholders, and ongoing efforts to improve it.

IKEA's implementation of green supply chain management methods sets a prominent standard for big retailers to adopt sustainability measures. By prioritizing sustainable sourcing, eco-friendly production, efficient logistics, and consumer interaction, IKEA showcases the compatibility of environmental responsibility with corporate success. This case study highlights the significance of dedication, ingenuity, and cooperation in facilitating the shift towards more environmentally friendly supply chains.

14.13 Future Development in Industrial Ecology

The future development of industrial ecology in the manufacturing industry is positioned to impact the direction of global sustainability initiatives substantially. Industries globally face increasing difficulties caused by environmental degradation, resource depletion, and climate change. The ideas of industrial ecology provide a plan for changing production processes. This transition focuses on optimizing resource utilization, minimizing waste, and establishing closed-loop systems that emulate natural ecosystems. This is a detailed analysis of the expected advancements in industrial ecology, specifically in the manufacturing industry. Initially, incorporating circular economy ideas into industrial processes is anticipated to gather speed. The circular economy concept prioritizes the extended use of resources by implementing measures including durable design, recycling, and remanufacturing. This strategy minimizes waste and the need for new materials, resulting in a more sustainable utilization of resources.

Manufacturers are becoming more aware of this approach's economic and environmental advantages, integrating circular economy methods into their operations to accomplish efficiency improvements and sustainability goals. Furthermore, the manufacturing sector will significantly benefit from technical innovation to advance industrial ecology. Emerging technologies, like additive manufacturing (3D printing), artificial intelligence (AI), and the Internet of Things (IoT), can fundamentally transform industrial processes. For instance, additive manufacturing reduces material waste by sequentially constructing products. At the same time, artificial intelligence (AI) and the Internet of Things (IoT) may enhance production processes to maximize energy and material efficiency. As these technologies progress, their implementation will enable the use of more environmentally friendly production methods that accord with the objectives of industrial ecology.

Furthermore, the growth of industrial symbiosis networks, in which waste or by-products from one operation are utilized as the input for another, is expected to increase. These networks facilitate the effective distribution of resources among various production organizations, reducing waste and decreasing environmental consequences. Industrial symbiosis facilitates collaboration across industries, leading to a comprehensive approach to sustainability, in which the products or resources generated by one sector may be utilized by another. Expanding these networks will be vital in attaining industrial ecology's waste reduction and resource efficiency goals. Additionally, there will be a heightened focus on life cycle assessment (LCA) and sustainable supply chain management. Life Cycle Assessment (LCA) offers a thorough structure for assessing the ecological consequences of products across their entire life cycle, enabling informed choices about materials, design, and production methods. Similarly, sustainable supply chain management incorporates environmental and social sustainability factors into supply chain operations. As corporations aim to decrease their environmental impact, these practices will become more and more essential to the manufacturing sector's adoption of industrial ecology.

Furthermore, it is anticipated that there will be advancements in regulatory frameworks and policy incentives to facilitate the adoption of industrial ecology concepts in the manufacturing sector. Governments and international organizations would probably implement more stringent environmental rules and provide incentives to encourage sustainable industrial practices. These regulatory initiatives will have a crucial impact on motivating firms to embrace industrial ecology concepts, thus pushing the industry towards increased sustainability.

Moreover, customer demand for sustainable products will influence the manufacturing sector's adoption of industrial ecology methods. With the increased awareness of environmental concerns, people actively seek items created sustainably and obtained ethically. This requirement will compel firms to embrace more environmentally friendly methods across their supply chains, encompassing everything from acquiring raw materials to the manufacturing process and proper waste disposal. Education and collaboration will play a crucial role in the future advancement of industrial ecology in the manufacturing sector. The effectiveness of industrial ecology relies on the broad comprehension and application of its fundamental concepts. Continual education and training for experts in the industry, along with cooperation among firms, academia, and governments, are necessary for this. By cultivating a mindset of sustainability and collaboration, the manufacturing industry can wholeheartedly seize the possibilities offered by industrial ecology. Ultimately, the future advancement of industrial ecology in the manufacturing industry presents a hopeful trajectory for attaining sustainable production and consumption. By incorporating circular economy ideas, technological innovation, industrial symbiosis, and sustainable supply chain management, and with the help of supporting legislative frameworks and customer demand, the industry may make substantial progress in minimizing its environmental footprint. Education and collaboration are crucial for achieving this objective since they will ensure widespread adoption and successful implementation of the concepts of industrial ecology.

References

Adidas and Parley. (2020). *Adidas Parley collection.*

Apple Inc. (2020a). *Apple supplier responsibility 2020 progress report.*

Apple Inc. (2020b). *2020 progress report. Supplier responsibility.*

Association for Supply Chain Management (ASCM). (2020). *Certified supply chain professional (CSCP).*

Auld, G., Balboa, C., Bernstein, S., & Cashore, B. (2008). The emergence of non-state market-driven (NSMD) global environmental governance: A cross-sectoral assessment. *Governance, 21*(4), 503–524.

BASF SE. (2020). *BASF report 2020: Integrated corporate report.*

Baumgartner, R. J., & Rauter, R. (2017). Strategic perspectives of corporate sustainability management to develop a sustainable organization. *Journal of Cleaner Production, 140*, 81–92. https://doi.org/10.1016/j.jclepro.2016.06.085

Boons, F., & Spekkink, W. (2012). Levels of institutional capacity and actor expectations about industrial symbiosis: Evidence from the Dutch stimulation program 1999–2004. *Journal of Industrial Ecology, 16*(1), 61–69 (The paper examines the importance of institutional capacity and stakeholder engagement in successfully implementing industrial symbiosis initiatives, drawing on evidence from a Dutch program).

Chertow, M. R. (2007). Uncovering' industrial symbiosis. *Journal of Industrial Ecology.*

Chertow, M. R., & Park, J. (2016). Scholarship and practice in industrial symbiosis: 1989–2014. In *Handbook of recycling: State-of-the-art for practitioners, analysts, and scientists.* Elsevier. https://doi.org/10.1016/B978-0-12-396459-5.00001-8

Chesbrough, H. (2003). *Open innovation: The new imperative for creating and profiting from technology.* Harvard Business School Press.

Cohen-Rosenthal, E., & Musnikow, J. (2017). *Eco-industrial strategies.* Taylor & Francis.

Coursera. (2020). *Sustainability courses.*

Dell Technologies. (2020). *Legacy of good annual report.*

Despeisse, M., Oates, M. R., & Ball, P. D. (2012). Modelling and mapping industrial systems to analyse and manage material flows and resource use. *Journal of Cleaner Production, 31*, 124–137 (The authors provide methodologies for mapping and modeling industrial systems to improve material flow management and resource use efficiency across different scales).

Despeisse, M., et al. (2017). The role of manufacturing in achieving sustainable development goals. *Journal of Industrial Ecology.*

DHL. (2020). *DHL GoGreen Solutions.*

Eccles, R. G., Perkins, K. M., & Serafeim, G. (2012). How to become a sustainable company. *MIT Sloan Management Review, 53*(4), 43–50. https://sloanreview.mit.edu/article/how-to-become-a-sustainable-company/

edX. (2020). *Supply chain management courses.*

Ehrenfeld, J., & Gertler, N. (1997). Industrial ecology in practice: The evolution of interdependence at Kalundborg. *Journal of Industrial Ecology, 1*(1), 67–79 (This seminal paper discusses the

Kalundborg Eco-Industrial Park in Denmark as a pioneering example of industrial symbiosis, showcasing the benefits of interconnected industrial processes).
Ehrenfeld, J. R. (2000). Industrial ecology: paradigm shift or normal science? *American Behavioral Scientist, 44*(2), 229–244.
Ellen MacArthur Foundation. (2020a). *Circular economy 100 (CE100).*
Ellen MacArthur Foundation. (2020b). *New plastics economy initiative.*
European Commission. (2018). *Circular economy package.*
European Commission. (2020a). *Horizon 2020 Work Programme.*
European Commission. (2020b). *Waste electrical and electronic equipment (WEEE).*
European Commission. (2020c). *EU emissions trading system (EU ETS).*
Fam, D., et al. (2017). *Transdisciplinary research and practice for sustainability outcomes.* Routledge.
General Electric. (2020). *GE reports: Additive manufacturing.*
Ghisellini, P., et al. (2016). A review on circular economy: The expected transition to a balanced interplay of environmental and economic systems. *Journal of Cleaner Production.*
Gibbs, D., & Deutz, P. (2007). Reflections on implementing industrial ecology through eco-industrial park development. *Journal of Cleaner Production, 15*(17), 1683–1695 (The authors reflect on the challenges and opportunities of developing eco-industrial parks, highlighting the role of policy and collaborative efforts in fostering sustainable industrial practices).
Global Reporting Initiative (GRI). (2020). *GRI standards.*
Green Purchasing Network. (2020). *About the GPN.*
Gregson, N., et al. (2015). *Interrogating the circular economy: The moral economy of resource recovery in the EU.* Economy and Society.
HP Inc. (2020). *Sustainable impact report.*
H&M. (2020). *H&M sustainability report 2020.*
IBM. (2020). *IBM food trust.*
IBM Food Trust. (2020). *Blockchain for supply chain transparency.*
IKEA. (2020). *IKEA sustainability report.*
Kollmuss, A., & Agyeman, J. (2002). Mind the gap: Why do people act environmentally and what are the barriers to pro-environmental behavior? *Environmental Education Research, 8*(3), 239–260. https://doi.org/10.1080/13504620220145401
Korhonen, J., Honkasalo, A., & Seppälä, J. (2018a). Circular economy: The concept and its limitations. *Ecological Economics.*
Korhonen, J., et al. (2018b). Circular economy as an essentially contested concept. *Journal of Cleaner Production.*
Li, X., Li, W., Sun, L., Cao, Y., & Wang, X. (2021). Big data in smart farming: A review. *Agricultural Systems, 189*, 103046. https://doi.org/10.1016/j.agsy.2021.103046
Lifset, R., & Graedel, T. E. (2002). Industrial ecology: goals and definitions. *A handbook of industrial ecology*, 3–15.
Lifset, R., Graedel, T. E., Allenby, B. R., & Linroth, M. T. (2013). Industrial ecology: The role of manufactured capital in sustainability. *Proceedings of the National Academy of Sciences, 110*(20), 6260–6264 (This article explores the role of technological innovation and manufactured capital in advancing the goals of industrial ecology and sustainability).
Lombardi, D. R., & Laybourn, P. (2012). Redefining industrial symbiosis: Crossing academic-practitioner boundaries. *Journal of Industrial Ecology.*
Lozano, R. (2013). The state of sustainability reporting in universities. *International Journal of Sustainability in Higher Education, 14*(5), 537–552. https://doi.org/10.1108/IJSHE-03-2012-0024
Martinez, S., et al. (2010). Local food systems: Concepts, impacts, and issues. *USDA Economic Research Service Report.*
Montreuil, B. (2011). Toward a Physical Internet: meeting the global logistics sustainability grand challenge. *Logistics Research, 3*, 71–87.
Nicholls, A., & Opal, C. (2005). *Fair trade: Market-driven ethical consumption.* Sage Publications.

Nike. (2020a). *Nike Flyknit innovation.*
Nike. (2020b). *Sustainable innovation.*
Patagonia. (2020a). *Environmental & social initiatives.*
Ormazabal, M., Prieto-Sandoval, V., Puga-Leal, R., & Jaca, C. (2018). Circular economy in Spanish SMEs: challenges and opportunities. *Journal of cleaner production, 185*, 157–167.
Patagonia. (2020b). Worn wear program.
Pedersen, E. R. G., & Gwozdz, W. (2014). From resistance to opportunity-seeking: Strategic responses to institutional pressures for corporate social responsibility in the Nordic fashion industry. *Journal of Business Ethics, 119*(2), 245–264.
Porter, M. E., & van der Linde, C. (1995). Toward a new conception of the environment-competitiveness relationship. *Journal of Economic Perspectives.*
Rajput, S., & Singh, S. P. (2019). Blockchain and IoT based cognitive edge framework for sharing economy services in a smart city. *Journal of Cleaner Production, 238*, 117977.
Schaltegger, S., & Wagner, M. (2011). Sustainable entrepreneurship and sustainability innovation: Categories and interactions. *Business Strategy and the Environment, 20*(4), 222–237. https://doi.org/10.1002/bse.682
Schaltegger, S., & Csutora, M. (2012). Eco-efficiency. *Journal of Industrial Ecology.*
Schneider Electric. (2020a). *Sustainability report.*
Schneider Electric. (2020b). *Energy University.*
Science Based Targets initiative (SBTi). (2020). *Science-based targets.*
Seuring, S., & Müller, M. (2008). From a literature review to a conceptual framework for sustainable supply chain management. *Journal of Cleaner Production, 16*(15), 1699–1710.
Siemens, A. G. (2020). *Siemens sustainability information.*
Sustainable Apparel Coalition. (2020). *The Higg index.*
Sustainable Supply Chain Foundation (SSCF). (2020). *About us.*
Tesla. (2020). *Gigafactory.*
Toyota Motor Corporation. (2020). *Toyota environmental challenge 2050.*
Tseng, M.-L., et al. (2018a). Circular economy meets Industry 4.0: Can big data drive industrial symbiosis? *Resources, Conservation and Recycling, 131*, 146–147.
Tseng, M.-L., Tan, R. R., Chiu, A. S. F., Chien, C.-F., & Kuo, T.-C. (2018b). Sustainable industrial production: Practices and circular economy. *Journal of Cleaner Production.*
Unilever. (2020a). *Sustainable living.*
Unilever. (2020b). *Unilever sustainable living plan.*
United Nations Environment Programme (UNEP). (2017). *Green public procurement.*
UPS. (2020). *UPS ORION.*
U.S. Department of Energy (2020). *Loan Programs Office.*
U.S. Environmental Protection Agency. (2020). *SmartWay transport partnership.*
Vollmer, I., Jenks, M. J. F., Roelands, M. C. P., White, R. J., van Harmelen, T., de Wild, P., van der Laan, G. P., Visser, W., & Crane, A. (2020). Corporate sustainability and the individual: Understanding what drives sustainability professionals as change agents. In *Working paper series.* Ashridge Business School.
Walmart. (2020). *Project Gigaton.*
World Bank. (2018). *Green finance.*
World Business Council for Sustainable Development (WBCSD). (2020). *Tire industry project.*
World Wildlife Fund (WWF). (2020). *Supply chain solutions.*
Zeng, S., et al. (2017). The influence of corporate culture on the successful implementation of environmental management systems. *Journal of Cleaner Production, 142*, 2875–2884.
Zheng, P., & Luo, Z. (2020). Energy efficiency and renewable energy integration in data centers. strategies and sustainability assessment. *Renewable and Sustainable Energy Reviews, 119*, 109595. https://doi.org/10.1016/j.rser.2019.109595

Zhu, Q., & Sarkis, J. (2004a). Relationships between operational practices and performance among early adopters of green supply chain management practices in Chinese manufacturing enterprises. *Journal of Operations Management, 22*(3), 265–289.

Zhu, Q., & Sarkis, J. (2004b). The moderating effects of institutional pressures on emergent green supply chain practices and performance. *International Journal of Production Research.*

Printed by Printforce, the Netherlands